ELECTRICAL WIRING
Industrial

BASED ON THE 2011
NATIONAL ELECTRICAL CODE®

ELECTRICAL WIRING
Industrial

14TH EDITION

STEPHEN L. HERMAN

DELMAR
CENGAGE Learning™   Australia Canada Mexico Singapore Spain United Kingdom United States

DELMAR
CENGAGE Learning

Electrical Wiring Industrial, 14th Edition
Stephen L. Herman

Vice President, Career and Professional Editorial: Dave Garza

Director of Learning Solutions: Sandy Clark

Acquisitions Editor: Stacy Masucci

Managing Editor: Larry Main

Senior Product Manager: John Fisher

Editorial Assistant: Andrea Timpano

Vice President, Career and Professional Marketing: Jennifer Baker

Marketing Director: Deborah Yarnell

Marketing Manager: Kathryn Hall

Associate Marketing Manager: Scott A. Chrysler

Production Director: Wendy Troeger

Production Manager: Mark Bernard

Content Project Manager: Barbara LeFleur

Senior Art Director: David Arsenault

Technology Project Manager: Christopher Catalina

Cover Images:
Electronic schematic: © Shane White/iStockphoto
Lightbulb illustration: © Joseph Villanova
Tower to distribute electricity from a power station: © JCVStock/Shutterstock

For product information and technology assistance, contact us at
Professional Group Cengage Learning Customer & Sales Support, 1-800-354-9706
For permission to use material from this text or product, submit all requests online at **cengage.com/permissions.**
Further permissions questions can be e-mailed to **permissionrequest@cengage.com.**

Library of Congress Control Number: 2010930412

ISBN-13: 978-1-1111-2489-2

ISBN-10: 1-1111-2489-2

Delmar
5 Maxwell Drive
Clifton Park, NY 12065-2919
USA

Cengage Learning is a leading provider of customized learning solutions with office locations around the globe, including Singapore, the United Kingdom, Australia, Mexico, Brazil and Japan. Locate your local office at: **international.cengage.com/region**

Cengage Learning products are represented in Canada by Nelson Education, Ltd.

For your lifelong learning solutions, visit **delmar.cengage.com**

Visit our corporate website at **cengage.com.**

Notice to the Reader

Publisher does not warrant or guarantee any of the products described herein or perform any independent analysis in connection with any of the product information contained herein. Publisher does not assume, and expressly disclaims, any obligation to obtain and include information other than that provided to it by the manufacturer. The reader is expressly warned to consider and adopt all safety precautions that might be indicated by the activities described herein and to avoid all potential hazards. By following the instructions contained herein, the reader willingly assumes all risks in connection with such instructions. The publisher makes no representations or warranties of any kind, including but not limited to, the warranties of fitness for particular purpose or merchantability, nor are any such representations implied with respect to the material set forth herein, and the publisher takes no responsibility with respect to such material. The publisher shall not be liable for any special, consequential, or exemplary damages resulting, in whole or part, from the readers' use of, or reliance upon, this material.

Printed in the U.S.A.
2 3 4 5 15 14 13 12 11

Contents

CHAPTER 9

CHAPTER 10

CHAPTER 11

Preface

INTENDED USE AND LEVEL

Electrical Wiring—Industrial is intended for use in industrial wiring courses at two-year community and technical colleges. The text walks the reader step by step through an industrial building, providing the basics on installing industrial wiring systems. An accompanying set of plans at the back of the book allows the student to proceed step by step through the wiring process by applying concepts learned in each chapter to an actual industrial building, in order to understand and meet requirements set forth by the *National Electrical Code®* (*NEC®*).

SUBJECT AND APPROACH

The fourteenth edition of *Electrical Wiring—Industrial* is based on the 2011 *NEC*. The *NEC* is used as the basic standard for the layout and construction of electrical systems. To gain the greatest benefit from this text, the learner must use the *NEC* on a continuing basis.

> In addition to the *NEC*, the instructor should provide the learner with applicable state and local wiring regulations as they may affect the industrial installation.

In addition to the accurate interpretation of the requirements of the *NEC,* the successful completion of any wiring installation requires the electrician to have a thorough understanding of basic electrical principles, a knowledge of the tools and materials used in installations, familiarity with commonly installed equipment and the specific wiring requirements of the equipment, the ability to interpret electrical construction drawings, and a constant awareness of safe wiring practices.

Electrical Wiring—Industrial builds upon the knowledge and experience gained from working with the other texts in the Delmar Cengage Learning electrical wiring series and related titles. The basic skills developed through previous applications are now directed to industrial installations. The industrial electrician is responsible for the installation of electrical service, power, lighting, and special systems in new construction; for the changeover from old systems to new in established industrial buildings; for the provision of additional electrical capacity to meet the growth requirements of an industrial building; and for periodic maintenance and repair of the various systems and components in the building.

FEATURES

An introduction to *plans and sitework* is the topic of the first chapter in the book, providing explanations of identifying symbols and interpreting the plans in order to help orient the student to the industrial job site. *Examples* are integrated into the text and take the student step by step through problems to illustrate how to derive solutions using newly introduced mathematical formulas and calculations. *Industrial building drawings* are included in the back of the book, offering students the opportunity to apply the concepts that they have learned in each chapter as they step through the wiring process. *Review questions* at the end of each chapter allow students to test what they have learned and to target any sections that require further review.

NEW TO THIS EDITION

- As with previous editions of this book, all *NEC* references have been updated to the 2011 *NEC*.
- The *NEC* use of the terms *compute, computed,* and *computation* is now replaced with *calculate, calculated,* and *calculations.*
- Chapter 1, Plans and Sitework, has a new section on Testing the Site for Grounding Requirement.
- Many of the line drawings have been replaced or updated to help the student better understand the concepts being presented in the text.
- Chapter 6, Using Wire Tables and Determining Conductor Sizes, and Chapter 8, Basic Motor Controls, also contain new coverage.

To access additional course materials including CourseMate, please visit www.cengagebrain.com. At the CengageBrain.com home page, search for the ISBN of your title (from the back cover of your book) using the search box at the top of the page. This will take you to the product page where these resources can be found.

This edition of *Electrical Wiring—Industrial* was completed after all normal steps of revising the *NEC* NFPA 70 were taken and before the actual issuance and publication of the 2011 edition of the *NEC*. These steps include: The National Fire Protection Association (NFPA) solicits proposals for 2011 *NEC;* interested parties submit proposals to NFPA; proposals are sent to Code-Making Panels (CMPs); proposals are reviewed by CMPs and Technical Correlating Committee, Report on Proposals document is published; interested parties submit comments on the proposals to NFPA; review of Comments is conducted by CMPs and Technical Correlating Committee; Report on Comments document is published; review of all Proposals and Comments is conducted at NFPA Annual Meeting; and new motions are permitted to be made at the NFPA Annual Meeting. Finally, the Standard Council meets to review actions made at the NFPA Annual Meeting and to authorize publication of the *NEC*.

Every effort has been made to be technically correct, but there is the possibility of typographical errors or appeals made to the NFPA board of directors after the normal review process that could result in reversal of previous decisions by the CMPs.

If changes in the *NEC* do occur after the printing of this book, these changes will be incorporated in the next printing.

The NFPA has a standard procedure to introduce changes between *Code* cycles after the actual *NEC* is printed. These are called *Tentative Interim Amendments,* or TIAs. TIAs and corrected typographical errors can be downloaded from the NFPA Web site, http://www.nfpa.org, to make your copy of the *Code* current.

SUPPLEMENTS

The Instructor Resources CD contains an Instructor Guide as a PDF with answers to all review questions included in the book, as well as an ExamView testbank, chapter presentations, and a topical presentation in PowerPoint. (order #: 1-1111-2490-6).

Visit us at http://www.delmarelectric.com, now LIVE for the 2011 Code cycle!

This newly designed Web site provides information on other learning materials offered by Delmar Cengage Learning, as well as industry links, career profiles, job opportunities, and more!

ABOUT THE AUTHOR

Stephen L. Herman has been both a teacher of industrial electricity and an industrial electrician for many years. He received his formal education at Catawba Valley Technical College in Hickory, North Carolina. After working as an industrial electrician for several years, he became the Electrical Installation and Maintenance instructor at Randolph Technical College in Asheboro, North Carolina. After nine years, he returned to industry as an electrician. Mr. Herman later became the lead Electrical Technology instructor at Lee College in Baytown, Texas. After serving 20 years at Lee College, he retired from teaching and now lives with his wife in Pittsburg, Texas. Mr. Herman has received the Halliburton Education Foundation's award for excellence in teaching. He has been a guest speaker at professional organizations and has twice been a judge for the national motor control competition at Skills USA.

Acknowledgments

The author and publisher wish to thank the following reviewers for their contributions:

DeWain Belote
Pinellas Tech Educational Center
St. Petersburg, Florida

Les Brinkley
Ashtabula County JVC
Ashtabula, Ohio

Al Clay
Pittsburg, Texas

Warren Dejardin
Northeast Wisconsin Technical
College
Oneida, Wisconsin

Thomas Lockett
Vatterott College
Quincy, Illinois

Special thanks to Mike Forister for his thorough technical review of the *Code* content.

The author also wishes to thank the following companies for their contributions of data, illustrations, and technical information:

Air-Temp Division, Chrysler Corporation
Allen-Bradley Co., Systems Division
Allen-Bradley Co., Drives Division
Allis Chalmers
American Standard Co.
ARCO Electrical Products Corp.
Audisone Inc.
Biddle Instruments
Bulldog Electrical Products Co.
Burndy Co.
Clarage Fan Co.
Crouse-Hinds ECM

Eaton Corp.Cutler-Hammer Products
Edwards Co., Inc
ESE
G&W Electrical Specialty Co.
General Electric Co.
Jensen Electric Company
Kellems Division, Harvey Hubble, Inc.
Wm. J. O'Connell and Stipes Publishing
    Company
Square D Company
Uticor Technology
Westinghouse Electric Corporation

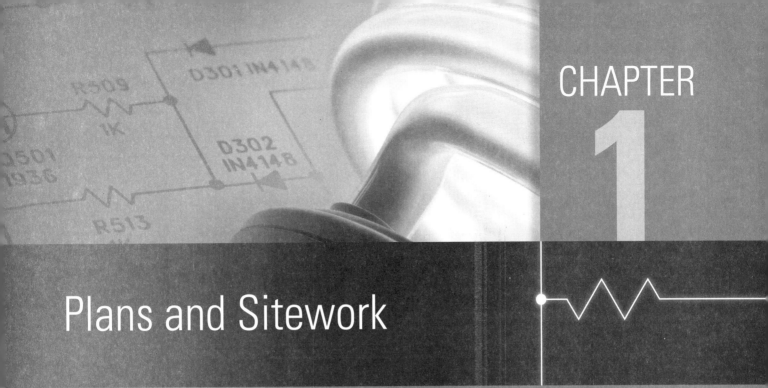

CHAPTER 1

Plans and Sitework

OBJECTIVES

After studying this chapter, the student should be able to

- read site plans to determine the location of the specific items.
- select materials for electrical sitework.
- identify underground wiring methods.
- perform International System of Units (SI) to English and English to SI conversions.
- calculate metric measurements.

⌐∿⊣ CONSTRUCTION PLANS

An electrician who has previously wired a residence or a commercial building is familiar with electrical floor plans and symbols. Although the electrical plans and symbols are basically similar for an industrial building project, additional emphasis is often placed on the sitework. The electrician must continually coordinate and work with the general foreman who is employed by the general contractor.

After the contract for the project is awarded, the electrical contractor must inspect the site plans to determine the approximate location of the industrial building on the site, as well as the locations of underground wiring, raceways, and manholes. The contractor then moves a trailer to the site and locates it so that it will require a minimal amount of relocation during construction. This trailer is used to store materials and tools during the construction of the building.

Building Location

The building location is given on the site plan by referring to existing points such as the centerline of a street. If the electrical contractor and the crew arrive on the site before the general contractor arrives, they are not required to "stake out" (locate) the building. However, they should be able to determine its approximate location. A site plan, such as the one given on Sheet Z-1 of the industrial building plans included in this text, shows the property lines and the centerlines of the street from which the electrician can locate the building and other site improvements.

⌐∿⊣ EXPLANATION OF PLAN SYMBOLS

Contour Lines

Contour lines are given on the site plan to indicate the existing and the new grading levels. If the required underground electrical work is to be installed before the grading is complete, trenches must be provided with enough depth to ensure that the installations have the proper cover after the final grading. The responsibility of who does the ditch-work (general contractor or electrician) is usually agreed upon before the contract is awarded.

Figure 1-1 gives the standard symbols used on construction site plans for contour lines and other features.

Bench Mark

The bench mark (BM), as given on the site plan, is the reference point from which all elevations are located. The bench mark elevation is established by the surveyor responsible for the preliminary survey of the industrial site. This BM elevation is related to a city datum or to the mean sea level value for the site. The elevation is usually given in feet and tenths of a foot. For example, an elevation of 123.4 ft is read as "one hundred twenty-three and four-tenths feet." Table 1-1 is used in making conversions from tenths of a foot to inches.

Elevations

The electrician must give careful attention to the elevations of the proposed building. These details are shown on Sheet Z-1 of the enclosed plans for the industrial building. These drawings provide valuable information concerning the building construction. Measurements on the elevations may be a plus or a minus reference to the BM elevation as given on the site plan.

TABLE 1-1

Conversions of tenths of a foot to inches.

Tenths	Decimal	Fractional
0.1 ft	1.2 in.	1³⁄₁₆ in.
0.2 ft	2.4 in.	2⅜ in.
0.3 ft	3.6 in.	3⅝ in.
0.4 ft	4.8 in.	4¹³⁄₁₆ in.
0.5 ft	6 in.	6 in.
0.6 ft	7.2 in.	7³⁄₁₆ in.
0.7 ft	8.4 in.	8⅜ in.
0.8 ft	9.6 in.	9⅝ in.
0.9 ft	10.8 in.	10¹³⁄₁₆ in.

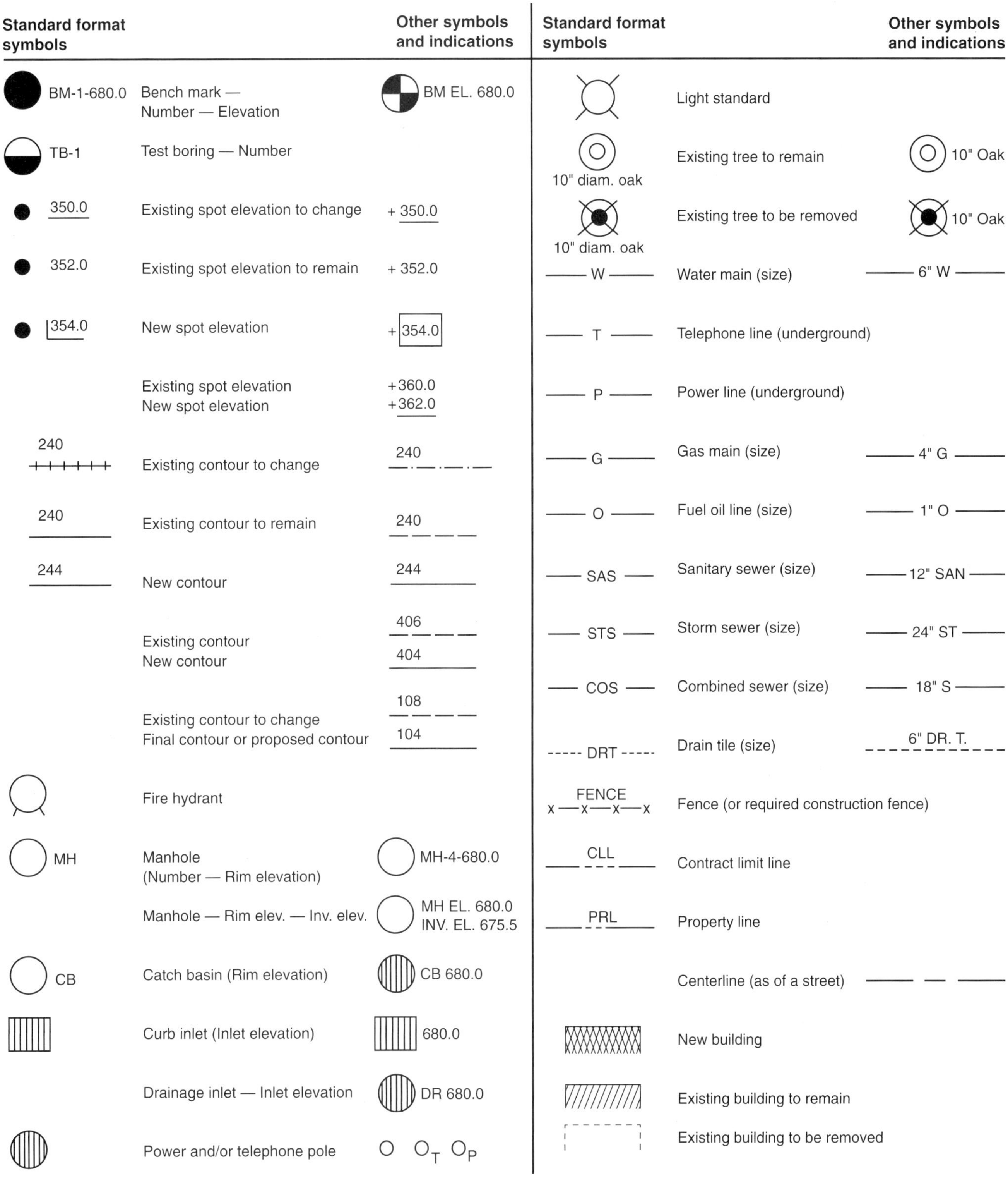

Standard format symbols		Other symbols and indications	Standard format symbols		Other symbols and indications
● BM-1-680.0	Bench mark — Number — Elevation	◑ BM EL. 680.0	⊗	Light standard	
◗ TB-1	Test boring — Number		◎ 10" diam. oak	Existing tree to remain	◎ 10" Oak
● 350.0	Existing spot elevation to change	+ 350.0	⊛ 10" diam. oak	Existing tree to be removed	⊗ 10" Oak
● 352.0	Existing spot elevation to remain	+ 352.0	— W —	Water main (size)	— 6" W —
● 354.0	New spot elevation	+ 354.0	— T —	Telephone line (underground)	
	Existing spot elevation / New spot elevation	+360.0 / +362.0	— P —	Power line (underground)	
240 +++++	Existing contour to change	240 —·—·—	— G —	Gas main (size)	— 4" G —
240 ———	Existing contour to remain	240 ———	— O —	Fuel oil line (size)	— 1" O —
244 ———	New contour	244 ———	— SAS —	Sanitary sewer (size)	— 12" SAN —
	Existing contour / New contour	406 / 404	— STS —	Storm sewer (size)	— 24" ST —
	Existing contour to change / Final contour or proposed contour	108 / 104	— COS —	Combined sewer (size)	— 18" S —
⌀	Fire hydrant		----- DRT -----	Drain tile (size)	6" DR. T. ----
◯ MH	Manhole (Number — Rim elevation)	◯ MH-4-680.0	FENCE x—x—x—x	Fence (or required construction fence)	
	Manhole — Rim elev. — Inv. elev.	◯ MH EL. 680.0 INV. EL. 675.5	CLL ---	Contract limit line	
◯ CB	Catch basin (Rim elevation)	◉ CB 680.0	PRL ---	Property line	
▦	Curb inlet (Inlet elevation)	▦ 680.0		Centerline (as of a street)	— — —
	Drainage inlet — Inlet elevation	◉ DR 680.0	▨	New building	
◍	Power and/or telephone pole	◯ O_T O_P	▨	Existing building to remain	
			⬚	Existing building to be removed	

FIGURE 1-1 Site plan symbols. (*Delmar/Cengage Learning*)

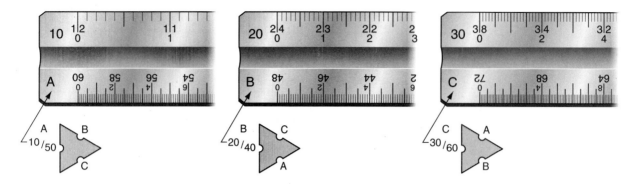

FIGURE 1-2 Scale. (*Delmar/Cengage Learning*)

Invert Elevation (INV)

When an invert elevation (INV) is given, this quantity indicates the level of the *lower* edge of the inside of a conduit entering the manhole (this conduit is usually the lower one in an installation). Refer ahead to Figure 1-13.

Site Plan Scales

Residential site plans generally are drawn to the same scale as is used on the building plans; that is, ⅛″ = 1′-0″ or ¼″ = 1′-0″. However, industrial building site plans typically use scales ranging from 1″ = 20′ and 1″ = 30′ up to 1″ = 60′. It is recommended that the electrician use a special measuring device, called a scale, to measure the site plans, Figure 1-2.

SITEWORK

There may be requirements for several different types of electrical systems to be installed on the site apart from the building itself. The electrician should review the plans and specifications carefully to be aware of all requirements. It is then the responsibility of the electrical contractor/ electrician to ensure that these requirements are met and that installations are made at the most advantageous time and in a fashion that will not conflict with sitework being carried out by other trades.

TESTING THE SITE FOR GROUNDING REQUIREMENTS

When determining the site for a building, one of the most important considerations is the system ground. Proper grounding helps protect against transient currents, electrical noise, and lightning strikes. Several methods can be used to test the electrical grounding system. The effectiveness of the grounding system greatly depends on the resistivity of the earth at the location of the system ground. The resistivity of the earth varies greatly throughout the world and even within small areas. Many factors affect the earth's resistivity such as soil type (clay, shell, sand, etc.), moisture content, electrolyte content, (acids, salts, etc.), and temperature.

In theory, the system ground is considered to have a resistance of zero because it is connected to system grounds everywhere, via the neutral conductor, Figure 1-3. In actual practice, however, the current carrying capacity of the grounding system can vary greatly from one area to another.

Testing

There are different methods for determining the resistivity of the grounding system. An old method used by electricians for many years is to connect a 100 watt lamp between the ungrounded (hot) conductor and the grounding conductor, Figure 1-4. To perform this test, the grounding conductor must be disconnected from the neutral bus in the panel. The brightness of the lamp gives an

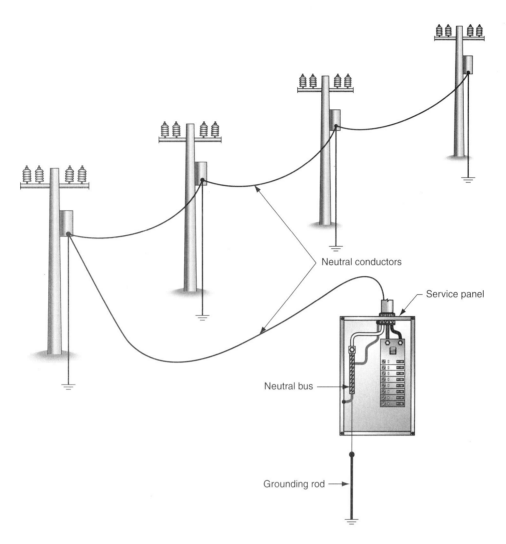

Neutral conductors

Service panel

Neutral bus

Grounding rod →

FIGURE 1-3   All neutral conductors are bonded together forming a continuous grounding system. (*Delmar/Cengage Learning*)

indication of the effectiveness of the grounding system. Although this test gives an indication that the grounding system either works or not, it does not give any indication of the actual resistance of the system. To measure the actual resistance of the grounding system requires the use of special equipment such as a ground resistance tester, Figure 1-5. There are three main tests used to measure ground resistance, the Wenner four point test, the three-point fall-of-potential test, and the clamp-on ground resistance test.

The Wenner Four Point Method

The Wenner four point test is generally performed before building construction begins. This method measures the ground resistance over a wide area. The results are used in designing the grounding system to ensure that it performs properly. This test requires the use of a four pole ground resistance meter, four metal rods, and conductors. The four rods are driven into the ground in a straight line, with equal space between each rod, Figure 1-6. To perform this test, the ground resistance tester produces a known amount of current between rods C1 and C2, producing a voltage drop across rods P1 and P2. The amount of voltage drop is proportional to the amount of current and ground resistance. Readings are generally taken with probes C1 and C2 spaced 5, 10, 15, 20, 30, 40, 60, 80, and 100 feet apart. If possible, it is recommended to perform the test with the probes spaced 150 feet apart.

The calculated soil resistance is the average of the soil resistance from the surface to a depth equal to the

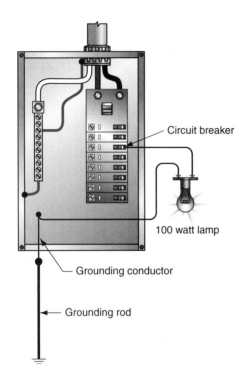

Circuit breaker

100 watt lamp

Grounding conductor

Grounding rod

FIGURE 1-4   A 100 watt lamp is used to test the grounding system. (*Delmar/Cengage Learning*)

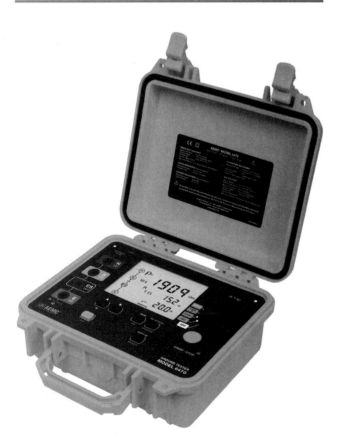

FIGURE 1-5   Ground resistance tester.
(*Photo Courtesy of AEMC Instruments*)

space between the probes. If the probes are set 30 feet apart, for example, each probe will provide an average resistance measurement from the surface to a depth of 30 feet. The tests should not only be made with the probes spaced different distances apart, but also with the probes in different directions from a central point. If the site is large enough, it is generally recommended to perform the test along at least two sides, generally from one corner to the other. It should be noted that underground structures such as metal water pipes can influence the readings. The best results will be obtained by gathering as much data as possible.

Three-Point Fall-of-Potential Test

The fall-of-potential test requires the use of a ground resistance meter. It is performed after the installation of the grounding system and should be done annually to assure the quality of the grounding system. Annual testing provides protection against the degradation of the system before damage to equipment and performance problems occur.

In the three-point fall-of-potential test, the three points of ground contact are

1. the system ground (grounding rod) (point A);

2. a current probe placed some distance from the grounding rod (point B); and

3. a voltage probe that is inserted at various distances between the grounding rod and the current probe (C). The voltage probe is placed in a straight line between the grounding rod and the current probe.

Ideally, the current probe (B) should be placed at a distance that is at least 10 times the length of the grounding rod (A), Figure 1-7. If the grounding rod is 8 feet in length, the current probe should be placed at least 80 feet from the grounding rod.

To perform this test, the grounding rod must be disconnected (electrically isolated) from the neutral bus in the service panel. Failure to do so will completely invalidate the test. The meter provides a known amount of current that flows from the current probe and back to the meter through the system grounding rod. The resistance of the earth causes a voltage drop that is measured between the current probe and the voltage probe. The amount of voltage drop is proportional to the amount of current flow and the ground resistance.

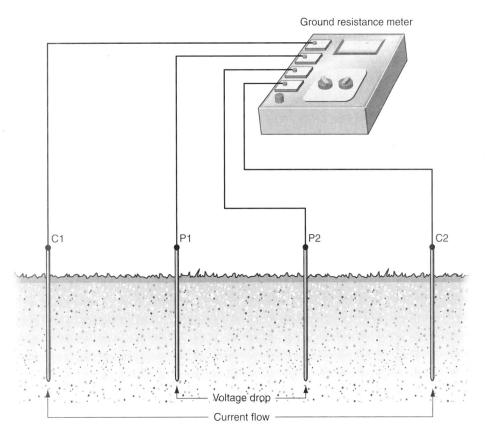

FIGURE 1-6 The Wenner four point test. (*Delmar/Cengage Learning*)

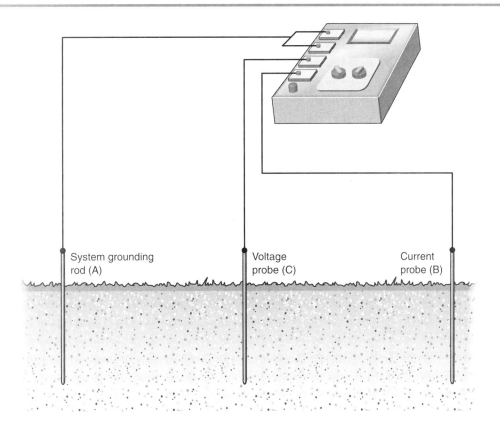

FIGURE 1-7 The three-point fall-of-potential test. (*Delmar/Cengage Learning*)

Resistance readings should be taken at several locations by moving the voltage probe a distance equal to 10% of the distance between the system grounding rod and the current probe. If performed properly, the three-point ground resistance test is the most accurate method of determining ground resistance.

The Clamp-On Ground Resistance Test

The clamp-on ground resistance test requires the use of a special clamp-on ground resistance meter. This test has several advantages over the three-point fall-of-potential test.

1. The service grounding system does not have to be disconnected and isolated from the neutral bus.

2. There are no probes that have to be driven into the ground or long connecting conductors.

3. The neutral conductor supplied by the utility company ties innumerable grounds together in parallel. The clamp-on ground tester measures the effective resistance of the entire grounding system.

4. Because this test is performed by a clamp-on meter, there are no connections that have to be broken or reconnected, resulting in a safer procedure, Figure 1-8.

The clamp-on ground resistance tester, Figure 1-9, contains two transformers. One transformer induces a small fixed voltage at approximately 2 kHz on the grounding conductor. If a path exists, the voltage will result in a current flow. The path is provided by the grounding system under test, the utility neutral, and the utility grounding system. The second transformer inside the meter senses the amount of current at the unique frequency provided by the first transformer. The amount of current is proportional to the induced

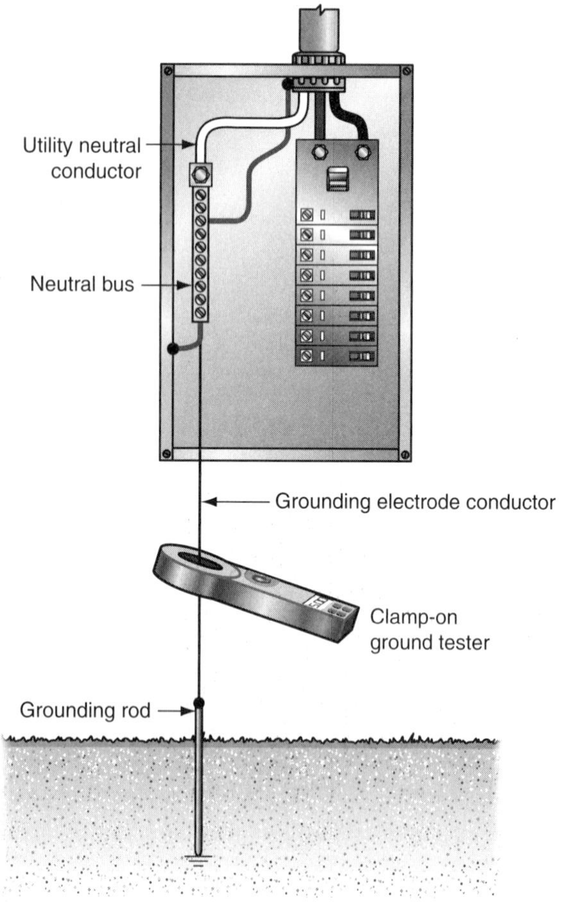

FIGURE 1-8   The clamp-on ground resistance test. (*Delmar/Cengage Learning*)

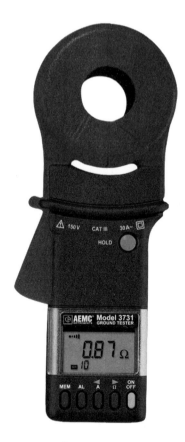

FIGURE 1-9   Ground resistance tester. (*Photo Courtesy of AEMC Instruments*)

voltage and the resistance of the grounding system. The meter uses the two known electrical quantities to calculate the resistance of the grounding system.

INTERPRETING THE SITE PLAN

Notations that do not normally appear on a site plan have been added to plan Z1 of the plans located in the back of the text. These notations are aids used to locate specific spots on the plan. The notations are identified by an asterisk followed by a number such as *1, *2, and so on.

Refer to the Composite Site Plan. Note the bench mark located in the southeast quadrant of the plan. This is the point at which the surveyor began measuring the elevations seen on the plan. Notice that some of the elevation lines have crossing hash marks. The hash marks indicate that that section of the elevation is to be changed. Locate the contour lines for 748 and 749. Parts of these lines have crossing hash marks and parts do not. Only the sections denoted with hash marks are to be changed.

The new elevations are shown with dark heavy lines. These dark heavy lines are shown to connect at some point with the existing contour lines. The elevation of the connecting contour line indicates what the new elevation is intended to be. At position *1, located in the upper southwest quadrant, a heavy dark line connects with the 749 elevation line. The area indicated by the new contour line is to be 749. Locate the new contour line connecting with the 749 contour line at *2. Trace this line to the point where it intersects with the layout of the building. Notice that the entire building is positioned in an area marked by these two new contour lines. This indicates that the building site is to be changed to a uniform 749 ft in preparation for pouring the concrete slab.

New spot elevations are used to indicate an elevation different from that marked by the plot plan. For example, locate the new contour line at *3. This new contour line connects to the 747 contour line. Now locate the new spot elevation at position *4. The arrow points to the curb inlet drain. The curb inlet drain is located in an area that is indicated to be 747 ft. The new spot elevation, however, shows that the curb inlet drain is to be 0.3 ft (90 mm) lower than the surrounding area.

Telephone Service

Telephone service is provided by conduit that runs from the telephone pole. The conduit runs underground at a minimum depth of 18 in. (450 mm) and then is run up the telephone pole for a distance of 8 ft (2.5 m), Figure 1-10. A temporary standard pole cap is installed to protect the equipment from water until the cables are pulled in. The telephone company later removes this cap and extends the conduit up the pole to the point of connection. The conduit is then sealed with a special telephone fitting or with a compound known as *gunk*. A long sweep conduit elbow or quarter bend is installed at the base of the pole. At the lowest point of this fitting, a small V-groove is cut or a ⅜-in. (9.5-mm) hole is drilled for moisture drainage. This drainage hole is known as a *weep hole*. A small dry well is then constructed below the weep hole and is filled with rocks. A pull wire (fish wire) is installed in the raceway from the pole to the junction box at the point where it enters the building. In general, 12-gauge galvanized wire is used as the fish wire, but a nylon string will do as well.

Direct Burial Wiring

The electrician may have a choice of several methods of installing underground wiring. The selection of the method to be used depends on the type of materials available and whether or not provisions are to be made for replacing the conductors. If direct burial cable is used, Figure 1-11, care must be taken to protect the cable from damage. For example, the cable can be installed in the ground to a greater depth than that at which normal digging takes place. Added protection is obtained by placing a treated board over the cable to provide a shield against digging and probing near the cable. The cable should also be surrounded by a layer of sand to prevent any abrasion of the cable by sharp stones and other objects in the soil.

Underground Raceways

Although underground raceways are more expensive to install, they provide many advantages that direct burial installations do not, such as permitting the removal of the original conductors and/or

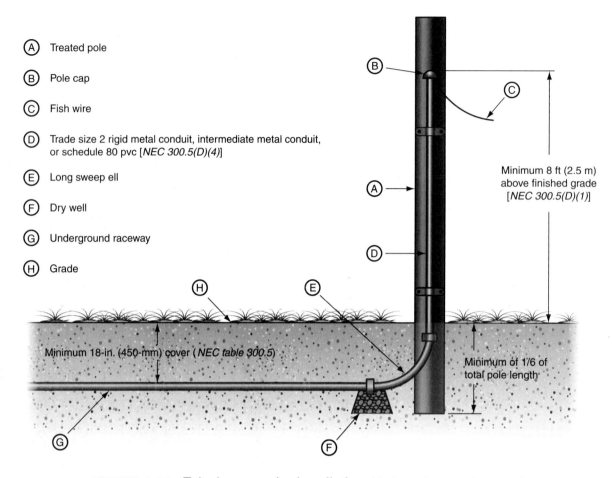

(A) Treated pole

(B) Pole cap

(C) Fish wire

(D) Trade size 2 rigid metal conduit, intermediate metal conduit, or schedule 80 pvc [*NEC 300.5(D)(4)*]

(E) Long sweep ell

(F) Dry well

(G) Underground raceway

(H) Grade

Minimum 8 ft (2.5 m) above finished grade [*NEC 300.5(D)(1)*]

Minimum 18-in. (450-mm) cover (*NEC table 300.5*)

Minimum of 1/6 of total pole length

FIGURE 1-10   Telephone service installation. (*Delmar/Cengage Learning*)

the installation of new conductors with higher current or voltage ratings. Underground raceways are available in a number of different materials, including rigid metal conduit and rigid nonmetallic conduit.

Rigid metal conduit can be installed directly in the soil if (*300.5 and 300.6 of the National Electrical Code* [*NEC*]):

• ferrous conduits (iron or steel) do not rely solely on enamel for corrosion protection;

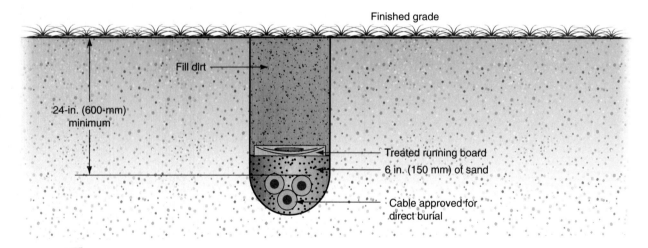

Finished grade

Fill dirt

24-in. (600-mm) minimum

Treated running board

6 in. (150 mm) of sand

Cable approved for direct burial

FIGURE 1-11   An installation of direct burial cable. (*Delmar/Cengage Learning*)

- the conduit is made of a material judged suitable for the condition; and
- the conduit is not placed in an excavation that contains large rocks, paving materials, cinders, large or sharply angular substances, or corrosive material.

Special precautions should be taken when using nonferrous conduit (aluminum) to prevent the conduit from contacting sodium chloride (salt) mixtures. Concrete mixes often use such mixtures to lower the freezing temperature of the green concrete. The chemical reaction between the aluminum and the salt may cause the concrete to fracture or spall (chip or fragment). When protection is desired or required for the type of raceway used, concrete is poured around the conduit, as shown in Figure 1-12, with at least 2 in. (50 mm) of cover in compliance with *NEC Table 300.5*.

The use of rigid polyvinyl chloride conduit type PVC is covered in *NEC Article 352*. These conduits may be used:

- concealed in walls, floors, and ceilings;

- under cinder fill;
- in locations subject to severe corrosive conditions;
- in dry and damp locations;
- exposed where not subject to physical damage; and
- underground.

If the electrical system to be installed operates at a potential higher than 600 volts, the nonmetallic conduit must be encased in not less than 2 in. (50 mm) of concrete.

NEC Article 344 gives the installation requirements for rigid metal conduit and *NEC Article 352* covers rigid polyvinyl chloride conduit type PVC.

The minimum requirements for the installation of conduit and cables underground are given in *NEC Table 300.5*. The general installation requirements are as follows.

For direct burial cables:

- the minimum burial depth is 24 in. (600 mm);

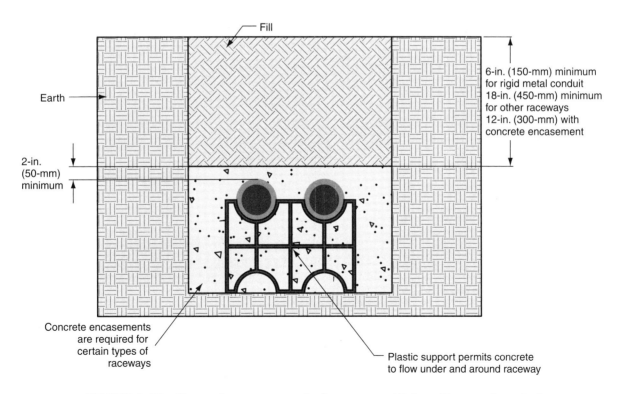

FIGURE 1-12 Concrete encasement of raceways. (*Delmar/Cengage Learning*)

- where necessary, additional protection is to be provided, such as sand, running boards, or sleeves;

- a residential exception permits cable burial to a depth of only 12 in. (300 mm) with GFCI protection; *NEC Table 300.5,* column 4.

For rigid polyvinyl chloride conduit type PVC:

- the minimum burial depth is 18 in. (450 mm);

- a 12-in. (300-mm) burial depth is permitted if a 2-in. (50-mm) concrete cover is provided over conduit;

- a 24-in. (600-mm) burial depth is required in areas subjected to heavy vehicular traffic.

For rigid conduit:

- the minimum burial depth is 6 in. (150 mm);

- a 24-in. (600-mm) burial depth is required in areas subjected to heavy vehicular traffic.

Manholes

Underground raceways terminate in underground manholes similar to the one shown in Figure 1-13. These manholes vary in size depending on the number and size of raceways and conductors that are to be installed. The drain is an important part of the installation because it removes moisture and allows

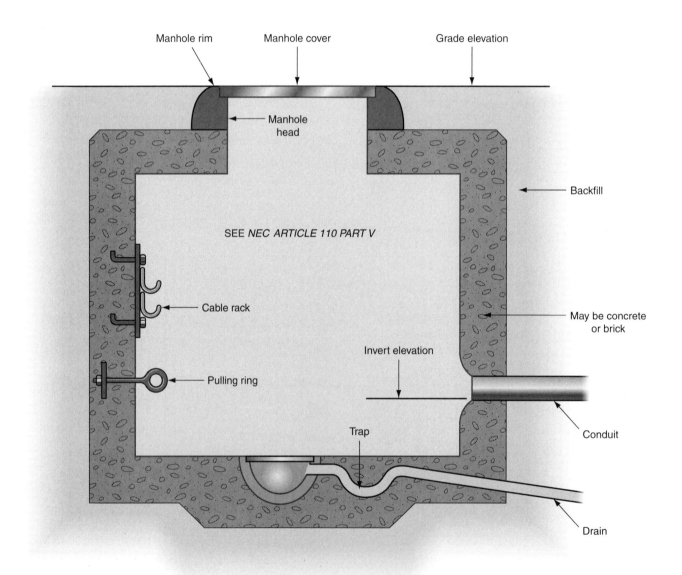

FIGURE 1-13 Typical manhole. (*Delmar/Cengage Learning*)

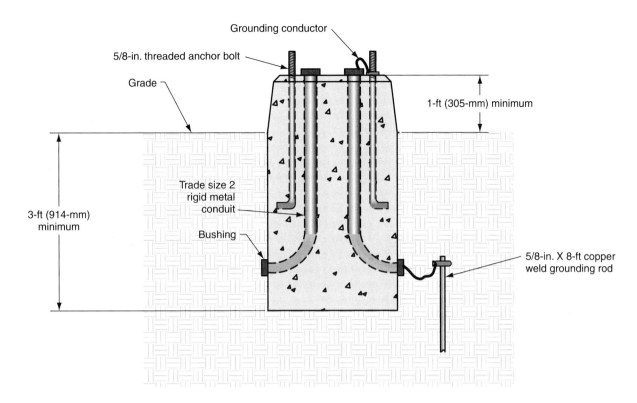

Grounding conductor

5/8-in. threaded anchor bolt

Grade

1-ft (305-mm) minimum

Trade size 2
rigid metal
conduit

3-ft (914-mm)
minimum

Bushing

5/8-in. X 8-ft copper
weld grounding rod

FIGURE 1-14 Typical concrete base for area lighting standard. (*Delmar/Cengage Learning*)

the manhole to remain relatively dry. If a storm sewer is not available for drainage, the installation of a dry well is an alternate choice.

Lighting Standards

Most types of area lighting standards require the installation of a concrete base, Figure 1-14. The manufacturer of the lighting standard should provide a template for the placement of the anchor bolts. If the manufacturer fails to provide a template for the placement of anchor bolts, the electrician should supply the general contractor with the template. The conduit installed in the base should be supplied with bushings on the ends to protect the cables. It is important that proper grounding be achieved at each lighting standard. A grounding conductor shall be installed with the supply conductors as the earth cannot be the sole grounding path; see *250.54*. This section also permits the installation of supplementary grounding electrodes as shown in Figure 1-14. It is mandatory that all conductive parts, including the grounding

electrode, base, bolts, and conduits, be bonded together to achieve comprehensive grounding. See *250.2*, *250.134*, and *250.54*.

METRICS (SI) AND THE *NEC*

The United States is the last major country in the world not using the metric system as the primary system. We have been very comfortable using English (U.S. Customary) values, but this is changing. Manufacturers are now showing both inch-pound and metric dimensions in their catalogs. Plans and specifications for governmental new construction and renovation projects started after January 1, 1994, have been using the metric system. You may not feel comfortable with metrics, but metrics are here to stay. You might just as well get familiar with the metric system.

Some common measurements of length in the English (Customary) system are shown with their metric (SI) equivalents in Table 1-2.

TABLE 1-2

Customary and metric comparisons.

Customary Units	*NEC* SI Units	SI Units
0.25 in.	6 mm	6.3500 mm
0.5 in.	12.7 mm	12.7000 mm
0.62 in.	15.87 mm	15.8750 mm
1.0 in.	25 mm	25.4000 mm
1.25 in.	32 mm	31.7500 mm
2 in.	50 mm	50.8000 mm
3 in.	75 mm	76.2000 mm
4 in.	100 mm	101.6000 mm
6 in.	150 mm	152.4000 mm
8 in.	200 mm	203.2000 mm
9 in.	225 mm	228.6000 mm
1 ft	300 mm	304.8000 mm
1.5 ft	450 mm	457.2000 mm
2 ft	600 mm	609.6000 mm
2.5 ft	750 mm	762.0000 mm
3 ft	900 mm	914.4000 mm
4 ft	1.2 m	1.2192 m
5 ft	1.5 m	1.5240 m
6 ft	1.8 m	1.8288 m
6.5 ft	2.0 m	1.9182 m
8 ft	2.5 m	2.4384 m
9 ft	2.7 m	2.7432 m
10 ft	3.0 m	3.0480 m
12 ft	3.7 m	3.6576 m
15 ft	4.5 m	4.5720 m
18 ft	5.5 m	5.4864 m
20 ft	6.0 m	6.0960 m
22 ft	6.7 m	6.7056 m
25 ft	7.5 m	7.6200 m
30 ft	9.0 m	9.1440 m
35 ft	11.0 m	10.6680 m
40 ft	12.0 m	12.1920 m
50 ft	15.0 m	15.2400 m
75 ft	23.0 m	22.8600 m
100 ft	30.0 m	30.4800 m

The *NEC* and other National Fire Protection Association (NFPA) Standards are becoming international standards. All measurements in the 2011 *NEC* are shown with metrics first, followed by the inch-pound value in parentheses—for example, 600 mm (24 in.).

In *Electrical Wiring—Industrial*, ease in understanding is of utmost importance. Therefore, inch-pound values are shown first, followed by metric values in parentheses—for example, 24 in. (600 mm).

A *soft metric conversion* is when the dimensions of a product already designed and manufactured to the inch-pound system have their dimensions converted to metric dimensions. The product does not change in size.

A *hard metric measurement* is where a product has been designed to SI metric dimensions. No conversion from inch-pound measurement units is involved. A *hard conversion* is where an existing product is redesigned into a new size.

In the 2011 edition of the *NEC*, existing inch-pound dimensions did not change. Metric conversions were made, then rounded off. Please note that when comparing calculations made by both English and metric systems, slight differences will occur as a result of the conversion method used. These differences are not significant, and calculations for both systems are therefore valid. Where rounding off would create a safety hazard, the metric conversions are mathematically identical.

For example, if a dimension is required to be 6 ft, it is shown in the *NEC* as 1.8 m (6 ft). Note that the 6 ft remains the same, and the metric value of 1.83 m has been rounded off to 1.8 m. This edition of *Electrical Wiring—Industrial* reflects these rounded-off changes. In this text, the inch-pound measurement is shown first—for example, 6 ft (1.8 m).

Trade Sizes

A unique situation exists. Strange as it may seem, what electricians have been referring to for years has not been correct!

Raceway sizes have always been an approximation. For example, there has never been a ½-in. raceway! Measurements taken from the *NEC* for a few types of raceways are shown in Table 1-3.

You can readily see that the cross-sectional areas, critical when determining conductor fill, are different. It makes sense to refer to conduit, raceway, and tubing sizes as trade sizes. The *NEC* in 90.9(C)(1) states that *where the actual measured size of a product is not the same as the nominal size, trade size designators shall be used rather*

TABLE 1-3

Trade size of raceways vs. actual inside diameter.

Trade Size	Inside Diameter (I.D.)
½ Electrical Metal Tubing	0.622 in.
½ Electrical Nonmetallic Tubing	0.560 in.
½ Flexible Metal Conduit	0.635 in.
½ Rigid Metal Conduit	0.632 in.
½ Intermediate Metal Conduit	0.660 in.

TABLE 1-5

This table compares the trade size of a knockout with the actual measurement of the knockout.

Trade Size Knockout	Actual Measurement
½	⅞ in.
¾	1³⁄₃₂ in.
1	1⅜ in.

than dimensions. Trade practices shall be followed in all cases. This edition of *Electrical Wiring—Industrial* uses the term *trade size* when referring to conduits, raceways, and tubing. For example, instead of ½-in. electrical metal tubing (EMT), it is referred to as trade size ½ EMT.

The *NEC* also uses the term *metric designator*. A ½-in. EMT is shown as *metric designator 16 (½)*. A 1-in. EMT is shown as *metric designator 27 (1)*. The numbers 16 and 27 are the metric designator values. The (½) and (1) are the trade sizes. The metric designator is the raceways' inside diameter—in rounded-off millimeters (mm). Table 1-4 shows some of the more common sizes of conduit, raceways, and tubing. A complete table is found in the *NEC, Table 300.1(C)*. Because of possible confu-

sion, this text uses only the term *trade size* when referring to conduit and raceway sizes.

Conduit knockouts in boxes do not measure up to what we call them. Table 1-5 shows trade size knockouts and their actual measurements.

Outlet boxes and device boxes use their nominal measurement as their *trade size*. For example, a 4 in. × 4 in. × 1½ in. does not have an internal cubic-inch area of 4 in. × 4 in. × 1½ in. = 24 cubic inches. *Table 314.16(A)* shows this size box as having an area of 21 in.³ This table shows *trade sizes* in two columns—millimeters and inches.

Table 1-6 provides the detailed dimensions of some typical sizes of outlet and device boxes in both metric and English units.

In practice, a square outlet box is referred to as 4 × 4 × 1½-inch square box, 4″ × 4″ × 1½″ square box, or trade size 4 × 4 × 1½ square box. Similarly, a single-gang device box might be referred to as a 3 × 2 × 3-inch device box, a 3″ × 2″ × 3″-deep device box, or a trade size 3 × 2 × 3 device box. The box type should always follow the trade size numbers.

Trade sizes for construction material will not change. A 2 × 4 is really a *name*, not an actual dimension. A 2 × 4 stud will still be referred to as a 2 × 4 stud. This is its *trade size*.

In this text, measurements directly related to the *NEC* are given in both inch-pound and metric units. In many instances, only the inch-pound units are shown. This is particularly true for the examples of raceway calculations, box fill calculations, and load calculations for square foot areas, and on the plans (drawings). To show both English and metric measurements on a plan would certainly be confusing and would really clutter up the plans, making them difficult to read.

TABLE 1-4

This table shows the metric designator for raceways through trade size 3.

METRIC DESIGNATOR AND TRADE SIZE

Metric Designator	Trade Size
12	⅜
16	½
21	¾
27	1
35	1¼
41	1½
53	2
63	2½
78	3

TABLE 1-6

Table 314.16(A) Metal Boxes

Box Trade Size			Minimum Volume		Maximum Number of Conductors* (arranged by AWG size)						
mm	in.		cm³	in.³	18	16	14	12	10	8	6
100 × 32	(4 × 1¼)	round/octagonal	205	12.5	8	7	6	5	5	5	2
100 × 38	(4 × 1½)	round/octagonal	254	15.5	10	8	7	6	6	5	3
100 × 54	(4 × 2⅛)	round/octagonal	353	21.5	14	12	10	9	8	7	4
100 × 32	(4× 1¼)	square	295	18.0	12	10	9	8	7	6	3
100 × 38	(4 × 1½)	square	344	21.0	14	12	10	9	8	7	4
100 × 54	(4 × 2⅛)	square	497	30.3	20	17	15	13	12	10	6
120 × 32	(4¹¹⁄₁₆ × 1¼)	square	418	25.5	17	14	12	11	10	8	5
120 × 38	(4¹¹⁄₁₆ × 1½)	square	484	29.5	19	16	14	13	11	9	5
120 × 54	(4¹¹⁄₁₆ × 2⅛)	square	689	42.0	28	24	21	18	16	14	8
75 × 50 × 38	(3 × 2 × 1½)	device	123	7.5	5	4	3	3	3	2	1
75 × 50 × 50	(3 × 2 × 2)	device	164	10.0	6	5	5	4	4	3	2
75× 50 × 57	(3× 2 × 2¼)	device	172	10.5	7	6	5	4	4	3	2
75 × 50 × 65	(3 × 2 × 2½)	device	205	12.5	8	7	6	5	5	4	2
75 × 50 × 70	(3 × 2 × 2¾)	device	230	14.0	9	8	7	6	5	4	2
75 × 50 × 90	(3 × 2 × 3½)	device	295	18.0	12	10	9	8	7	6	3
100 × 54 × 38	(4 × 2⅛ × 1½)	device	169	10.3	6	5	5	4	4	3	2
100 × 54 × 48	(4 × 2⅛ × 1⅞)	device	213	13.0	8	7	6	5	5	4	2
100 × 54 × 54	(4 · 2⅛ × 2⅛)	device	238	14.5	9	8	7	6	5	4	2
95 × 50 × 65	(3¾ × 2 × 2½)	masonry box/gang	230	14.0	9	8	7	6	5	4	2
95 × 50 × 90	(3¾ × 2 × 3½)	masonry box/gang	344	21.0	14	12	10	9	8	7	4
min. 44.5 depth	FS — single cover/gang (1¾)		221	13.5	9	7	6	6	5	4	2
min. 60.3 depth	FD — single cover/gang (2⅜)		295	18.0	12	10	9	8	7	6	3
min. 44.5 depth	FS — multiple cover/gang (1¾)		295	18.0	12	10	9	8	7	6	3
min. 60.3 depth	FD — multiple cover/gang (2⅜)		395	24.0	16	13	12	10	9	8	4

*Where no volume allowances are required by 314.16(B)(2) through (B)(5).

Reprinted with permission from NFPA 70-2011.

Because the *NEC* rounded off most metric conversion values, a calculation using metrics results in a different answer when compared with the same calculation done using inch-pounds. For example, load calculations for a residence are based on 3 volt-amperes per square foot or 33 volt-amperes per square meter.

For a 40 ft × 50 ft dwelling:

3 VA × 40 ft × 50 ft = 6000 volt-amperes.

In metrics, using the rounded-off values in the *NEC*:

33 VA × 12 m × 15 m = 5940 volt-amperes.

The difference is small, but nevertheless, there is a difference.

To show calculations in both units throughout this text would be very difficult to understand and would take up too much space. Calculations in either metrics or inch-pounds are in compliance with *90.9(D)*. In *90.9(C)(3)* we find that metric units are not required if the industry practice is to use inch-pound units.

It is interesting to note that the examples in *Chapter 9* of the *NEC* use inch-pound units, not metrics.

Guide to Metric Usage

The metric system is a *base-10* or *decimal* system in that values can be easily multiplied or divided by

TABLE 1-7

Numerical system prefixes.

Name	Exponential	Metric (SI)	Script	Customary
mega	(10^6)	1 000 000	one million	1,000,000
kilo	(10^3)	1 000	one thousand	1000
hecto	(10^2)	100	one hundred	100
deka		10	ten	10
unit		1	one	1
deci	(10^{-1})	0.1	one-tenth	1/10 or 0.1
centi	(10^{-2})	0.01	one-hundredth	1/100 or 0.01
milli	(10^{-3})	0.001	one-thousandth	1/1000 or 0.001
micro	(10^{-6})	0.000 001	one-millionth	1/1,000,000 or 0.000,001
nano	(10^{-9})	0.000 000 001	one-billionth	1/1,000,000,000 or 0.000,000,001

10 or powers of 10. The metric system as we know it today is known as the International System of Units (SI) derived from the French term *le Système International d'Unités.*

In the United States, it is the practice to use a period as the decimal marker and a comma to separate a string of numbers into groups of three for easier reading. In many countries, the comma has been used in lieu of the decimal marker and spaces are left to separate a string of numbers into groups of three. The SI system, taking something from both, uses the period as the decimal marker and the space to separate a string of numbers into groups of three, starting from the decimal point and counting in either direction. For example, 12 345.789 99. An exception to this is when there are four numbers on either side of the decimal point. In this case, the third and fourth numbers from the decimal point are not separated. For example, 2015.1415.

In the metric system, the units increase or decrease in multiples of 10, 100, 1000, and so on. For instance, one megawatt (1,000,000 watts) is 1000 times greater than one kilowatt (1000 watts).

By assigning a name to a measurement, such as a watt, the name becomes the unit. Adding a prefix to the unit, such as kilo-, forms the new name *kilowatt*, meaning 1000 watts. Refer to Table 1-7 for prefixes used in the numerical systems.

Certain prefixes shown in Table 1-7 have a preference in usage. These prefixes are *mega-, kilo-*, the unit itself, *centi-, milli-, micro-*, and *nano-*. Consider that the basic metric unit is a meter (one). Therefore, a kilometer is 1000 meters, a centimeter is 0.01 meter, and a millimeter is 0.001 meter.

The advantage of the SI metric system is that recognizing the meaning of the proper prefix lessens the possibility of confusion.

In this text, when writing numbers, the names are often spelled in full, but when used in calculations, they are abbreviated. For example: m for meter, mm for millimeter, in. for inch, and ft for foot. It is interesting to note that the abbreviation for inch is followed by a period (12 in.), but the abbreviation for foot is not followed by a period (6 ft). Why? Because ft. is the abbreviation for fort.

SUMMARY

As time passes, there is no doubt that metrics will be commonly used in this country. In the meantime, we need to take it slow and easy. The transition will take time. Table 1-8 shows useful conversion factors for converting English units to metric units.

TABLE 1-8

Useful conversions and their abbreviations.

inches (in.) × 0.0254 = meter (m)	square centimeters (cm²) × 0.155 = square inches (in.²)
inches (in.) × 0.254 = decimeters (dm)	square feet (ft²) × 0.093 = square meters (m²)
inches (in.) × 2.54 = centimeters (cm)	square meters (m²) × 10.764 = square feet (ft²)
centimeters (cm) × 0.3937 = inches (in.)	square yards (yd²) × 0.8361 = square meters (m²)
millimeters (mm) = inches (in.) × 25.4	square meters (m²) × 1.196 = square yards (yd²)
millimeters (mm) × 0.039 37 = inches (in.)	kilometers (km) × 1 000 = meters (m)
feet (ft) × 0.3048 = meters (m)	kilometers (km) × 0.621 = miles (mi)
meters (m) × 3.2802 = feet (ft)	miles (mi) × 1.609 = kilometers (km)
square inches (in.²) × 6.452 = square centimeters (cm²)	

REVIEW QUESTIONS

All answers should be written in complete sentences, and calculations should be shown in detail.

1. In a set of construction drawings, where would an electrician find information about the location and placement of a building? _____

   _____

   _____

2. From information on the Composite Site Plan, where is the lowest area on the site and what is the elevation? _____

   _____

   _____

3. Using a scale, what is the length, in feet, of the "footprint" of the industrial building?

   _____

   _____

4. What would be the elevation, at the pole, of the bottom of a trench being dug to install the telephone service using rigid nonmetallic conduit? _____

   _____

   _____

5. Raceways are specified for use under sidewalks and drives for the installation of the underground cable serving the site lighting. How many feet of raceway should be ordered for this installation? _____

   _____

   _____

6. What is the difference, in SI units, between the lowest and the highest contours? _____

   _____

   _____

To answer the following questions, examine the composite site plan, the north and west elevations, the site plan symbols, and the *NEC*.

7. What is the elevation of the manhole rim where the bench mark is established? _____

    _____

    _____

8. What is the elevation of the first floor of the industrial building? _____

    _____

    _____

9. What is the vertical distance from the manhole rim to the first floor of the industrial building? (Measure in decimal feet.) _____

    _____

    _____

10. What is the vertical distance from the manhole rim to the first floor of the industrial building, measured in feet and inches accurate to $\frac{1}{16}$ inch?_____

    _____

    _____

11. Where is the preferred area for location of the construction trailer? Why did you choose that area? _____

    _____

    _____

12. It was determined that the rigid nonmetallic conduit for the telephone service could be installed in a trench with a bottom elevation of 743.65 ft. If the conduit is allowed to rise a distance of 1 ft, how deep is the trench at the building? _____

    _____

    _____

13. The distance from the first floor of the office wing to the second floor is how many meters? _____

    _____

14. A cable containing two insulated conductors and a bare grounding wire is installed to a lighting standard mounted on a base similar to the one shown in Figure 1-14. Assume you are the electrician in charge. What instruction would you give to a first-year apprentice who will make up the grounding connection? _____

    _____

    _____

15. Why is it necessary to have a good grounding system for the building? _____

    _____

    _____

16. What are the three most common methods of determining ground resistance? _____

    _____

    _____

CHAPTER

2

The Unit Substation

OBJECTIVES

After studying this chapter, the student should be able to

- define the functions of the components of a unit substation.
- select the proper size of high-voltage fuse.
- explain how to set transformer taps.
- describe how a ground detector operates.
- identify the proper metering connections.

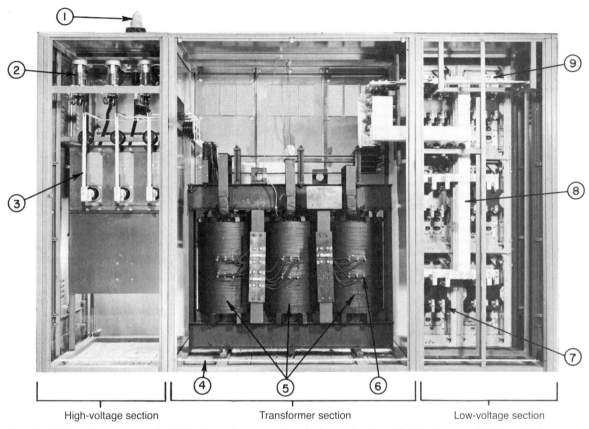

FIGURE 2-1 Unit substation. (*Courtesy ABB*)

① Pothead ② Lightning arrester ③ High-voltage fused switch ④ Grounding bus ⑤ Transformer ⑥ Taps ⑦ Load side terminals ⑧ Secondary bus ⑨ Neutral connections

High-voltage section   Transformer section   Low-voltage section

Power companies commonly supply high voltage service to large commercial or industrial buildings and complexes. The customer owns the step-down transformers, metering, and switching equipment necessary to supply the low voltage loads. This equipment is housed in a *unit substation*, Figure 2-1. The unit substation consists of three compartments: the high-voltage section, the transformer section, and the low-voltage section.

THE HIGH-VOLTAGE SECTION

The Pothead

The high-voltage section must include a means by which the incoming line can be terminated. A device called a *pothead* provides a reliable method of terminating a high-voltage cable, Figure 2-1 and Figure 2-2. To connect the incoming lead-covered cable at the pothead, the cable is opened and the conductors are bared for several inches. The wiping sleeve of the pothead is cut off until the opening is the correct size to receive the cable. The cable is then inserted until the lead sheath is inside the sleeve. The following steps are then completed in the order given: (1) The cable conductors are connected to the terminals at the end of the porcelain insulators; (2) the lead cable is wiped (soldered) to the wiping sleeve; and (3) the pothead is filled with a protective and insulating compound (usually made from an asphalt or resin base). The pothead installation is now ready for the external connections. Several precautions should be observed when the pothead is filled with the selected compound. First, the correct compound is heated to a specified temperature (usually between 250°F and 450°F). The pothead is then filled according to the manufacturer's instructions. Extreme care must be taken to ensure that voids do not occur within the pothead where moisture can accumulate.

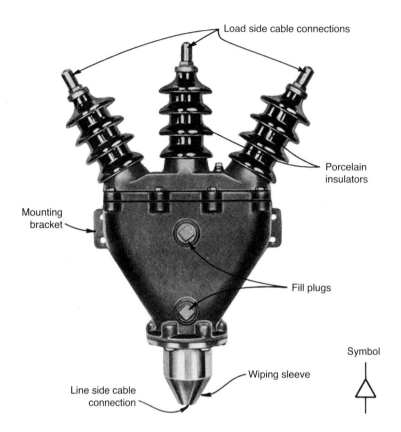

Load side cable connections

Porcelain
insulators

Mounting
bracket

Fill plugs

Symbol

Wiping sleeve

Line side cable
connection

FIGURE 2-2   Pothead. (*Courtesy ABB*)

Lightning Arresters

Lightning arresters, Figure 2-3, are installed on buildings in areas where lightning storms are common. These devices are designed to provide a *low-impedance* path to ground for any surge currents such as those resulting from a lightning strike. Surge arresters installed in accordance with the requirements of *NEC Article 280* shall be installed on each ungrounded overhead service conductor. See *230.209.* The internal components of the arrester vary according to the type of arrester and the specific application. The electrician must ensure that a good ground connection is made to the arrester.

⎯⎓⎯ **TIP:** If the transformer section of the unit substation is to be given a megohm-meter test, the line connection to the arrester must be disconnected during the test to prevent a false ground reading. ●

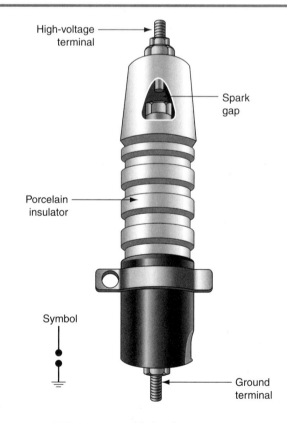

High-voltage
terminal

Spark
gap

Porcelain
insulator

Symbol

Ground
terminal

FIGURE 2-3   Lightning arrester.
(*Delmar/Cengage Learning*)

High-Voltage, Current-Limiting Fuses

High-voltage, current-limiting fuses are installed as protective devices in power distribution systems such as the one installed in the industrial building. The selection of the proper fuse is based on several factors, including the continuous current rating, voltage rating, frequency rating, interrupt rating, and coordination. The fuse selected for a particular installation must meet the predetermined voltage and frequency requirements listed. Fuses are available for both 25- and 60-hertz systems and for voltage ratings of 2400 volts and up, Figure 2-4.

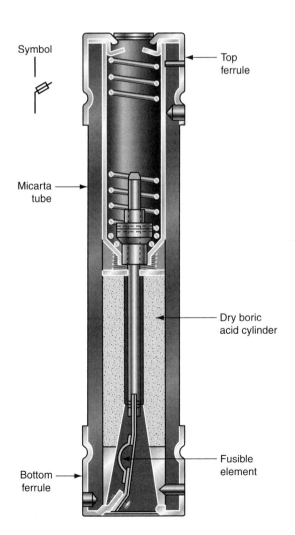

FIGURE 2-4 Cutaway view of high-voltage fuse. (*Delmar/Cengage Learning*)

Symbol

Top ferrule

Micarta tube

Dry boric acid cylinder

Fusible element

Bottom ferrule

Continuous Current Rating

High-voltage fuses are available with either an N or an E rating. These ratings indicate that certain standards established by the Institute of Electrical and Electronic Engineers, Inc. (IEEE) and National Electrical Manufacturers Association (NEMA) have been met. The N rating represents an older set of standards and indicates that a cable-type fuse link will open in less than 300 seconds at a load of 220 percent of its rated current.

An E-type fuse rated at 100 amperes or less will open in 300 seconds at a current of 200 to 240 percent of its rating. Above 100 amperes, an E-rated fuse will open in 600 seconds at a current of 220 to 264 percent of its rated current. The electrician should note, however, that an E-rated fuse does not provide protection in the range of one to two times the continuous load current rating.

The selection of the fuse with the correct continuous current rating to provide transformer protection is based on the following recommendations:

- select a fuse with the lowest rating that has a minimum melting time of 0.1 second at 12 times the continuous current rating of the transformer;
- select a fuse with a continuous current rating of 1.6 times the continuous current rating of the transformer;
- select a fuse that complies with *NEC Article 450.*

⎍ TRANSFORMER PROTECTION

In general, fuses are selected for high-voltage protection because they are less expensive than other types of protection, are extremely reliable, and do not require as much maintenance as do circuit breakers. The protection will be further enhanced if the protective device has the proper interrupt rating.

The minimum interrupt rating permitted for a fuse in a specific installation is the maximum symmetrical fault current available at the fuse location. Power companies will provide the information when requested and will recommend a fuse rating in excess of this value.

OVERCURRENT PROTECTION

Interrupt Rating

As stated earlier, the maximum rating of overcurrent devices for transformers rated at 600 volts or higher is set forth in *NEC Table 450.3(A)*. To use this table, the *percent impedance (%Z)* of the transformer must be known. This value is stamped on the nameplate of transformers rated 25 kVA and larger. See *450.11*. The actual impedance of a transformer is determined by its physical construction, such as the gauge of the wire in the winding, the number of turns, the type of core material, and the magnetic efficiency of the core construction. Percent impedance is an empirical value that can be used to predict transformer performance. It is common practice to use the symbol %Z to represent the percent impedance. Percentages must be converted to a decimal form before they can be used in a mathematical formula.

When this conversion has been made, the symbol .Z will be used to represent the decimal impedance, that is, the percent impedance in decimal form. The percent value is converted to a numerical value by moving the decimal point two places to the left, thus, 5.75 percent becomes 0.0575. This value has no units, as it represents a ratio.

When working with any transformer, it is important to keep in mind the full meaning of the terms *primary* and *secondary* and *high-voltage* and *low-voltage*. The primary is the winding that is connected to a voltage source; the secondary is the winding that is connected to an electrical load. The source may be connected to either the low-voltage or the high-voltage terminals of the transformer. If a person inadvertently connects a high-voltage source to the low-voltage terminals, the transformer would increase the voltage by the ratio of the turns. A 600-volt to 200-volt transformer would become a 600-volt to 1800-volt transformer if the connections were reversed. This would not only create a very dangerous situation but could also result in permanent damage to the transformer because of excessive current flow in the winding. Always be careful when working with transformers, and never touch a terminal unless the power source has been disconnected.

The percent impedance is measured by connecting an ammeter across the low-voltage terminals and a variable voltage source across the high-voltage terminals. This arrangement is shown in Figure 2-5. The connection of the ammeter is short-circuiting the secondary of the transformer. An ammeter should be chosen that has a scale with about twice the range of the value to be measured so that the reading will be taken in the middle of the range. If the current to be measured is expected to be about 30 amperes, a meter with a 0- to 60-ampere range would be ideal. Using a meter with a range under 40 amperes or over 100 amperes may not permit an accurate reading.

After the connections have been made, the voltage is increased until the ammeter indicates the rated full-load current of the secondary (low-voltage winding). The value of the source voltage is then used to calculate the decimal impedance (.Z). The .Z is found

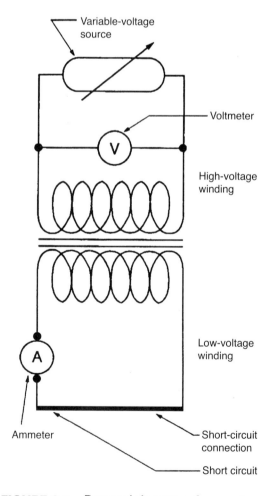

FIGURE 2-5 Determining transformer impedance. (*Delmar/Cengage Learning*)

by determining the ratio of the source voltage as compared to the rated voltage of the high-voltage winding.

 EXAMPLE ────────────

Assume that the transformer shown in Figure 2-5 is a 2400/480-volt, 15-kVA transformer. To determine the impedance of the transformer, first calculate the full-load current rating of the secondary winding. Given the transformer rating in VA, and the secondary voltage E, the secondary current I can be calculated:

$$I = \frac{VA}{E} = \frac{15,000}{480} = 31.25 \text{ amperes}$$

Next, increase the source voltage connected to the high-voltage winding until there is a current of 31.25 amperes in the low-voltage winding. For the purpose of this example, assume that voltage value to be 138 volts. Finally, determine the ratio of source voltage as compared to the rated voltage.

$$.Z = \frac{\text{Source voltage}}{\text{Rated voltage}}$$

$$= \frac{138}{2400} = 0.0575$$

To change the decimal value to %Z, move the decimal point two places to the right and add a % sign. This is the same as multiplying the decimal value by 100.

$$\%Z = 5.75\%$$

Transformer impedance is a major factor in determining the amount of voltage drop a transformer will exhibit between no load and full load and in determining the current in a short-circuit condition. When the transformer impedance is known, it is possible to calculate the maximum possible short-circuit current. This would be a worst-case scenario, and the available short-circuit current would decrease as the length of the connecting wires increases the impedance. The following formulas can be used to calculate the short-circuit current value when the transformer impedance is known.

$$\text{(Single-phase) } I_{sc} = \frac{VA}{E \times .Z}$$

$$\text{(3-phase) } I_{sc} = \frac{VA}{E \times \sqrt{3} \times .Z}$$

The equation $I = VA/.Z$ is read "amperes equals volt-amperes divided by the decimal impedance." This equation is not an application of Ohm's law because decimal impedance is not measured in ohms. The purpose of the equation is to determine the current in a circuit when the transformer capacity (volt-amperes) and the percent impedance are given.

The equation for calculating the rated current for a single-phase transformer is

$$I = \frac{VA}{E}$$

The equation for calculating the rated current for a 3-phase transformer is

$$I = \frac{VA}{E \times \sqrt{3}}$$

The short-circuit current can be determined by dividing the rated secondary current by the decimal impedance of the transformer:

$$I_{sc} = \frac{I_{SECONDARY}}{.Z}$$

The short-circuit current for the transformer in the previous example would be

$$I_{sc} = \frac{31.25}{0.0575}$$

$$= 543.5 \text{ amperes}$$

DETERMINING TRANSFORMER FUSE SIZE

The transformer impedance value is also used to determine the fuse size for the primary and secondary windings. It will be assumed that the transformer shown in Figure 2-5 is a step-down transformer and the 2400-volt winding is used as the primary and the 480-volt winding is used as the secondary. *NEC Table 450.3(A)* indicates that the fuse size for a primary over 600 volts and having an impedance of 6 percent or less is 300 percent of the rated current. The rated current for the primary winding in this example is

$$I = \frac{15,000}{2400}$$

$$= 6.25 \text{ amperes}$$

The fuse size will be

$$6.25 \times 3.00 = 18.75 \text{ amperes}$$

NEC Table 450.3(A) Note 1 permits the next higher fuse rating to be used if the calculated value does not correspond to one of the standard fuse sizes listed in *240.6*. The next higher standard fuse size is 20 amperes.

NEC Table 450.3(A) indicates that if the secondary voltage is 600 volts or less, the fuse size will be set at 125 percent of the rated secondary current. In this example, the fuse size will be

$$31.25 \times 1.25 = 39.06 \text{ amperes}$$

A 40-ampere fuse will be used as the secondary short-circuit protective device, Figure 2-6.

Transformers Rated 600 Volts or Less

Fuse protection for transformers rated 600 volts or less is stipulated by *450.3(B)*. If the rated primary current is less than 9 amperes, the overcurrent protective device can be set at **not more than** 167 percent of this value. If the primary current is less than 2 amperes, the short-circuit protective device can be set at **not more than** 300 percent of this value.

Notice that if the primary current is 9 amperes or more, it is permissible to increase the fuse size to the next highest standard rating. If the primary current is less than 9 amperes, the next lowest fuse size must be used.

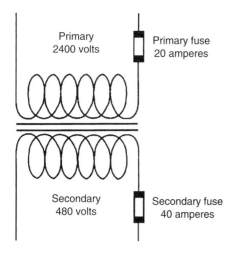

Primary
2400 volts

Primary fuse
20 amperes

Secondary
480 volts

Secondary fuse
40 amperes

FIGURE 2-6 Transformer fusing.
(Delmar/Cengage Learning)

NEC 450.3(B) addresses only transformers that have short-circuit protection in both the primary and secondary windings. If the secondary winding is protected with an overcurrent protective device that is not rated more than 125 percent of the rated secondary current, the primary winding does not have to be provided with separate overcurrent protection if the feeder it is connected to is protected with an overcurrent protective device that is not rated more than 250 percent of the primary current.

EXAMPLE

Assume that a transformer is rated 480/120 volts, and the secondary winding is protected with a fuse that is not greater than 125 percent of its rated current. Now assume that the rated primary current of this transformer is 8 amperes. If the feeder supplying the primary of the transformer is protected with an overcurrent protective device rated at 20 amperes or less (8 × 2.50 = 20), the primary of the transformer does not require separate overcurrent protection, Figure 2-7. The *Code* further states in *450.3(B)* that if the transformer is rated at 600 volts or less and has been provided with a thermal overload device in the primary winding by the manufacturer, no further primary protection is required if the feeder overcurrent protective device is not greater than six times the primary current for a transformer with a rated impedance of not more than 6 percent, and not more than four times the primary rated current for a transformer with a rated impedance greater than 6 percent but not more than 10 percent.

EXAMPLE

Assume that a transformer has a primary winding rated at 240 volts and is provided with a thermal overload device by the manufacturer. Also assume that the primary has a rated current of 3 amperes and an impedance of 4 percent. To determine whether separate overcurrent protection is needed for the primary, multiply the rated primary current [*NEC Table 450.3(B), Note 3*] by 6 (3 × 6 = 18 amperes). If the branch-circuit protective device supplying power to

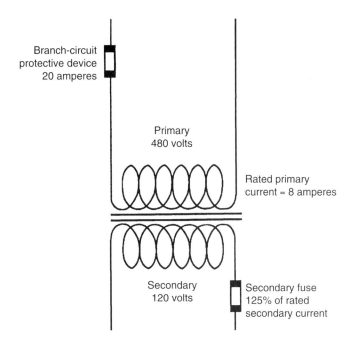

Branch-circuit
protective device
20 amperes

Primary
480 volts

Rated primary
current = 8 amperes

Secondary
120 volts

Secondary fuse
125% of rated
secondary current

FIGURE 2-7 Transformer overcurrent protection. (*Delmar/Cengage Learning*)

the transformer primary has an overcurrent protective device rated at 18 amperes or less, no additional protection is required. If the branch-circuit overcurrent protective device is calculated greater than 18 amperes, a separate overcurrent protective device for the primary will be required.

If separate overcurrent protection is required, it will be calculated at 167 percent of the primary current rating because the primary current is less than 9 amperes but greater than 2 amperes. In this example, the primary overcurrent protective device should be rated at

$$3 \times 1.67 = 5.01 \text{ amperes}$$

A 5-ampere fuse would be used to provide primary overcurrent protection.

Coordination

Coordination is the process of selecting protective devices so that there is a minimum of power interruption in case of a fault or overload. In other words, for a particular situation, a value of high-voltage fuse should be selected that ensures that other protective devices between it and the loads can react to a given condition in less time.

Coordination studies require that the time-current characteristic of the different protective devices be compared and that the selection of the proper devices be made accordingly. Problems in the coordination of high-voltage fusing occur most frequently when

1. circuit breakers are used as secondary protective devices, and

2. a single main protective device is installed on the secondary side of the transformer.

THE TRANSFORMER SECTION

There is little difference between the transformer in a unit substation and any other power transformer (see the connection diagram in Figure 2-8). However, the topic of transformer taps should be explained in some detail.

Taps

Although voltage systems are generally classified by a voltage value, such as a 2300-volt or a 4160-volt system, this exact value is rarely the voltage provided at the transformer. To compensate for this probable voltage difference, taps are built into the transformer, Figure 2-9. These taps are

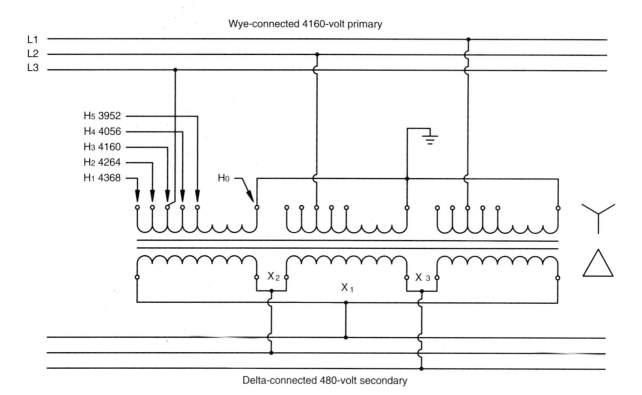

FIGURE 2-8   Three-phase power transformer with a wye connected primary and delta connected secondary. (*Delmar/Cengage Learning*)

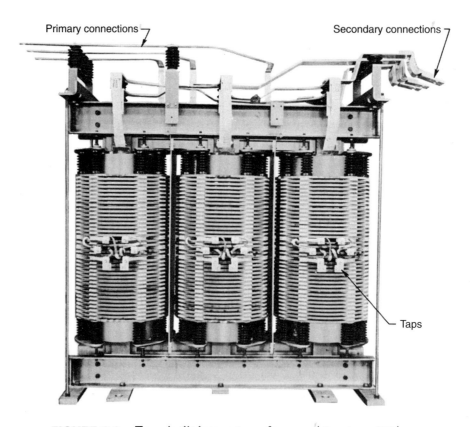

FIGURE 2-9   Taps built into a transformer. (*Courtesy ABB*)

usually provided at 2½-percent increments above and below the standard rated voltage. For example, taps on a 4160/480-volt transformer may provide for voltages of 3952, 4056, 4160, 4264, and 4368 volts. Connections at the proper voltage levels will provide the desired 480 volts on the secondary.

THE LOW-VOLTAGE SECTION

After the incoming voltage is reduced to the desired value, it is taken by busbars into the low-voltage section. Here, protective devices are installed to distribute the voltage throughout the area to be served. Numerous variations in the arrangements of these devices are possible depending upon the needs of the installation. A main device can be installed to interrupt the total power, or any combination of main and feeder devices can be used.

Grounding

The majority of the connections to ground are made in the low-voltage section. However, the electrician should be aware that a grounding bus usually runs the entire length of the unit substation. This bus provides the means for a positive ground connection between the compartments, as well as a convenient place to make other ground connections. Two types of grounding connections are of special interest. The system grounding connection is used to connect a phase or the neutral of the transformer secondary to ground. This grounding electrode conductor is sized according to *NEC 250.66* and *Table 250.66*.

The second grounding connection of special interest is the connection of all the incoming metal raceways to the grounding system. There are no problems in grounding when the raceways enter the substation through the metal structure. However, when a raceway enters through the base of the unit, a grounding connection must be installed between the conduit and the grounding system. This conductor is sized according to *NEC 250.122* and *Table 250.122*.

Ground Detectors

A careful inspection of Figure 2-8 reveals that there are two grounding connections, one on the center tap of the 3-phase wye, high-voltage connection and another to X2 on the 3-phase delta secondary. The decision as to whether the high side is to be grounded is made by the utility company. The general rule, however, is to ground the wye system as shown in this figure. The grounding of the secondary system is optional when the system is a 480-volt delta-connected type, according to *250.20(B)*. If the phase is grounded, then special attention should be given to *240.22*. This section says that if fuses are used for overcurrent protection, they should be installed only in the ungrounded conductors.

An alternative to grounding the secondary is to let it *float*; that is, the secondary remains ungrounded. If this design is selected, ground detectors should be installed, Figure 2-10, to detect any unintentional system grounding. See *250.21(B)*. It should be noted that if a conductor makes contact with ground at any point, the entire system is grounded. However, such a ground may not be an effective ground connection,

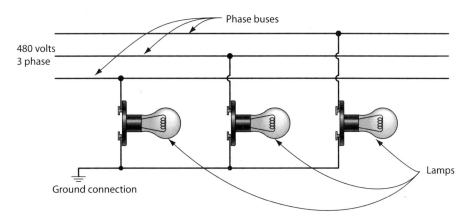

FIGURE 2-10   Ground detectors for ungrounded systems. (*Delmar/Cengage Learning*)

and serious equipment damage may result when a second ground connection occurs on another phase.

The ground detector system in Figure 2-10 consists of three lamps connected as shown. The lamps used have the same voltage ratings as for the line-to-line voltage. The lamps light dimly when there are no grounded conductors. If any phase becomes grounded, however, the lamp connected to that phase dims even more or goes out entirely, while the other two lamps become brighter. Thus, a quick visual check by maintenance personnel can determine whether a ground has developed. Ground detectors are shown on the riser diagram on Sheet E-1 of the plans for the industrial building.

THE HIGH-VOLTAGE METERING EQUIPMENT

The specifications for the industrial building indicate that to provide for energy use measurements, two ¾-inch conduits must be run from the high-voltage section of the unit substation to a cabinet located in a caged section of the loading platform behind the unit substation, Figure 2-11. Current and potential transformers located in the high-voltage section (4160 volts) are an integral part of the substation as assembled at the factory. (The ratio of the potential transformers is 40:1, and the ratio of the current transformers is 400:1.) The metering cabinet is provided with meter test blocks and an instrument autotransformer.

A double 3-phase meter socket trough must be installed above the metering cabinet. This socket trough is connected to the cabinet with conduit nipples. Connections between the current and potential transformers in the high-voltage section of the unit substation and the autotransformer and meter sockets in the cabinet are made with size 12 American Wire Gauge (AWG) wire. The autotransformer is designed to provide voltage components of the proper magnitude and at the correct phase angles to the potential coils of the reactive meter (to be described shortly). These voltage components are

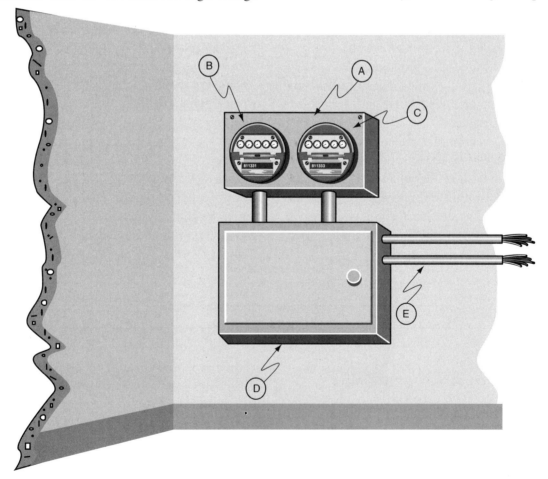

FIGURE 2-11 Meter installation in a caged section of an alcove at the end of the loading platform. (*Delmar/Cengage Learning*)

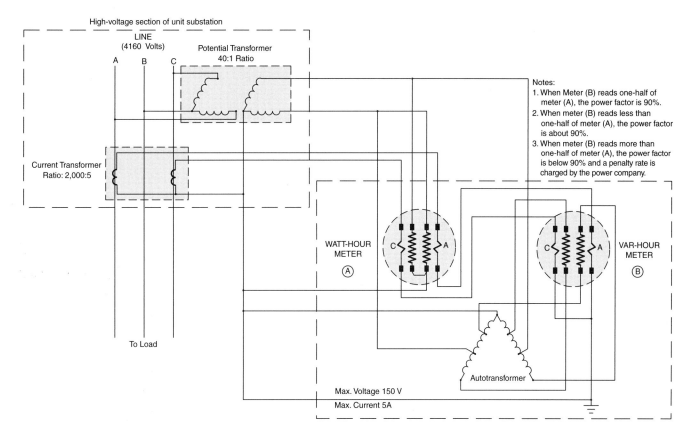

High-voltage section of unit substation

LINE
(4160 Volts)

Potential Transformer
40:1 Ratio

A   B   C

Current Transformer
Ratio: 2,000:5

To Load

Notes:
1. When Meter (B) reads one-half of meter (A), the power factor is 90%.
2. When meter (B) reads less than one-half of meter (A), the power factor is about 90%.
3. When meter (B) reads more than one-half of meter (A), the power factor is below 90% and a penalty rate is charged by the power company.

WATT-HOUR
METER
(A)

C   A

C   A

VAR-HOUR
METER
(B)

Autotransformer

Max. Voltage 150 V
Max. Current 5A

FIGURE 2-12   Connections for high-voltage metering of watt-hours and var-hours.
(*Delmar/Cengage Learning*)

90° out of phase with the line voltage. The left-hand meter socket is wired to receive a standard socket-type watt-hour meter that measures active kilowatt hours. The right-hand meter socket will receive a reactive var-hour (volt-amperes-reactive) meter, Figure 2-12. This second meter measures reactive kilovar-hours.

The two meters are provided with 15-minute demand attachments. The local power company furnishes these meters, which are installed by power company personnel. The meters each have two elements and maximum ratings of 150 volts and 5 amperes. The two demand attachments (not shown) register the demand in kilowatts and kilovars for the respective meters if the demand is sustained for a period of more than 15 minutes at any one time.

Industrial Power Rates

The rates charged by the power company for the energy used are based on the readings of the meter registers and the maximum demand indicators.

Some power companies charge a penalty if the power factor falls below a certain level, as indicated in the example shown in Figure 2-12, Note 3. Assume that the reactive meter reading is one-half the kilowatt-hour reading. Thus, the tangent of the phase angle is $\frac{5}{10}$, or 0.5, and the cosine of the phase angle is 0.9. As a result, no penalty is imposed by the power company because the power factor is 90 percent.

Preferential rates are given when the power transformer is owned by the customer. Further rate reductions are made when the metering measurements are taken on the high-voltage side of the transformer. Both of these conditions for preferential rates are present in the industrial installation being considered in this text.

If the power factor is unity (1.00), it is evident that the reactive meter (var-hour meter) indicating disk is stationary. However, if the power factor falls below unity, then the var-hour meter disk will rotate in one direction for a lagging current and in the opposite direction for a leading current.

In the industrial building, two 350-kVAR synchronous condensers furnish leading current as desired to raise and correct the power factor. (Synchronous condensers are described in Chapter 10.) Simple adjustments of these machines minimize the kVAR-hours, as registered on the reactive meter.

REVIEW QUESTIONS

All answers should be written in complete sentences, and calculations should be shown in detail.

The three main components of a unit substation follow. For each component, name the principal parts and identify their function(s). The parts are listed in Figure 2-1.

1. High-voltage section _____

_____

_____

2. Transformer section _____

_____

_____

3. Low-voltage section _____

_____

_____

The remaining questions are related to the equipment that is associated with the unit substation.

4. Explain the operation of ground detectors and identify the situation that would require their use. _____

_____

5. Explain the reasoning for installation of the two meters. _____

_____

_____

6. Two current coils are installed. What is their function and why was it not necessary to install three coils? _____

_____

_____

7. If the secondary is ungrounded (such as that shown in Figure 2-5), what connections would likely be made to the grounding bus? _____

_____

_____

8. If the secondary is grounded, what connection(s), in addition to those listed in Question 7, would be made to the grounding bus? _____

_____

_____

Feeder Bus System

OBJECTIVES

After studying this chapter, the student should be able to

- set forth the benefits of using busways.
- identify common applications of busways.
- list the components of busways.
- describe various support systems.

FEEDER DUCTS

Modern industrial electrical systems use several methods to transport electrical energy from the source of supply to the points within the plant where panelboards or switchboards are located. These methods may include the use of heavy feeder conductors or cables run in troughs or trays, or heavy busbars enclosed in ventilated ducts. For the industrial building covered in this text, busbars in a ventilated enclosure are specified and shown on the plans. The proper name for this assembly is a busway; however, most electricians and others call the assembly a bus duct. *NEC Article 368* contains the provisions for the installation of busways.

The source of electrical energy in this case is the unit substation located at the rear of the industrial building. Two ventilated feeder busways originate at the unit substation.

Feeder busway sections are available in standard 10-ft (3-m) lengths and in other lengths on special order. Numerous fittings can be used to make branches, turn corners (in both the edgewise and the flat types of installation), and in general, follow the contours of a building.

The enclosure containing the buses is constructed of two identical ventilated steel halves. When these halves are bolted together, they form a complete housing for the busbars. The copper busbars are supported on insulators inside the enclosure, Figure 3-1.

The enclosure contains six busbars that are connected together in pairs to form a 3-conductor system. The busbars are machine-wrapped with varnished cambric insulating tape, except where connections are to be made.

The connection of the busway to enclosures such as unit substations is accomplished with flanged end connections. In addition, these connections are used to transpose the positions of the buses connected to the same phase, Figure 3-2. This transposition reduces the impedance of the total length of the busway. Because each phase is located at two places (or more in larger busways), the effects of the magnetic field are reduced and the opposition to the current flow is also reduced. See *300.20* in the *NEC* for installation requirements related to induced currents.

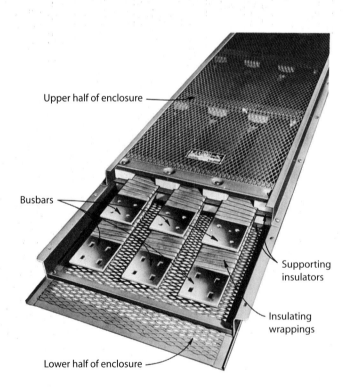

FIGURE 3-1   Feeder busway section. (*Courtesy Siemens*)

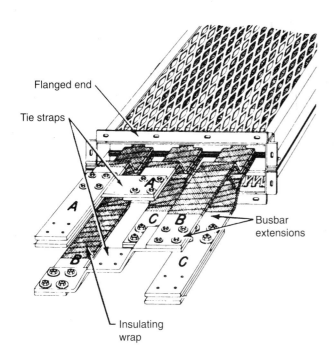

FIGURE 3-2   Busway showing cross connections of phases. (*Courtesy Siemens*)

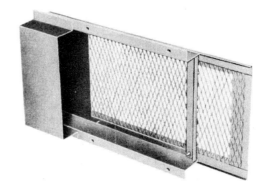

FIGURE 3-3   End closure for ventilated busway. (*Courtesy Siemens*)

Feeder Busway No. 1

Feeder busway No. 1 of the industrial building has a 600-volt and 1000-ampere rating. It starts at the low-voltage section of the unit substation and rises vertically for almost 8 ft (2.5 m). At this point, a tee section is installed to carry the busway in an edgewise, double-branch formation in both directions along the east wall of the main structure (toward the north and south walls of the building).

When the two busway branches meet the north and south walls of the building, edgewise ells change the direction of the branches. The branches of feeder busway No. 1 continue in an edgewise installation along the north and south walls of the manufacturing area at a height of about 16 ft (4.9 m) above the floor. The feeder busway running along the south ends at approximately the midpoint of the wall.

The branch of the feeder busway running along the north wall extends to the west wall of the manufacturing area. An edgewise ell installed at this point changes the direction of the busway once again. It now continues along the west wall of the building and ends before reaching the southwest corner of the manufacturing area. End closing sections, Figure 3-3, are used at the termination of each of the busway runs, *NEC 368.10.*

Power can be tapped from the feeder busway at any *handhole opening.* The handhole openings are located at every joint in the enclosure. For standard lengths, the joints and the handhole openings are 10 ft (3 m) apart. Cable tap boxes are used for cable or conduit tapoff or feed-ins at any handhole opening, Figure 3-4A. Tap box cable lugs and straps, Figure 3-4B, are provided with each tap box. Fusible switch adapters (cubicles) and circuit-breaker cubicles are available for use when it is necessary to connect loads to the feeder, *NEC 368.17(C).*

Feeder Busway No. 2

Much of the information presented for the ventilated feeder busway No. 1 also applies to feeder busway No. 2. This feeder also begins at the low-voltage section of the unit substation and rises vertically to a point slightly below the overhead roof structure of the building. An elbow section is installed at this point to change the direction of the busway while positioning it so that it runs horizontally to form a flat type of installation. The busway runs in a westerly direction down the center of the manufacturing area until it reaches the approximate center of the area where a tee section is installed. As a result, branches of the same feeder duct run north and south and extend as far as the two outside plug-in busways, Figure 3-5.

Feeder busway No. 2 is designed to carry large current values with a minimum power loss and at a low operating temperature. The busway is rated at 1600 amperes and 600 volts.

FIGURE 3-4A   Busway with cable tap box. (*Courtesy Siemens*)

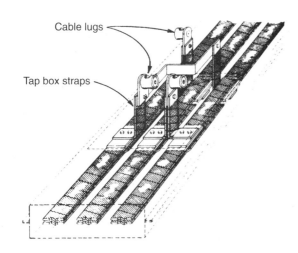

Cable lugs

Tap box straps

FIGURE 3-4B   Bus extension to facilitate cable connections. (*Delmar/Cengage Learning*)

Feeder busway No. 2 is constructed of flat, closely spaced, completely insulated, paired-phase busbars enclosed in a ventilated steel casing similar to that of busway No. 1. Straight sections, elbows, tees, and crosses are standard components available for use so that the duct can be installed horizontally or vertically, edgewise or flat, and can meet any turn or elevation requirements. The casing ends of adjacent sections overlap and are bolted together to form a rigid scarf-lap joint, Figure 3-6.

The flat busbars overlap in the same manner as for the casing. The busbars are bolted together with spring washers, cap screws, and splined nuts furnished with the sections. Vinyl plastic snap-on covers insulate the bolted busbar sections. There are two busbars per phase for a total of six bars. Each bar measures ¼ in. by 2⁷⁄₁₆ in. (6 mm by 58 mm).

Each busbar has a cross-sectional area of 0.61 in.² (394 mm²); thus, the total area per phase is 1.22 in.² (788 mm²). Because the assembly is rated for 1600 amperes, the current density in the busbars is

$$\frac{1600 \text{ A}}{1.22 \text{ in.}^2} = \begin{array}{l} 1311 \text{ amperes/in.}^2 \\ \text{(of cross-sectional area)} \end{array}$$

$$\frac{1600 \text{ A}}{788 \text{ mm}^2} = 2 \text{ amperes/mm}^2$$

When this is compared with a standard density value of 1000 amperes per square inch (1.55 amperes per square millimeter), the value of transposing the buses to reduce the impedance is evident.

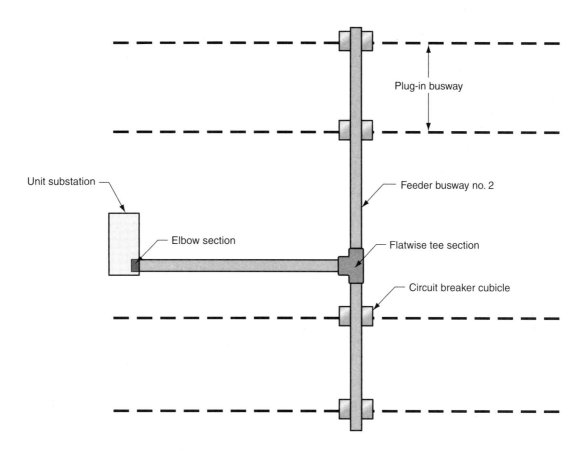

Plug-in busway

Unit substation

Feeder busway no. 2

Elbow section

Flatwise tee section

Circuit breaker cubicle

FIGURE 3-5 Feeder busway No. 2. (*Delmar/Cengage Learning*)

Busbar joined with spring washer and cap screw

Hex head cap screw

Spring cup washer

Splined nut

Silvered contact surfaces

Offset end—$^{7}/_{16}$" × $^{3}/_{4}$" slotted holes

Straight end—with splined nut insert

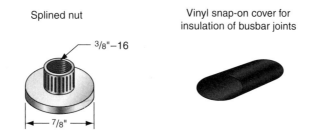

Splined nut

Vinyl snap-on cover for insulation of busbar joints

$^{3}/_{8}$"–16

$^{7}/_{8}$"

FIGURE 3-6 Busbar accessories. (*Delmar/Cengage Learning*)

The pairs of busbars are joined together at the flanged ends located at the substation. End closers are installed at the two dead-end sections of the feeder.

THE CIRCUIT-BREAKER CUBICLES

Circuit-breaker cubicles are used to connect the No. 2 feeder busway to the 225-ampere, plug-in busway runs, Figure 3-7. There are ten of these runs in the industrial building; thus, five double circuit-breaker cubicles are required.

The circuit-breaker cubicle consists of a cube-shaped steel housing. This housing can be attached to the lower side of the ventilated feeder duct. Two 225-ampere circuit breakers are provided in a single housing. Openings in the sides of the cubicle permit the attachment of a plug-in busway. This busway runs in opposite directions at right angles to the feeder (review Figure 3-5). The circuit breakers protect the plug-in duct from overloads as required by *368.17(C)*. This section of the *Code* requires that a means such as chains, ropes, or sticks be provided so that the disconnecting means can be operated from the floor. In this installation, a rope is to be connected to each operating handle and extended to within 7 ft (2.1 m) of the floor.

FIGURE 3-7 Circuit-breaker cubicle. (*Courtesy Siemens*)

PLUG-IN BUSWAY

The plug-in busway is actually a subfeeder taken from the No. 2 ventilated feeder. According to the layout shown on the plans, the plug-in busway makes 3-phase, 480-volt power available to all parts and at any point of the manufacturing area of the plant.

The busbars in the busway are coated with silver at each connection point. Silver is unequaled as an electrical conductor. In addition, silver is less subject to pitting (corrosion) than is copper. Thus, when the bus plug fingers contact the silver coating of the busbars, a high-conductivity connection is ensured. Standard plug-in sections are 10 ft (3 m) in length and consist of two identical formed steel halves that are bolted together to form the complete outside housing, Figure 3-8. This housing also provides the scarf-lap feature, which permits two adjacent duct sections to overlap each other by 12 in. (300 mm). The resulting lap simulates an interlocked joint and provides high rigidity and strength to the assembly, Figure 3-9.

The busway specified for the industrial building is 3-phase duct and is rated at 225 amperes and 480 volts. Although ells, tees, and cross sections (or fittings) are not required for the industrial building installation, such fittings are available for use when specified by the design or layout. Some of these fittings are shown in Figure 3-10.

Power takeoff plug-in openings are spaced at convenient intervals on alternate sides of the enclosure. (Each side has the same number of openings.) Bus plugs can be inserted into any one of these openings. In this manner, branch circuits can be dropped to any item of equipment requiring electric power. The design of the bus plugs is such that they ground against the enclosure before the plug fingers contact the busbars. Additional safety is provided by this design during plug insertion, Figure 3-11.

A plug-in busway is used to provide a flexible tapoff means for motor branch circuits. In other words, the plug-in busway transports electrical energy from the ventilated feeder to the locations of the production machines. Tapoff openings every 10 in. (250 mm) along the busway mean that there is always a convenient location to connect a machine.

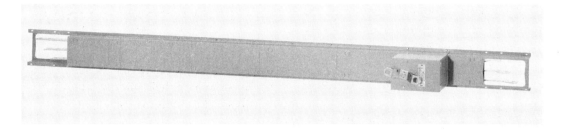

FIGURE 3-8 Plug-in busway. (*Courtesy Siemens*)

The plug-in busway is much like a panelboard extending through a complete load area. However, the busway system is much more flexible. If a machine is to be moved from one location to another, it is a simple matter to unplug the circuit-protective device, move it and the machine to the new location, and plug the protective device back into the busway system. A move of this type can be made without shutting off power to the system or disrupting production in any way. Figure 3-12 shows a typical power distribution system.

The ten separate runs of the plug-in busway in the industrial building start at the circuit-breaker cubicles and extend for a distance of approximately 96 ft (29 m) in either direction (east-west). The runs are about 32 ft (9.6 m) apart on centers with a lesser distance between the outside runs and the walls of the structure. The ends of each run are fitted with end closer fittings.

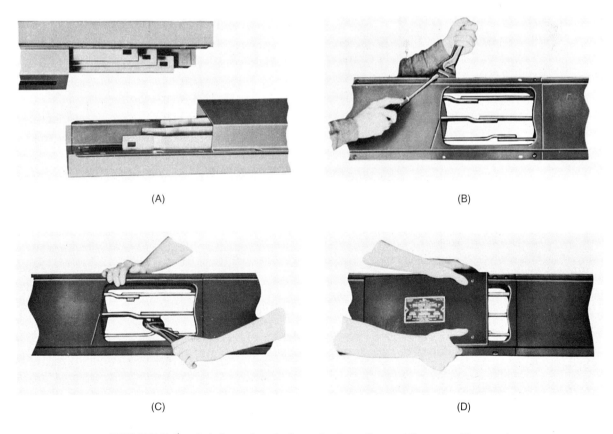

(A)

(B)

(C)

(D)

FIGURE 3-9 Joining plug-in bus duct sections. (*Courtesy Siemens*)

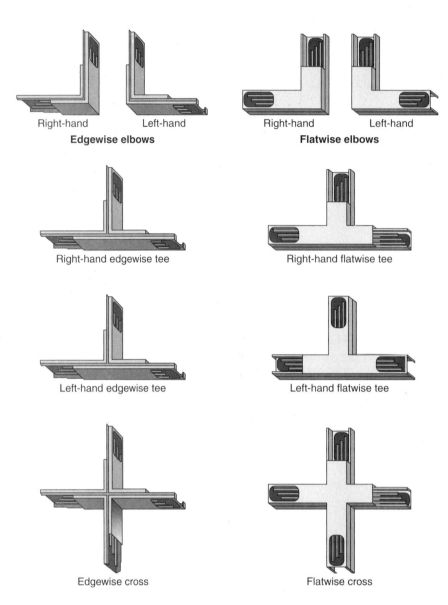

Right-hand   Left-hand
Edgewise elbows

Right-hand   Left-hand
Flatwise elbows

Right-hand edgewise tee

Right-hand flatwise tee

Left-hand edgewise tee

Left-hand flatwise tee

Edgewise cross

Flatwise cross

FIGURE 3-10 Plug-in busway fittings. (*Delmar/Cengage Learning*)

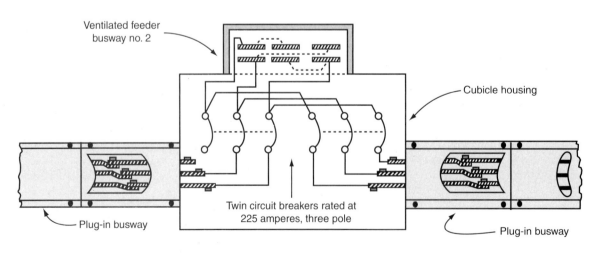

Ventilated feeder
busway no. 2

Cubicle housing

Plug-in busway

Twin circuit breakers rated at
225 amperes, three pole

Plug-in busway

FIGURE 3-11 Plug-in busway. (*Delmar/Cengage Learning*)

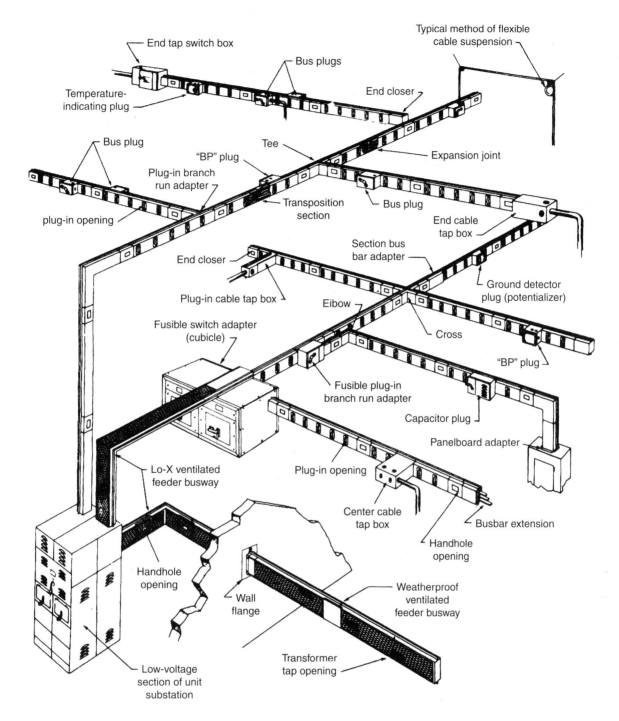

FIGURE 3-12 Power distribution system. (*Delmar/Cengage Learning*)

Method of Suspension

There is an almost unlimited selection of methods for hanging or supporting the plug-in busway. Support arrangements are shown in Figure 3-13 to illustrate some of the more common methods of hanging sections using clamp hangers. Prefabricated clamp hangers eliminate the drilling, cutting, or bend-

ing generally associated with hangers constructed on the job. Clamp hanger halves are slipped over the duct casing and are bolted together. Support arrangements shown include bracket supports, strap hangers, rod hangers, and messenger cable suspension.

The busway used in the industrial building is supported by rods and messenger cables. These cables, in turn, are supported from the overhead

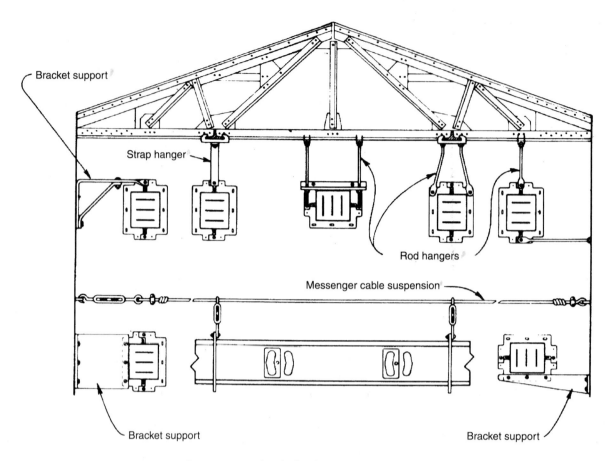

FIGURE 3-13 Support methods for busway. (*Delmar/Cengage Learning*)

structure. The busways are all supported at intervals of 5 ft (1.5 m) or less in accordance with *368.30*.

 BUS PLUGS

One bus plug, Figure 3-14, must be furnished for each machine in the manufacturing area of the plant. According to the plans and specifications for the industrial building, there are 111 machines to be supplied with power from the plug-in system. The number and size of the bus plugs required are summarized in Table 3-1.

The bus plugs provide branch-circuit protection for each of the machines and must be selected according to the specific requirements of the individual machines. The plug-in devices are identified on Sheet E-2 of the plans with regard to the type of machine tool to be supplied. More detailed information is given in the specifications. One advantage of the fusible plug-in unit is that a minimum number

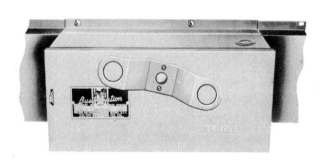

FIGURE 3-14 Bus plug. (*Courtesy Siemens*)

TABLE 3-1

Number and size of required bus plugs.

Number of Bus Plugs Required	Protective Devices Rating, Amperes	Switch Rating, Amperes
47	15	30
26	20	30
30	30	30
3	60	60
5	90	100

of sizes are needed. The plug-in unit size is based on the switch ampere rating. In addition, the protective device rating can be easily changed. Where such devices are located out of reach of the machine operators, suitable means must be provided for operating the disconnect means. See *368.17(C)*.

REVIEW QUESTIONS

All answers should be written in complete sentences, calculations should be shown in detail, and *Code* references should be cited when appropriate.

1. The current density of the 1600-ampere busway was calculated to be 1313 amperes per square inch. Compare this with the allowable current density of a 500-kcmil (thousand circular mils)–type THWN conductor. _____

   _____

   _____

   _____

2. Would it be permissible to cut six openings in the top of the unit substation, with each busbar installed through an individual opening? Why or why not? _____

   _____

   _____

   _____

3. Describe what is meant by transposing the buses and what is achieved. _____

   _____

   _____

   _____

4. Describe when it is appropriate to use busways and when plug-in busways are preferred.

   _____

   _____

   _____

5. Describe at least four support methods and give examples of when their use would be appropriate. _____

   _____

   _____

   _____

   _____

   _____

4

Panelboards

OBJECTIVES

After studying this chapter, the student should be able to

- identify panelboard types.
- select and adjust circuit breakers.
- make feeder connections to panelboards.

──◦◦◦── PANELBOARDS

Circuit control and overcurrent protection must be provided for all circuits and the power-consuming devices connected to these circuits. Lighting and power panelboards located throughout the building being supplied with electrical energy provide this control and protection. Fifteen panelboards are provided in the industrial building to feed electrical energy to the various circuits, Table 4-1.

All of the required panelboards are listed in the specifications and are shown on the plans or are referred to on the riser diagram. These panelboards distribute the electrical energy and protect the circuits supplying outlets throughout the building. The schedule in Table 4-1 shows that eleven of the fifteen panelboards listed supply lighting and receptacle circuits. As a general rule, a panelboard for which more than 10 percent of its overcurrent devices are rated at 30 amperes or less, and for which neutral connections are provided, is defined as a lighting and appliance branch-circuit panelboard. Throughout this text, this type of panelboard will be called a *lighting panelboard*. Panelboards not meeting these requirements are known as power or distribution panelboards.

Lighting and Appliance Panelboards

The basic requirements for panelboards are given in *408.36*. Panelboards P-1 through P-10 and P-12 are considered as lighting and appliance panelboards (Table 4-1).

For these panelboards, follow these requirements:

- *The panelboard shall have a rating of not less than the minimum feeder capacity* as calculated according to *NEC Article 408.*

- *The panelboard must be protected using an overcurrent protective device with a trip setting not exceeding the rating of the panelboard.*

- The total 3-hour load shall not exceed 80 percent of the panelboard rating except when specifically rated for 100 percent continuous duty.

Panelboards are available with standard main ratings of 100, 225, 400, 600, 800, and 1200 amperes.

FIGURE 4-1 Lighting and appliance panelboard with main breaker. (*Courtesy Square D*)

These panelboards may be installed without a main protective device and can be connected directly to a feeder protected at not more than the rating of the panelboard. Individual protection is required on lighting and appliance panelboards when these panelboards are connected to the secondary of a transformer having only primary protection, Figure 4-1.

When a subfeeder, such as the one from transformer TA (as shown on Sheet E-1 of the working drawings), serves more than one panelboard, then connections must be made in the subfeeder for each of the panelboards. These connections can be made either by tapping the conductor or by using subfeed lugs in the panelboards, Figure 4-2.

If subfeed lugs are used, the electrician must ensure that the lugs are suitable for making multiple connections, as required by *110.14(A)*. In general, this means that a separate lug is to be provided for each conductor being connected, Figure 4-3.

TABLE 4-1

Schedule of electric panelboards for the industrial building.

Panelboard No.	Location	Mains	Voltage Rating	No. of Circuits	Breaker Ratings	Poles	Purpose
P-1	Basement N. Corridor	Breaker 100 A	208/120 V 3 φ, 4 W	19 2 5	20 A 20 A 20 A	1 2 1	Lighting and Receptacles Spares
P-2	1st Floor N. Corridor	Breaker 100 A	208/120 V 3 φ, 4 W	24 2 0	20 A 20 A	1 2	Lighting and Receptacles Spares
P-3	2nd Floor N. Corridor	Breaker 100 A	208/120 V 3 φ, 4 W	24 2 0	20 A 20 A	1 2	Lighting and Receptacles Spares
P-4	Basement S. Corridor	Breaker 100 A	208/120 V 3 φ, 4 W	24 2 0	20 A 20 A	1 2	Lighting and Receptacles Spares
P-5	1st Floor S. Corridor	Breaker 100 A	208/120 V 3 φ, 4 W	23 2 1	20 A 20 A 20 A	1 2 1	Lighting and Receptacles Spares
P-6	2nd Floor S. Corridor	Breaker 100 A	208/120 V 3 φ, 4 W	22 2 2	20 A 20 A 20 A	1 2 1	Lighting and Receptacles Spares
P-7	Mfg. Area N. Wall E.	Breaker 100 A	208/120 V 3 φ, 4 W	5 7 2	50 A 20 A 20 A	1 1 1	Lighting and Receptacles Spares
P-8	Mfg. Area N. Wall W.	Breaker 100 A	208/120 V 3 φ, 4 W	5 7 2	50 A 20 A 20 A	1 1 1	Lighting and Receptacles Spares
P-9	Mfg. Area S. Wall E.	Breaker 100 A	208/120 V 3 φ, 4 W	5 7 2	50 A 20 A 20 A	1 1 1	Lighting and Receptacles Spares
P-10	Mfg. Area S. Wall W.	Breaker 100 A	208/120 V 3 φ, 4 W	5 7 2	50 A 20 A 20 A	1 1 1	Lighting and Receptacles Spares
P-11	Mfg. Area East Wall	Lugs only 225 A	208 V 3 φ, 3 W	6	20 A	3	Blowers and Ventilators
P-12	Boiler Room	Breaker 100 A	208/120 V 3 φ, 4 W	10 4	20 A 20 A	1 1	Lighting and Receptacles Spares
P-13	Boiler Room	Lugs only 225 A	208 V 3 φ, 3 W	6	20 A	3	Oil Burners and Pumps
P-14	Mfg. Area East Wall	Lugs only 400 A	208 V 3 φ, 3 W	3 2 1	175 A 70 A 40 A	3 3 3	Chillers Fan Coil Units Fan Coil Units
P-15	Mfg. Area West Wall	Lugs only 600 A	208 V 3 φ, 3 W	5	100 A	3	Trolley Busway and Elevator

Note: Where a 2-pole circuit breaker is used, the space required is the same as for two single-pole breakers.

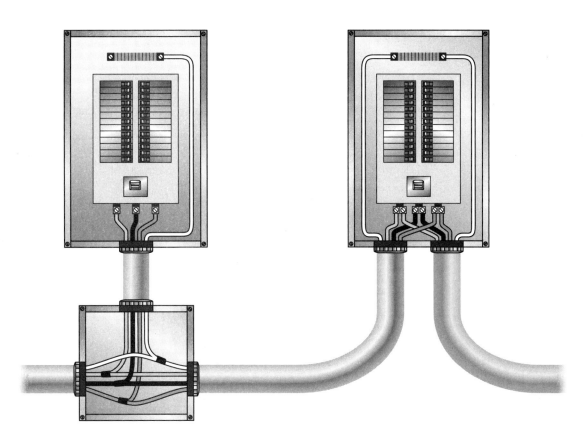

FIGURE 4-2   Methods of connecting panelboards. (*Delmar/Cengage Learning*)

If taps are made to the subfeeder, they may be reduced in size according to *240.21*. This specification is very useful in cases such as that of panelboard P-12. For this panelboard, a 100-ampere main breaker is fed by a 350-kcmil conductor. Within the distances given in the section, a conductor with a 100-ampere rating may be tapped to the subfeeder and connected to the 100-ampere main breaker in the panelboard.

The temperature rating of conductors must be selected and coordinated so as not to exceed the lowest temperature rating of any connected termination, conductor, or device [*110.14(C)*].

FIGURE 4-3   Cable connectors. (*Delmar/Cengage Learning*)

┳┳┳┫ BRANCH-CIRCUIT PROTECTIVE DEVICES

The schedule of panelboards for the industrial building, Table 4-1, shows that lighting panelboards P-1 through P-6 have 20-ampere circuit breakers, including two double-pole breakers (to supply special receptacle outlets). Two single-pole breakers, Figure 4-4, require the same installation space as for a double-pole breaker, Figure 4-5.

FIGURE 4-4   Single-pole breaker. (*Courtesy Square D*)

FIGURE 4-5   Double-pole breaker. (*Courtesy Square D*)

FIGURE 4-6   A 3-pole breaker. (*Courtesy Square D*)

For a 3-pole circuit breaker such as the one shown in Figure 4-6, three poles are required for each breaker used. When the panelboards are purchased, the interiors are specified by the total number of poles required; the circuit breakers are ordered separately.

┳┳┳┫ PANELBOARD PROTECTIVE DEVICE

The main for a panelboard may be either a fuse or a circuit breaker. Because the *Electrical Wiring–Commercial* text discusses the use of fuses in detail, this text will concentrate on the use of circuit breakers. The selection of the circuit breaker should be based on the necessity to

- provide the proper overload protection;
- ensure a suitable voltage rating;
- provide a sufficient interrupt current rating;
- provide short-circuit protection; and
- coordinate the breaker(s) with other protective devices.

The choice of the overload protection is based on the rating of the panelboard. The trip rating of the circuit breaker cannot exceed the ampacity of the busbars in the panelboard. The number of branch-circuit breakers generally is not a factor in the selection of the main protective device except in a practical sense. It is a common practice to have the total amperage of the branch breakers exceed the rating of the main breaker by several times.

The voltage ratings of the breakers must be higher than that of the system. Breakers are usually rated at 250 or 600 volts.

The importance of the interrupt rating is covered in detail in *Electrical Wiring–Commercial*. The student should recall that if there is any question as to the exact value of the short-circuit current available at a point, a circuit breaker with a high interrupt rating is to be installed.

Many circuit breakers used as the main protective device are provided with an adjustable magnetic trip, Figure 4-7. Adjustments of this trip determine the degree of protection provided by the circuit breaker if a short circuit occurs. The manufacturer of this device provides exact information about the adjustments to be made. In general, a low setting may be ten or twelve times the overload trip rating. Two rules should be followed whenever the magnetic trip is set:

- The lower setting provides the greater protection.
- The setting should be lower than the value of the short-circuit current available at that point.

POWER PANELBOARDS

The panelboard schedule in Table 4-1 shows that four panelboards in the industrial building are power panelboards. A typical power panelboard is shown in Figure 4-8. A common interior arrangement for a 3-wire, fusible panelboard is shown in Figure 4-9. The panelboard is supplied from a major source such as a transformer. The panelboard then provides circuits to individual loads, as shown in Figure 4-10.

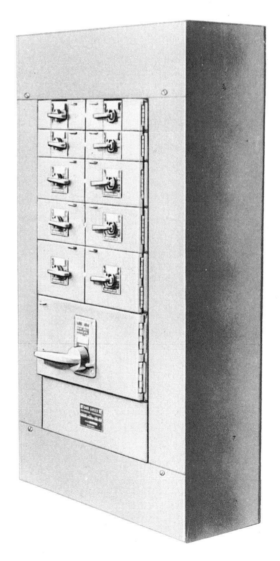

FIGURE 4-8 A typical power panelboard.
(*Courtesy Square D*)

FIGURE 4-7 Circuit breaker with adjustable magnetic trip. (*Courtesy Square D*)

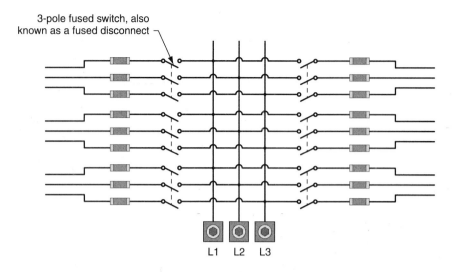

3-pole fused switch, also known as a fused disconnect

L1 L2 L3

FIGURE 4-9 Fusible power panelboard. (*Delmar/Cengage Learning*)

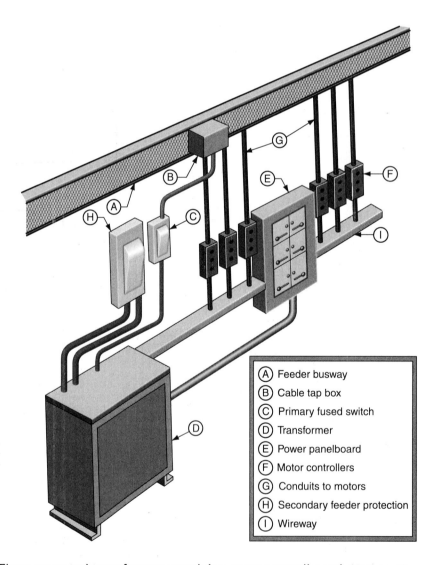

Ⓐ	Feeder busway
Ⓑ	Cable tap box
Ⓒ	Primary fused switch
Ⓓ	Transformer
Ⓔ	Power panelboard
Ⓕ	Motor controllers
Ⓖ	Conduits to motors
Ⓗ	Secondary feeder protection
Ⓘ	Wireway

FIGURE 4-10 Floor-mounted transformer supplying power panelboard. (*Delmar/Cengage Learning*)

All answers should be written in complete sentences, calculations should be shown in detail, and *Code* references should be cited when appropriate.

1. The schedule of panelboards is given in Table 4-1. How many of these are power panelboards and how are they different from the others? _____

    _____

    _____

2. Three-phase, 4-wire panelboards are usually constructed with an even number of spaces available for each phase; thus, the total number of spaces would be in increments of 6, such as 12, 18, 24, 30, 36, or 42. How many spaces would be available for the later addition of circuit breakers in panelboard P-1 after the panelboard has been equipped as scheduled? _____

    _____

    _____

3. Figure 4-2 illustrates two methods of feeding a panelboard from a feeder that continues on to serve other loads. Compare the two methods and indicate your preference.

    _____

    _____

    _____

4. Describe, in detail, what is illustrated in Figure 4-10. _____

    _____

    _____

    _____

    _____

    _____

5. How would you adjust the magnetic trip on a circuit breaker? _____

    _____

    _____

    _____

CHAPTER 5

Trolley Busways

OBJECTIVES

After studying this chapter, the student should be able to

- identify features of a trolley busway installation.
- identify features of a lighting trolley busway and installation.
- select components to support cord drops.

THREE-PHASE TROLLEY BUSWAY

Many modern industrial plants use systems of mobile trolley outlets that move along specially constructed busways. The industrial trolley bus is a 100-ampere, enclosed busbar electrical system. Such a trolley bus provides a continuous outlet system for feeding electrical energy to portable electric tools, cranes, hoists, and other electrical loads.

When the trolley system is installed over production and assembly lines, it provides current to equipment through trolleys that move along with the particular object being assembled. Because the busbars are totally enclosed in a steel casing, there are no exposed live parts to provide hazards to worker safety. This system eliminates the need for and the hazards of portable cords plugged into fixed outlets at the floor level.

THE TROLLEY BUSWAY RUNS

Sheet E-2 of the industrial building plans shows the layout of the four trolley busway runs to be installed. The specifications provide more detailed information about the trolley busway system. The four runs as shown on Sheet E-2 are labeled A, B, C, and D. These runs are 68 ft (20.7 m), 131 ft (40 m), 96 ft (29 m), and 106 ft (32.3 m) long, respectively.

The trolley systems are constructed of straight sections joined end to end, Figure 5-1. The standard section is 10 ft (3 m) long, but sections of less than 10 ft (3 m) in length are available so that a run can be made to exact dimensions. Curved sections and other fittings are also available.

Trolley Busway Run A

Trolley run A consists of straight busway extending for 68 ft (20.7 m). One 3-phase drop-out section is installed at the approximate midpoint of the run. A drop-out section provides the means for removing or inserting the trolleys, Figure 5-2. As shown in Figure 5-2, the drop-out section contains two hinged doors that open when a lever is raised. When the lever is in the down position, the doors are firmly closed and the trolleys move past this section smoothly. Blocking straps ensure that a trolley cannot be placed in the duct incorrectly. This feature also ensures that the polarity is always correct after the trolleys are inserted. Drop-out sections are available in lengths of 10 ft (3 m), and one drop section must be installed in each run.

The trolley busway is to be suspended 8 ft (2.5 m) above the floor according to the specifications and is supported by standard hangers. These hangers also serve as a means for joining adjacent sections and automatically aligning the busbars. The hangers are formed from 12-gauge steel. However, the only tool needed to join the industrial-type trolley duct is a screwdriver.

Intermediate-type hangers are used at the midpoint of each standard 10-ft (3-m) section to give extra support. The intermediate hangers fit snugly around the duct sections but do not interfere with the free passage of the trolleys. The combination of the standard hangers and the intermediate hangers supports the busway at 5-ft (1.5-m) intervals, resulting in a very rigid and secure installation, Figure 5-3.

FIGURE 5-1 Standard 10-ft (3-m) section of trolley busway. (*Delmar/Cengage Learning*)

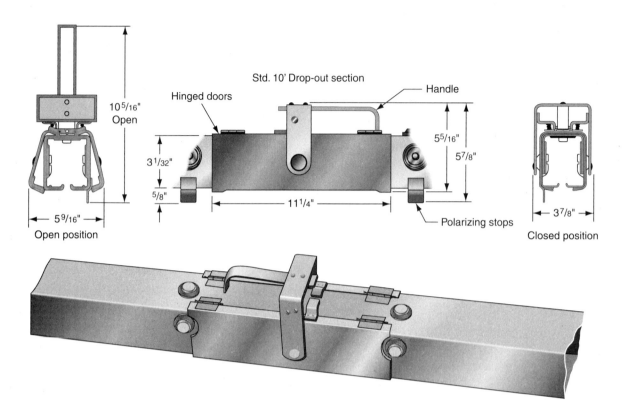

FIGURE 5-2 Standard drop-out section. (*Delmar/Cengage Learning*)

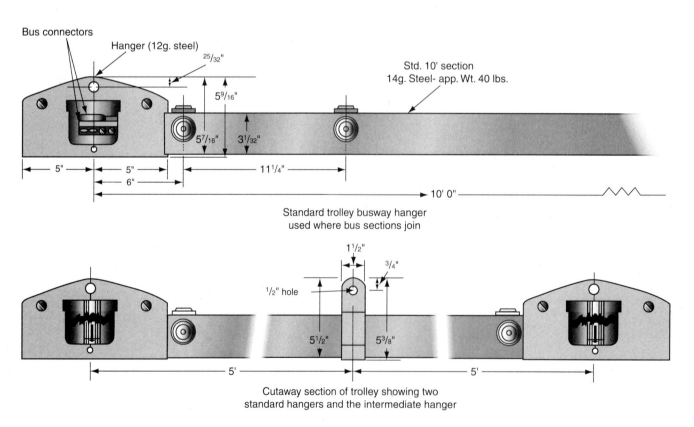

FIGURE 5-3 Supporting trolley busway. (*Delmar/Cengage Learning*)

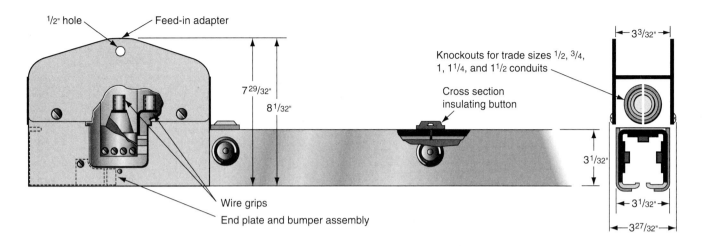

FIGURE 5-4   Trolley busway section with feed-in adapter and end plate with bumper.
(*Delmar/Cengage Learning*)

The standard and intermediate hangers are attached to the overhead structure by rod or strap-type supports, Figure 5-4.

Each end of the busway is capped with an end plate and bumper assembly. This device closes the ends of the busway run and acts as a bumper for any trolley reaching the end of the busway. The bumper absorbs shock and protects the trolley from damage.

Feed-in Adapter

The power supply from the panelboard is fed into the busway run through a *feed-in adapter*, Figure 5-4. The feed-in adapter has pressure-type wire terminals. This adapter can be used at the end of a run or it can be installed in the center of the busway to provide center feed for thc connection of the conduit or cable from the panelboard. Conduit must be installed between the power panelboard and the feed-in adapter to bring power to the trolley busway. A 100-ampere circuit will run in this conduit from panelboard P-15.

The Trolleys

Several types of both fused and unfused trolleys are available. The trolley specified for the industrial building is a fusible, box-type tool hanger with a heavy-duty rating. The box tool hanger, Figure 5-5, has a hinged cover and is provided with puller-type

fuse cutouts, plug receptacles, and cord clamps. For the industrial building, cutouts for 0- to 30-ampere fuses are provided. The trolley has eight wheels and four side thrust rollers to ensure smooth movement along the busway.

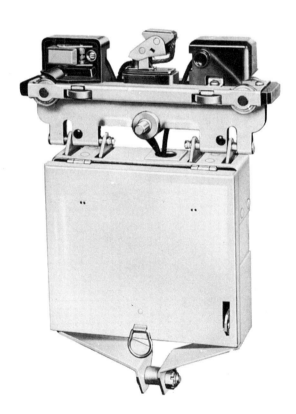

FIGURE 5-5   Trolley with box tool hanger.
(*Courtesy Siemens*)

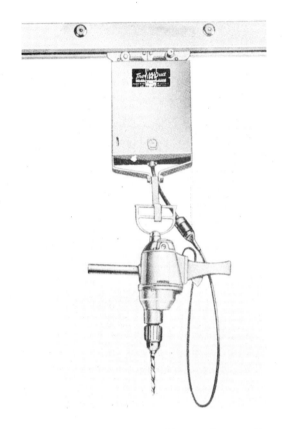
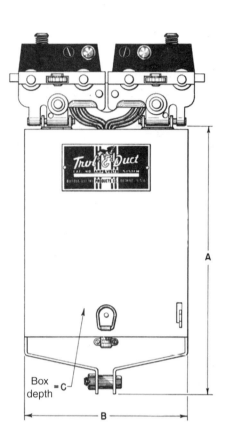

Box depth = C

FIGURE 5-6 Heavy-duty trolley with box tool hanger. (*Courtesy Siemens*)

The trolleys have six graphite bronze shoes. These shoes make contact with the busbars and provide a path that continues through the fuse cutout and receptacle to the heavy-duty, 4-wire rubber cord. This cord is used to attach the various portable tools such as electric drills, buffers, grinders, and other equipment to the busway system, Figure 5-6.

The 4-wire rubber cord provides three conductors to operate the 3-phase portable tools used on the job. The fourth conductor (green in color) is used to ground the equipment (*400.23* and *400.24*).

One end of the grounding conductor must be attached securely to the trolley, and the other end to the housing of the portable tool. The cord must be approved for heavy-duty usage and may be a type SJ containing 12 AWG conductors. The fuses in the trolley are rated at 20 amperes, Figure 5-7.

The specifications for the industrial building call for the use of one trolley for each 15 ft (4.5 m) or fraction thereof of trolley busway. Thus, for run A [68 ft (20.7 m) long], the contractor must furnish five trolleys. Figure 5-8 shows a typical installation of trolley busway.

The Conduit Run

Conduit must be installed from the feed-in adapter to the power panelboard to bring electrical energy to the duct runs. A 100-ampere feeder circuit will run in this conduit from panelboard P-15.

Trolley Busway Run B

As shown on the plans, trolley busway run B extends 131 ft (40 m) in length. The information supplied for busway run A also applies to trolley

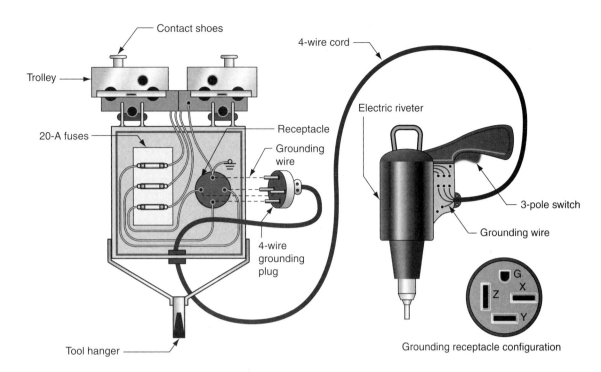

FIGURE 5-7 Details of trolley box-type tool hanger showing grounding connections. (*Delmar/Cengage Learning*)

FIGURE 5-8 Trolley busway may be used to feed power to hoisting equipment. (*Courtesy Siemens*)

runs B, C, and D. The location of run B is shown on Sheet E-2. As is the case for run A, conduit must be installed to power panelboard No. 15. The conduit runs must be routed to conform with the structure of the building. Nine trolleys are required for trolley run B.

Trolley Run C

Trolley run C is a straight run extending 96 ft (29 m) (see Sheet E-2 of the plans). This run is connected to power panelboard No. 15 by conduit from the feed-in adapter located at the end of the run. Six trolleys are required for trolley run C.

Trolley Run D

Trolley run D is 106 ft (32.3 m) long. The installation methods and parts used for this run are the same as those used in the other runs. Panelboard P-15 (located on the west wall of the manufacturing area) again supplies electrical energy to the run. According to the specifications, run D requires eight trolleys including tool hangers to be installed, one for every 15 ft (4.5 m) or fraction thereof of busway.

Each of the four trolley systems (A through D) is a 3-phase system and is rated at 208 volts and 100 amperes. The equipment attachment plugs used with the duct system are polarized. This feature eliminates several problems when portable tools having 3-phase motors are used. For example, reversed phases and the resulting reversal in the direction of rotation of the portable tools are eliminated.

These trolley systems have several advantages. The runs follow the production and assembly lines of the plant. This convenience tends to increase the amount of work that can be completed. A neater and safer production area is maintained because the tools the worker commonly uses are not scattered over the floor but rather are suspended directly over the working area.

LIGHTING IN THE MANUFACTURING AREA

The general illumination system for the manufacturing area of the plant consists of 180 fluorescent luminaires suspended from a system of 50-ampere lighting trolley busways. This type of lighting system is in wide use in industrial applications because of the mobility provided by the system, Figure 5-9. For example, in modern industry, production lines and machine layouts can be changed when so required by the introduction of new products or manufacturing methods. A lighting system composed of fixed outlets and a fixed conduit system is not easily adaptable to such changing requirements.

On the other hand, a system of luminaires suspended from a trolley system can be readily shifted

FIGURE 5-9 Typical trolley busway lighting system. (*Courtesy Siemens*)

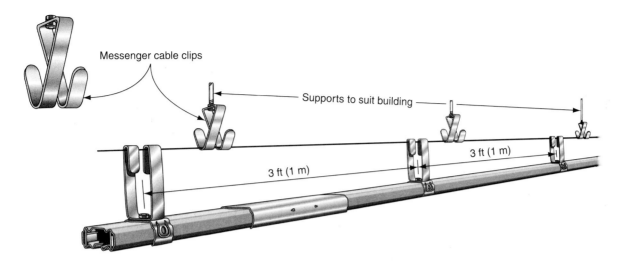

Messenger cable clips

Supports to suit building

3 ft (1 m)

3 ft (1 m)

FIGURE 5-10   Messenger cable suspension. (*Delmar/Cengage Learning*)

from one location to another as desired. It is not necessary to take down or replace heavy conduit systems as would be required to change a fixed lighting outlet system. If extensive alterations or a major plant changeover requires the removal of the lighting trolley duct, the duct sections are completely reusable.

The lighting trolley busways are rated at 50 amperes. Thus, 50-ampere lighting circuits are available, as compared with the 15- or 20-ampere branch circuits used in conventional lighting systems. The 50-ampere lighting circuits used in the trolley systems are approved by the Underwriters Laboratories, Inc. (UL).

The trolley lighting circuits in the industrial building consist of twenty 50-ampere branch circuits. These branch circuits, in turn, are composed of twenty trolley runs constructed of standard 10-ft (3-m) lengths and special lengths as necessary. Each run is about 96 ft (29.3 m) long. The runs are suspended by special clips from messenger wires stretched tightly below the roof structure trusswork. These messenger wires are adjusted for tension by turnbuckles located at the ends of the runs. Intermediate supports for the messenger cables, Figure 5-10, must be attached to the overhead structure at appropriate intervals to prevent any sagging of the lighting system.

The lighting trolley is available in 5-ft (1.5-m) and 10-ft (3-m) sections. These sections are joined by plain couplings. The ends of the sections contain trolley entrance end caps. The plain couplings make a positive connection, electrically and mechanically, between two duct sections and permit free passage of the trolleys along the duct runs. The trolley entrance and caps serve two purposes: They close the ends of the duct runs and they provide an entrance point for the insertion or removal of the trolleys, Figure 5-11.

Other trolley entrance couplings are used at the midpoint (approximately) of each duct run. This arrangement is an additional convenience when removing or inserting trolleys. To prevent arcing, trolleys should not be inserted or removed while they are under load.

Center feed-in boxes are used to bring the electrical supply cables to the trolley busway. Each box has two adjustable couplings (one on each end of the box) connected by removable flexible jumper wires. In addition, the feed-in box has two sets of concentric knockouts that provide a means of bringing the feeding conduits into the duct system, Figure 5-12.

The Busbars

The trolley busway is equipped with two copper busbars. Each bar has a cross-sectional area of 30,557 circular mils and is rated by UL at 50 amperes and 250 volts, Figure 5-13.

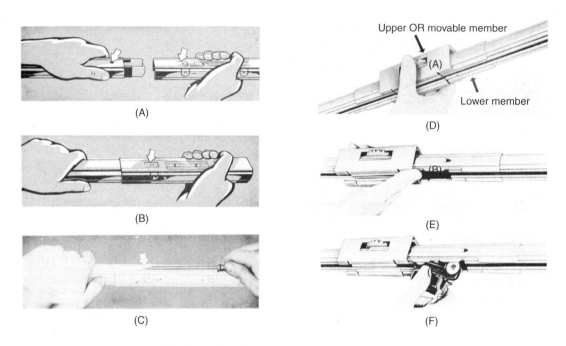

FIGURE 5-11 Plain and trolley entrance couplings. (*Courtesy Siemens*)

FIGURE 5-12 Feed-in box.
(*Delmar/Cengage Learning*)

The Trolleys

Several types of trolleys are available for use with the lighting trolley busway. The trolley specified for the industrial building is equipped with a cord clamp and grounding screw. The trolley has metal wheels and rolls freely along the busway from one position to another. Two heavy-duty, weight-supporting devices are used with the trolley when heavy luminaires are to be suspended

from it as is the case with the industrial building, Figure 5-14.

The twenty 50-ampere trolley busway branch lighting circuits in the manufacturing area run east and west from the twenty center feed-in boxes (see Sheet E-3 of the plans). These circuits form ten lines of trolley busway suspended lighting. Each circuit has nine trolleys, and an industrial-type, 96-in.-long (2.5-m-long) fluorescent luminaire is suspended from each trolley.

The luminaires used in the manufacturing area are industrial-type fluorescent luminaires. Each luminaire uses two F96T12/CW/VHO lamps rated at 215 watts each. However, the power losses in the ballasts increase this value to 450 volt-ampere per luminaire. The fluorescent ballasts used are all of the high power factor type with individual fusing provided in accordance with *368.17*.

The luminaires are also equipped with *lead-lag adders*, which cause the lamps to fire at different times. In other words, the current wave crest of one lamp occurs nearly one-half cycle before the second lamp receives the wave crest. This arrangement eliminates most of the stroboscopic effect that

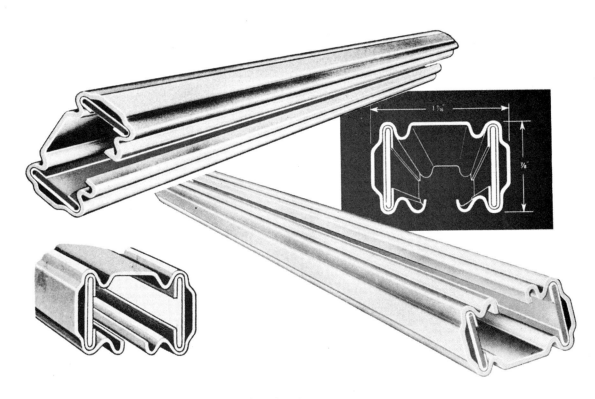

FIGURE 5-13   Busbar section. (*Delmar/Cengage Learning*)

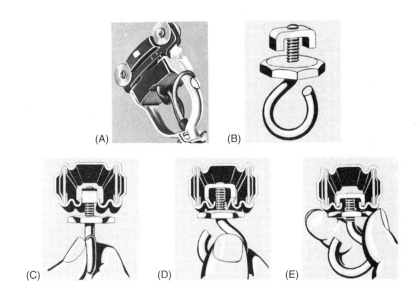

(A)   (B)

(C)   (D)   (E)

FIGURE 5-14   Heavy-duty weight supports. (*Delmar/Cengage Learning*)

occurs if individual lamps are used or if the lead-lag system is not used.

The luminaires are supported at the center by the trolley attachment and by two additional supports near the ends of each luminaire. Of the 180 lumi-naires installed in the industrial building, 45 are fed from each of the four panelboards.

In addition to the lighting, several receptacle outlets are also supplied with power from each of the panelboards. According to the plans and the Schedule of Receptacle Outlets included in the specifications, each of the four lighting panelboards in the manufacturing area of the plant contains seven 20-ampere circuit breakers feeding fifteen duplex grounding receptacles. Two or three receptacle outlets are assigned to each of these 120-volt circuits.

LIGHTING IN THE BOILER ROOM

Panelboard P-12 is located in the boiler room of the plant. This panelboard supplies the lighting and receptacle needs in this area. There are sixteen lighting outlets in the boiler room connected to four separate circuits so that there are four outlets on each circuit. The luminaires used in the boiler room are the same as those used in the manufacturing area. However, these sixteen luminaires are suspended by chains from fixed outlet boxes.

The Cord Drops (*NEC Article 400*)

The specifications call for the use of 4-conductor type SJ rubber cord to connect the various machines in the manufacturing area to the busway system. These cords are rated for heavy-duty usage. The colors of the individual conductors of the cord are black, white, red, and green.

The green conductor is reserved for equipment grounding. One end of this conductor is connected to the steel housing of the bus plug. The other end of the conductor is connected to the steel housing of the motor control equipment on the machine.

The drop from the overhead busway system to supply power to the machines located at various points on the floor is usually made by either of two methods.

One method involves the use of rigid or thin-wall conduit to extend from the bus plug to the machine that it will serve. The conduit may be run horizontally with or without bends to a point directly over the machine to be supplied and then dropped vertically. The resulting system is a rigid raceway assembly that must be supported by appropriate hangers. The ungrounded conductors are pulled into the conduit, which serves as the equipment ground.

One disadvantage of this method is its inflexibility when the layout of the machines being served must be rearranged. In such a case, the conduit assemblies must be taken down, the wire removed, and the conduit disassembled. Then, the entire run must be rebuilt to fit the new location using new wire and new conduit for part or all of the assembly.

The second method is to use rubber cord drops from the bus plug to the machine being served, Figure 5-15. This method is flexible in terms of making changes and thus is commonly used. The industrial building uses the rubber cord drop method.

Strain Reliefs

Strain relief grips are used in the cord drop method of supplying equipment to comply with *400.10*. The strain relief type of grip is designed for use at the terminals or ends of the rubber cord drop where it enters or leaves a knockout opening in the bus plug, in the disconnecting switch, or in a motor controller.

The bus drop grips are used at or near the ends of the rubber cord runs and also where the cord changes direction from the horizontal to the vertical, Figure 5-15 and Figure 5-16.

Figure 5-17 shows that the cord grips are constructed in a basketweave pattern. The grips are tubular in shape and are made from strands of galvanized plow steel wire. Grips are available in a variety of sizes to fit most cables or cords. When strain or tension is placed on the rubber cord so that it is pulled taut, the basket structure of the cord grip contracts to apply a stronger grip on the cord.

Bus drop safety springs are used to maintain the proper tension on the horizontal and vertical cord runs. These springs are available with 40-, 80-, and 150-pound (18.14-, 36.28-, and 68-kilogram) ratings. The selection of the proper spring depends upon the weight and length of the cord being supported, Figure 5-18.

Several different ways of using cord grips are shown in Figure 5-19.

FIGURE 5-15   Machines supplied by rubber cords from overhead busway. (*Delmar/Cengage Learning*)

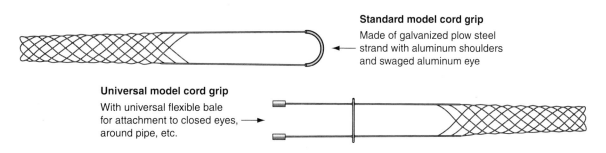

Standard model cord grip
Made of galvanized plow steel strand with aluminum shoulders and swaged aluminum eye

Universal model cord grip
With universal flexible bale for attachment to closed eyes, around pipe, etc.

FIGURE 5-16   Cord grip models. (*Delmar/Cengage Learning*)

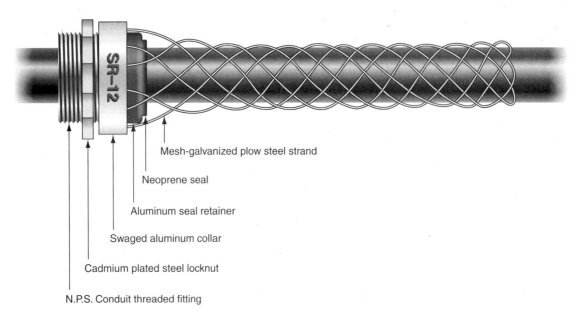

Mesh-galvanized plow steel strand

Neoprene seal

Aluminum seal retainer

Swaged aluminum collar

Cadmium plated steel locknut

N.P.S. Conduit threaded fitting

FIGURE 5-17 Bus drop and strain relief cord grip. (*Delmar/Cengage Learning*)

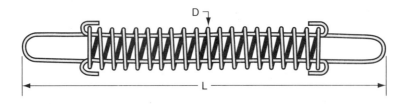

Catalog number	Maximum deflection	Breaking strength	No load length	Diameter
40 lb.	2.875 in. @ 45 lb. 73.0 mm @ 20.4 kg	500 lb. 227 kg	7.25 in. 184 mm	0.75 in. 19 mm
80 lb.	2.62 in. @ 110 lb. 66.5 mm @ 50 kg	850 lb. 385 kg	8.25 in. 209 mm	1 in. 25 mm
150 lb.	2.38 in. @175 lb. 60.5 mm @ 79 kg	850 lb. 385 kg	8.25 in. 209 mm	1.125 in. 28 mm

FIGURE 5-18 Bus drop safety spring. (*Delmar/Cengage Learning*)

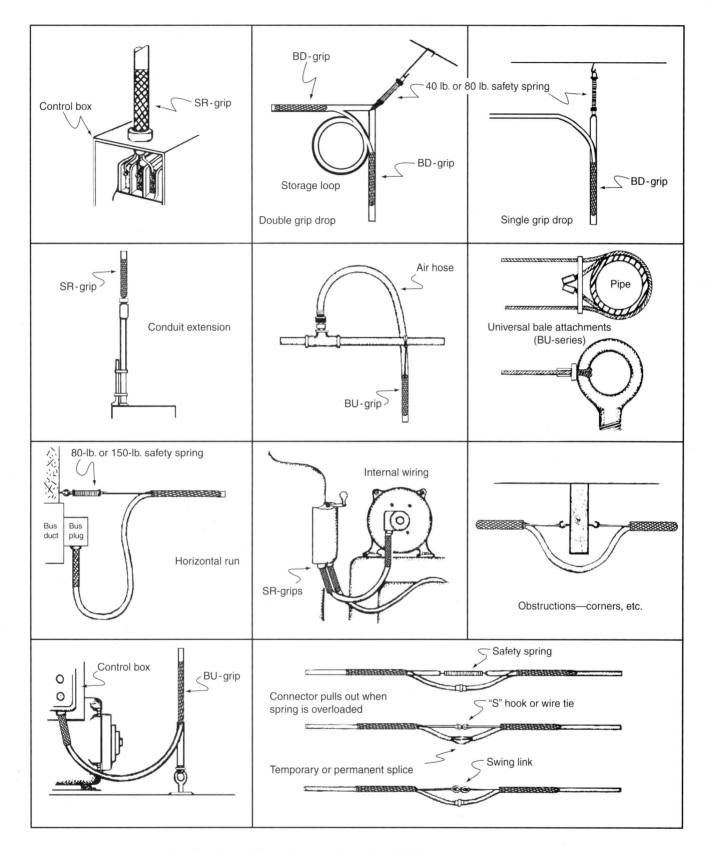

FIGURE 5-19   Applications of bus drop and strain relief cord grips. (*Delmar/Cengage Learning*)

All answers should be written in complete sentences, calculations should be shown in detail, and *Code* references should be cited when appropriate.

1. What advantages are there to having a trolley busway installed? _____

_____

_____

_____

2. Is the trolley busway considered to be a feeder or a branch circuit? _____

_____

_____

_____

3. What features does a trolley box–type hanger provide? _____

_____

_____

_____

4. Describe the operation of strain relief grips and identify their basic function. _____

_____

_____

_____

5. What are some of the advantages of installing the lighting on a busway? _____

_____

_____

_____

Using Wire Tables and Determining Conductor Sizes

OBJECTIVES

After studying this chapter, the student should be able to

- select a conductor from the proper wire table.
- discuss the different types of wire insulation.
- determine insulation characteristics.
- use correction and adjustment factors to determine the ampacity of conductors.
- determine the resistance of long lengths of conductors.
- determine the proper wire sizes for loads located long distances from the power source.
- list the requirements for using parallel conductors.
- discuss the use of a megohmmeter for testing insulation.

CONDUCTORS

- *NEC Article 310* addresses conductors for general wiring.
- *NEC Table 310.104(A)*, reproduced in this chapter as Table 6-1, lists the conductor application and gives specific information about each insulation type.

- *NEC Tables 310.15(B)(16)* through *310.15(B)(19)* are used to determine a conductor size according to the requirements of a circuit. *NEC Tables 310.15(B)(16)* and *310.15(B)(17)* are reproduced in this chapter as Tables 6-2 and 6-3, respectively.
- Table 6-2 (*NEC Table 310.15(B)(16)*) lists allowable ampacities for not more than three insulated, copper conductors in a raceway, based on an ambient air temperature of 86°F (30°C).

TABLE 6-1

Table 310.104(A) Conductor Applications and Insulations Rated 600 Volts

Trade Name	Type Letter	Maximum Operating Temperature	Application Provisions	Insulation	Thickness of Insulation					Outer Covering[1]
					AWG or kcmil	mm		mils		
Fluorinated ethylene propylene	FEP or FEPB	90°C 194°F	Dry and damp locations	Fluorinated ethylene propylene	14–10 8–2	0.51 0.76		20 30		None
		200°C 392°F	Dry locations — special applications[2]	Fluorinated ethylene propylene	14–8	0.36		14		Glass braid
					6–2	0.36		14		Glass or other suitable braid material
Mineral insulation (metal sheathed)	MI	90°C 194°F 250°C 482°F	Dry and wet locations For special applications[2]	Magnesium oxide	18–16[3] 16–10 9–4 3–500	0.58 0.91 1.27 1.40		23 36 50 55		Copper or alloy steel
Moisture-, heat-, and oil-resistant thermoplastic	MTW	60°C 140°F 90°C 194°F	Machine tool wiring in wet locations Machine tool wiring in dry locations. Informational Note: See NFPA 79.	Flame-retardant, moisture-, heat-, and oil-resistant thermoplastic	(A) (B) 22–12 10 8 6 4–2 1–4/0 213–500 501–1000	(A) 0.76 0.76 1.14 1.52 1.52 2.03 2.41 2.79	(B) 0.38 0.51 0.76 0.76 1.02 1.27 1.52 1.78	(A) 30 30 45 60 60 80 95 110	(B) 15 20 30 30 40 50 60 70	(A) None (B) Nylon jacket or equivalent
Paper		85°C 185°F	For underground service conductors, or by special permission	Paper						Lead sheath
Perfluoro-alkoxy	PFA	90°C 194°F 200°C 392°F	Dry and damp locations Dry locations — special applications[2]	Perfluoro-alkoxy	14–10 8–2 1–4/0	0.51 0.76 1.14		20 30 45		None
Perfluoro-alkoxy	PFAH	250°C 482°F	Dry locations only. Only for leads within apparatus or within raceways connected to apparatus (nickel or nickel-coated copper only)	Perfluoro-alkoxy	14–10 8–2 1–4/0	0.51 0.76 1.14		20 30 45		None
Thermoset	RHH	90°C 194°F	Dry and damp locations		14–10 8–2 1–4/0 213–500 501–1000 1001–2000	1.14 1.52 2.03 2.41 2.79 3.18		45 60 80 95 110 125		Moisture-resistant, flame-retardant, nonmetallic covering[1]
Moisture-resistant thermoset	RHW	75°C 167°F	Dry and wet locations	Flame-retardant, moisture-resistant thermoset	14–10 8–2 1–4/0 213–500 501–1000 1001–2000	1.14 1.52 2.03 2.41 2.79 3.18		45 60 80 95 110 125		Moisture-resistant, flame-retardant, nonmetallic covering
	RHW-2	90°C 194°F								
Silicone	SA	90°C 194°F 200°C 392°F	Dry and damp locations For special application[2]	Silicone rubber	14–10 8–2 1–4/0 213–500 501–1000 1001–2000	1.14 1.52 2.03 2.41 2.79 3.18		45 60 80 95 110 125		Glass or other suitable braid material

Reprinted with permission from NFPA 70-2011.

TABLE 6-1

Table 310.104(A)   *Continued*

Trade Name	Type Letter	Maximum Operating Temperature	Application Provisions	Insulation	Thickness of Insulation			Outer Covering[1]
					AWG or kcmil	mm	mils	
Thermoset	SIS	90°C 194°F	Switchboard wiring only	Flame-retardant thermoset	14–10 8–2 1–4/0	0.76 1.14 2.41	30 45 55	None
Thermoplastic and fibrous outer braid	TBS	90°C 194°F	Switchboard wiring only	Thermoplastic	14–10 8 6–2 1–4/0	0.76 1.14 1.52 2.03	30 45 60 80	Flame-retardant, nonmetallic covering
Extended polytetra-fluoro-ethylene	TFE	250°C 482°F	Dry locations only. Only for leads within apparatus or within raceways connected to apparatus, or as open wiring (nickel or nickel-coated copper only)	Extruded polytetra-fluoroethylene	14–10 8–2 1–4/0	0.51 0.76 1.14	20 30 45	None
Heat-resistant thermoplastic	THHN	90°C 194°F	Dry and damp locations	Flame-retardant, heat-resistant thermoplastic	14–12 10 8–6 4–2 1–4/0 250–500 501–1000	0.38 0.51 0.76 1.02 1.27 1.52 1.78	15 20 30 40 50 60 70	Nylon jacket or equivalent
Moisture- and heat-resistant thermoplastic	THHW	75°C 167°F 90°C 194°F	Wet location Dry location	Flame-retardant, moisture- and heat-resistant thermoplastic	14–10 8 6–2 1–4/0 213–500 501–1000 1001–2000	0.76 1.14 1.52 2.03 2.41 2.79 3.18	30 45 60 80 95 110 125	None
Moisture- and heat-resistant thermoplastic	THW	75°C 167°F 90°C 194°F	Dry and wet locations Special applications within electric discharge lighting equipment. Limited to 1000 open-circuit volts or less. (size 14-8 only as permitted in 410.68)	Flame-retardant, moisture- and heat-resistant thermoplastic	14–10 8 6–2 1–4/0 213–500 501–1000 1001–2000	0.76 1.14 1.52 2.03 2.41 2.79 3.18	30 45 60 80 95 110 125	None
	THW-2	90°C 194°F	Dry and wet locations					
Moisture- and heat-resistant thermoplastic	THWN	75°C 167°F	Dry and wet locations	Flame-retardant, moisture- and heat-resistant thermoplastic	14–12 10 8–6 4–2 1–4/0 250–500 501–1000	0.38 0.51 0.76 1.02 1.27 1.52 1.78	15 20 30 40 50 60 70	Nylon jacket or equivalent
	THWN-2	90°C 194°F						
Moisture-resistant thermoplastic	TW	60°C 140°F	Dry and wet locations	Flame-retardant, moisture-resistant thermoplastic	14–10 8 6–2 1–4/0 213–500 501–1000 1001–2000	0.76 1.14 1.52 2.03 2.41 2.79 3.18	30 45 60 80 95 110 125	None
Underground feeder and branch-circuit cable — single conductor (for Type UF cable employing more than one conductor, see Article 340.)	UF	60°C 140°F 75°C 167°F[5]	See Article 340.	Moisture-resistant Moisture- and heat-resistant	14–10 8–2 1–4/0	1.52 2.03 2.41	60[4] 80[4] 95[4]	Integral with insulation

Reprinted with permission from NFPA 70-2011.

(*continues*)

TABLE 6-1

Table 310.104(A)   *Continued*

Trade Name	Type Letter	Maximum Operating Temperature	Application Provisions	Insulation	Thickness of Insulation			Outer Covering[1]
					AWG or kcmil	mm	mils	
Underground service-entrance cable — single conductor (for Type USE cable employing more than one conductor, see Article 338.)	USE	75°C 167°F[5]	See Article 338.	Heat- and moisture-resistant	14–10 8–2 1–4/0 213–500 501–1000 1001–2000	1.14 1.52 2.03 2.41 2.79 3.18	45 60 80 95[6] 110 125	Moisture-resistant nonmetallic covering (See 338.2.)
	USE-2	90°C 194°F	Dry and wet locations					
Thermoset	XHH	90°C 194°F	Dry and damp locations	Flame-retardant thermoset	14–10 8–2 1–4/0 213–500 501–1000 1001–2000	0.76 1.14 1.40 1.65 2.03 2.41	30 45 55 65 80 95	None
Moisture-resistant thermoset	XHHW	90°C 194°F 75°C 167°F	Dry and damp locations Wet locations	Flame-retardant, moisture-resistant thermoset	14–10 8–2 1–4/0 213–500 501–1000 1001–2000	0.76 1.14 1.40 1.65 2.03 2.41	30 45 55 65 80 95	None
Moisture-resistant thermoset	XHHW-2	90°C 194°F	Dry and wet locations	Flame-retardant, moisture-resistant thermoset	14–10 8–2 1–4/0 213–500 501–1000 1001–2000	0.76 1.14 1.40 1.65 2.03 2.41	30 45 55 65 80 95	None
Modified ethylene tetrafluoro-ethylene	Z	90°C 194°F 150°C 302°F	Dry and damp locations Dry locations — special applications[2]	Modified ethylene tetrafluoro-ethylene	14–12 10 8–4 3–1 1/0–4/0	0.38 0.51 0.64 0.89 1.14	15 20 25 35 45	None
Modified ethylene tetrafluoro-ethylene	ZW	75°C 167°F 90°C 194°F 150°C 302°F	Wet locations Dry and damp locations Dry locations — special applications[2]	Modified ethylene tetrafluoro-ethylene	14–10 8–2	0.76 1.14	30 45	None
	ZW-2	90°C 194°F	Dry and wet locations					

[1] Some insulations do not require an outer covering.
[2] Where design conditions require maximum conductor operating temperatures above 90°C (194°F).
[3] For signaling circuits permitting 300-volt insulation.
•
[4] Includes integral jacket.
[5] For ampacity limitation, see 340.80.
[6] Insulation thickness shall be permitted to be 2.03 mm (80 mils) for listed Type USE conductors that have been subjected to special investigations. The nonmetallic covering over individual rubber-covered conductors of aluminum-sheathed cable and of lead-sheathed or multiconductor cable shall not be required to be flame retardant. For Type MC cable, see 330.104. For nonmetallic-sheathed cable, see Article 334, Part III. For Type UF cable, see Article 340, Part III.

Reprinted with permission from NFPA 70-2011.

TABLE 6-2

Table 310.15(B)(16) (formerly Table 310.16) Allowable Ampacities of Insulated Conductors Rated Up to and Including 2000 Volts, 60°C Through 90°C (140°F Through 194°F), Not More Than Three Current-Carrying Conductors in Raceway, Cable, or Earth (Directly Buried), Based on Ambient Temperature of 30°C (86°F)*

Size AWG or kcmil	Temperature Rating of Conductor [See Table 310.104(A).]						Size AWG or kcmil
	60°C (140°F)	75°C (167°F)	90°C (194°F)	60°C (140°F)	75°C (167°F)	90°C (194°F)	
	Types TW, UF	Types RHW, THHW, THW, THWN, XHHW, USE, ZW	Types TBS, SA, SIS, FEP, FEPB, MI, RHH, RHW-2, THHN, THHW, THW-2, THWN-2, USE-2, XHH, XHHW, XHHW-2, ZW-2	Types TW, UF	Types RHW, THHW, THW, THWN, XHHW, USE	Types TBS, SA, SIS, THHN, THHW, THW-2, THWN-2, RHH, RHW-2, USE-2, XHH, XHHW, XHHW-2, ZW-2	
	COPPER			ALUMINUM OR COPPER-CLAD ALUMINUM			
18	—	—	14	—	—	—	—
16	—	—	18	—	—	—	—
14**	15	20	25	—	—	—	—
12**	20	25	30	15	20	25	12**
10**	30	35	40	25	30	35	10**
8	40	50	55	35	40	45	8
6	55	65	75	40	50	55	6
4	70	85	95	55	65	75	4
3	85	100	115	65	75	85	3
2	95	115	130	75	90	100	2
1	110	130	145	85	100	115	1
1/0	125	150	170	100	120	135	1/0
2/0	145	175	195	115	135	150	2/0
3/0	165	200	225	130	155	175	3/0
4/0	195	230	260	150	180	205	4/0
250	215	255	290	170	205	230	250
300	240	285	320	195	230	260	300
350	260	310	350	210	250	280	350
400	280	335	380	225	270	305	400
500	320	380	430	260	310	350	500
600	350	420	475	285	340	385	600
700	385	460	520	315	375	425	700
750	400	475	535	320	385	435	750
800	410	490	555	330	395	445	800
900	435	520	585	355	425	480	900
1000	455	545	615	375	445	500	1000
1250	495	590	665	405	485	545	1250
1500	525	625	705	435	520	585	1500
1750	545	650	735	455	545	615	1750
2000	555	665	750	470	560	630	2000

*Refer to 310.15(B)(2) for the ampacity correction factors where the ambient temperature is other than 30°C (86°F).
**Refer to 240.4(D) for conductor overcurrent protection limitations.

- If the ambient temperature is above 86°F (30°C), a correction factor must be applied. These factors are given in table 310.15(B)(2)(a) reproduced in this chapter as Table 6-4.

- If there are four or more conductors in the raceway, an adjustment factor shall be applied. These factors are given in *NEC Table 310.15(B) (3)(a)*, reproduced in this chapter as Table 6-5.

Table 310.15(B)(17) (formerly Table 310.17) Allowable Ampacities of Single-Insulated Conductors Rated Up to and Including 2000 Volts in Free Air, Based on Ambient Temperature of 30°C (86°F)*

Size AWG or kcmil	Temperature Rating of Conductor [See Table 310.104(A).]						Size AWG or kcmil
	60°C (140°F)	75°C (167°F)	90°C (194°F)	60°C (140°F)	75°C (167°F)	90°C (194°F)	
	Types TW, UF	Types RHW, THHW, THW, THWN, XHHW, ZW	Types TBS, SA, SIS, FEP, FEPB, MI, RHH, RHW-2, THHN, THHW, THW-2, THWN-2, USE-2, XHH, XHHW, XHHW-2, ZW-2	Types TW, UF	Types RHW, THHW, THW, THWN, XHHW	Types TBS, SA, SIS, THHN, THHW, THW-2, THWN-2, RHH, RHW-2, USE-2, XHH, XHHW, XHHW-2, ZW-2	
	COPPER			ALUMINUM OR COPPER-CLAD ALUMINUM			
18	—	—	18	—	—	—	—
16	—	—	24	—	—	—	—
14**	25	30	35	—	—	—	—
12**	30	35	40	25	30	35	12**
10**	40	50	55	35	40	45	10**
8	60	70	80	45	55	60	8
6	80	95	105	60	75	85	6
4	105	125	140	80	100	115	4
3	120	145	165	95	115	130	3
2	140	170	190	110	135	150	2
1	165	195	220	130	155	175	1
1/0	195	230	260	150	180	205	1/0
2/0	225	265	300	175	210	235	2/0
3/0	260	310	350	200	240	270	3/0
4/0	300	360	405	235	280	315	4/0
250	340	405	455	265	315	355	250
300	375	445	500	290	350	395	300
350	420	505	570	330	395	445	350
400	455	545	615	355	425	480	400
500	515	620	700	405	485	545	500
600	575	690	780	455	545	615	600
700	630	755	850	500	595	670	700
750	655	785	885	515	620	700	750
800	680	815	920	535	645	725	800
900	730	870	980	580	700	790	900
1000	780	935	1055	625	750	845	1000
1250	890	1065	1200	710	855	965	1250
1500	980	1175	1325	795	950	1070	1500
1750	1070	1280	1445	875	1050	1185	1750
2000	1155	1385	1560	960	1150	1295	2000

*Refer to 310.15(B)(2) for the ampacity correction factors where the ambient temperature is other than 30°C (86°F).
**Refer to 240.4(D) for conductor overcurrent protection limitations.

Reprinted with permission from NFPA 70-2011.

INSULATION TYPE

A factor that determines the amount of current a conductor is permitted to carry is the type of insulation used. The insulation is the nonconductive covering around the wire, as shown in Figure 6-1. Different types of insulation can withstand more heat than other types. The voltage rating of the conductor is also determined by the type of insulation. The amount of voltage a particular type of insulation can withstand without breaking down is determined by the type of material from which it is made and its thickness. Table 6-1 [*NEC Table 310.104(A)*] lists different types of insulation and certain specifications about each one.

The table is divided into seven main columns. The first column lists the trade name of the insulation;

TABLE 6-4

Table 310.15(B)(2)(a) Ambient Temperature Correction Factors Based on 30°C (86°F)

For ambient temperatures other than 30°C (86°F), multiply the allowable ampacities specified in the ampacity tables by the appropriate correction factor shown below.

Ambient Temperature (°C)	Temperature Rating of Conductor			Ambient Temperature (°F)
	60°C	75°C	90°C	
10 or less	1.29	1.20	1.15	50 or less
11–15	1.22	1.15	1.12	51–59
16–20	1.15	1.11	1.08	60–68
21–25	1.08	1.05	1.04	69–77
26–30	1.00	1.00	1.00	78–86
31–35	0.91	0.94	0.96	87–95
36–40	0.82	0.88	0.91	96–104
41–45	0.71	0.82	0.87	105–113
46–50	0.58	0.75	0.82	114–122
51–55	0.41	0.67	0.76	123–131
56–60	—	0.58	0.71	132–140
61–65	—	0.47	0.65	141–149
66–70	—	0.33	0.58	150–158
71–75	—	—	0.50	159–167
76–80	—	—	0.41	168–176
81–85	—	—	0.29	177–185

Reprinted with permission from NFPA 70-2011.

TABLE 6-5

Table 310.15(B)(3)(a) Adjustment Factors for More Than Three Current-Carrying Conductors in a Raceway or Cable

Number of Conductors[1]	Percent of Values in Table 310.15(B)(16) through Table 310.15(B)(19) as Adjusted for Ambient Temperature if Necessary
4–6	80
7–9	70
10–20	50
21–30	45
31–40	40
41 and above	35

[1]Number of conductors is the total number of conductors in the raceway or cable adjusted in accordance with 310.15(B)(5) and (6).

Reprinted with permission from NFPA 70-2011.

the second lists its identification code letter; the third column lists its maximum operating temperature; and the fourth shows its applications and where it is permitted to be used. The fifth column lists the material from which the insulation is made; the sixth states its thickness; and the last column lists the type of outer covering over the insulation.

PROBLEM 1: Find the maximum operating temperature of type TW insulation. Refer to Table 6-1.

Solution: The eighteenth data row of Table 6-1 gives the specifications for type TW conductors.

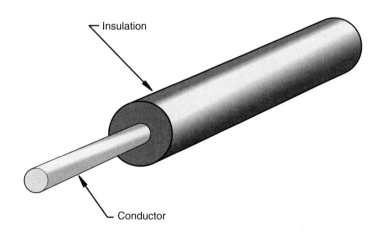

FIGURE 6-1   An insulated conductor. (*Delmar/Cengage Learning*)

The third column gives the maximum operating temperature as 60°C, or the equivalent 140°F.

PROBLEM 2: Can type THHN insulation be used in wet locations?

Solution: Locate type THHN insulation in Table 6-1. The fourth column indicates that this insulation can be used in dry and damp locations. This type of insulation cannot be used in wet locations. For an explanation of the difference between damp and wet locations, consult "locations" in *NEC Article 100*.

Conductor Metals

Another factor that determines the allowable ampacity of the conductor is the type of metal used for the wire. Table 6-2 (*NEC Table 310.15(B)(16)*) lists the current-carrying capacity of both copper and aluminum or copper-clad aluminum conductors. A study of the table reveals that a copper conductor is permitted to carry more current than an aluminum conductor of the same size and insulation type. An 8 AWG copper conductor with type TW insulation has an allowable ampacity of 40 amperes. An 8 AWG aluminum conductor with type TW insulation has an allowable ampacity of 30 amperes.

⎍⎍⎍ CORRECTION FACTORS

One of the main conditions that determines the current a conductor is permitted to carry is the ambient, or surrounding, air temperature. Table 6-2 lists the allowable ampacity of not more than three conductors in a raceway. These allowable ampacities are based on an ambient air temperature of 86°F or 30°C. If these conductors are to be used in a location with a higher ambient temperature, the ampacity of the conductor must be reduced.

The correction factor chart located in *Table 310.15(B)(2)(a)*, Table 6-4, provides the necessary factors for ambient temperatures from 50°F to 185°F (10°C to 85°C). This table is divided into the same number of columns as in the wire table. The

> ⎍⎍⎍ **NOTE:** After reduction, the current-carrying capacity of a conductor is referred to as the ampacity, not the allowable ampacity. ⬤

correction factors in each column are used for the conductors listed in the same column of the wire table.

PROBLEM 1: What is the ampacity of a 4 AWG copper conductor with type THWN insulation that will be used in an area with an ambient temperature of 43°C?

Solution: Determine the allowable ampacity of a 4 AWG copper conductor with type THWN insulation. Type THWN insulation is located in the second data column of Table 6-2. The table lists an allowable ampacity of 85 amperes. Refer to the Correction Factors shown in Table 6-4. In the left-hand column, select a temperature range that includes 43°C. The table lists a correction factor of 0.82. The ampacity is to be multiplied by the correction factor.

$$85 \times 0.82 = 69.7 \text{ amperes}$$

PROBLEM 2: What is the ampacity of a 1/0 AWG copper-clad aluminum conductor with type RHH insulation when the conductor is installed on insulators in free air, in an area with an ambient air temperature of 100°F?

Solution: In Table 6-3, locate the column that contains type RHH copper-clad aluminum. The table indicates an allowable ampacity of 205 amperes. Determine the correction factor from Table 6-4. Fahrenheit degrees are located in the far right hand column of *Table 310.15(B)(2)(a)*. The 100°F temperature is between 97°F and 104°F. The correction factor for this temperature is 0.91. Multiply the ampacity of the conductor by this factor.

$$205 \times 0.91 = 187 \text{ amperes}$$

⎍⎍⎍ MORE THAN THREE CONDUCTORS IN RACEWAY

Table 6-2 (*NEC Table 310.15.(B)(16)*) and *NEC Table 310.15(B)(18)* list allowable ampacities for three conductors in a raceway. If a raceway is to contain more than three conductors, the allowable ampacity of the conductors must be derated. This is because the heat from each conductor combines with the heat dissipated by the other conductors to produce a higher temperature inside the raceway. Table 6-5 [*NEC Table 310.15(B)(3)(a)*] lists the adjustment factors. If the raceway is used in a space with a greater ambient temperature than that listed in the appropriate wire table, the temperature correction formula also shall be applied.

- When conductors are of different systems, or if installing the conductors in a cable tray, *310.15(B)(3)* should be reviewed.

- Adjustment is not required for raceways size 24 (600 mm) or less in length.

PROBLEM: Twelve 14 AWG copper conductors with type RHW insulation are to be installed in conduit in an area with an ambient temperature of 110°F. What will be the ampacity of these conductors?

Solution: First, determine the allowable ampacity of a 14 AWG copper conductor with type RHW insulation. Type RHW insulation is located in the second data column of Table 6-2. A 14 AWG copper conductor has an allowable ampacity of 20 amperes. The next step is to use the correction factor for ambient temperature. A correction factor of 0.82 is appropriate.

$$20 \times 0.82 = 16.4 \text{ amperes}$$

Next, an adjustment factor located in Table 6-5 shall be applied. The table indicates a factor of 50 percent when 10 through 20 conductors are installed in a raceway.

$$16.4 \times 0.50 = 8.2 \text{ amperes}$$

A 14 AWG, Type RHW conductor when installed in a raceway, with a group of twelve conductors, in a 110°F ambient has an ampacity of 8.2 amperes.

UNDERGROUND CONDUCTORS

NEC Tables 310.60(C)(81) through *310.60(C)(84)* list ampacities and temperature correction factors for conductors intended for direct burial. *NEC Tables 310.60(C)(77)*, *310.60(C)(78)*, and *310.60(C)(79)* list conductors that are to be buried in electrical duct banks. An electrical duct can be a single metal or nonmetallic conduit. An electrical duct bank is a group of electrical ducts buried together, as shown in *NEC Figure 310.60*. When a duct bank is used, the center point of individual ducts should be separated by a distance of no less than 7.5 in. (190 mm).

CALCULATING CONDUCTOR SIZES AND RESISTANCE

Although the wire tables in the *NEC* are used to determine the proper wire size for most installations, there are instances in which these tables are not used. One example of this is the formula shown in *310.60(D)*. This formula is to be used for ampacities not listed in the wire tables. The following formula may be used under engineering supervision.

$$I = \sqrt{\frac{TC - (TA + \Delta TD)}{RDC(1 + YC)RCA}}$$

where
- I = current
- TC = conductor temperature in °C
- TA = ambient temperature in °C
- ΔTD = dielectric loss temperature rise
- RDC = direct-current (dc) resistance of conductor at temperature TC
- YC = component alternating-current (ac) resistance resulting from skin effect and proximity effect
- RCA = effective thermal resistance between conductor and surrounding conditions

LONG WIRE LENGTHS

It also becomes necessary to calculate wire sizes instead of using the tables in the *Code* when the length of the conductor is excessively long. The listed ampacities in the *Code* tables assume that the length of the conductor will not significantly increase the resistance of the circuit. When the wire length becomes extremely long, however, it is necessary to calculate the size of wire needed.

All wire contains resistance. As the length of wire is increased, it has the effect of adding resistance in series with the load. There are four factors that determine the resistance of a length of wire:

1. The material from which the wire is made. Different types of material have different wire resistances. A copper conductor will have less resistance than an aluminum conductor of the same size and length. An aluminum conductor

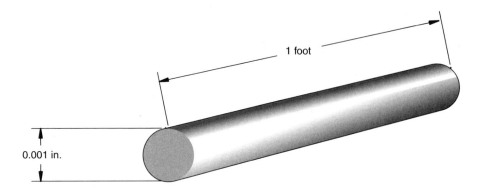

FIGURE 6-2 Mil foot. (*Delmar/Cengage Learning*)

will have less resistance than a piece of iron wire the same size and length.

2. The diameter of the conductor. The larger the diameter, the less resistance it will have. The diameter of a wire is measured in mils. One mil equals 0.001 inch. The circular mil area of a wire is the diameter of the wire in mils squared.

EXAMPLE

Assume a wire has a diameter of 0.064 inch. Converting to mils:

0.064 in. × 1000 mil per in. = 64 mils

The area in circular mils is

$64^2 = (64 \times 64) = 4096$ cmil

3. The length of the conductor. The longer the conductor, the more resistance it will have. Adding length to a conductor has the same effect as connecting resistors in series.

4. The temperature of the conductor. As a general rule, most conductive materials will increase their resistance with an increase of temperature. Some exceptions to this rule are carbon, silicon, and germanium. If the coefficient of temperature for a particular material is known, its resistance at different temperatures can be calculated. Materials that increase their resistance with an increase of temperature have a *positive* coefficient of temperature. Materials

that decrease their resistance with an increase of temperature have a *negative* coefficient of temperature.

In the customary system of measurement, a standard value of resistance is the mil foot. It is used to determine the resistance of different lengths and sizes of wire. A mil foot is a piece of wire 1 foot long and 1 mil in diameter, Figure 6-2. The resistances of a mil foot of wire at 20°C for different materials are shown in Table 6-6. Notice the wide range of resistances for different materials. The temperature coefficient of the different types of conductors

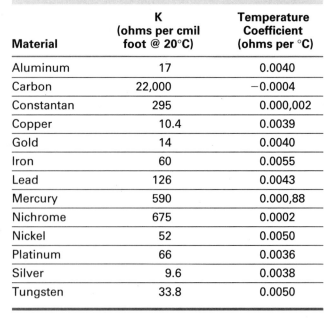

TABLE 6-6

Resistivity of materials.

Material	K (ohms per cmil foot @ 20°C)	Temperature Coefficient (ohms per °C)
Aluminum	17	0.0040
Carbon	22,000	−0.0004
Constantan	295	0.000,002
Copper	10.4	0.0039
Gold	14	0.0040
Iron	60	0.0055
Lead	126	0.0043
Mercury	590	0.000,88
Nichrome	675	0.0002
Nickel	52	0.0050
Platinum	66	0.0036
Silver	9.6	0.0038
Tungsten	33.8	0.0050

is also listed. The temperature of a conductor can greatly affect its resistance. Table 6-6 lists the ohms-per-mil foot at 20°C. The resistance of material is generally given at 20°C because it is the standard used in the *American Engineers Handbook*. The temperature coefficient of the material can be used to determine the resistance of a material at different temperatures.

PROBLEM: What is the ohms-per-mil foot at 75°C?

Solution: Use the formula:

$$R = R_{ref} [1 + \alpha(T - T_{ref})]$$

where
- R = conductor resistance at temperature "T"
- R_{ref} = Conductor resistance at reference temperature (20°C in this example)
- α = Coefficient of resistance for the conductor material
- T = Conductor temperature in °C
- T_{ref} = Reference temperature at which α is specified for the conductor material

$$R = 10.4[1 + 0.0039(75 - 20)]$$
$$R = 10.4[1 + 0.0039(55)]$$
$$R = 10.4[1 + 0.2145]$$
$$R = 10.4[1.2145]$$
$$R = 12.63$$

At a temperature of 75°C, copper would have a resistance of 12.63 ohms-per-mil foot.

CALCULATING RESISTANCE

Now that a standard measure of resistance for different types of materials is known, the resistance of different lengths and sizes of these materials can be calculated. The formula for calculating resistance of a certain length, size, and type of wire is

$$R = \frac{K \times L}{cmil}$$

where
- R = resistance of the wire
- K = ohms per mil foot
- L = length of wire in feet
- $cmil$ = circular mil area of the wire

This formula can be converted to calculate other values in the formula such as size, length, and area of wire used.

To find the size of wire, use

$$cmil = \frac{K \times L}{R}$$

To find the length of wire, use

$$L = \frac{K \times cmil}{K}$$

To find the type of wire, use

$$K = \frac{R \times cmil}{L}$$

PROBLEM 1: Find the resistance of 6 AWG copper wire 550 feet long. Assume a temperature of 20°C. The formula to be used is

$$R \text{ (ohms)} = \frac{K \text{ (ohms per mil ft)} \times L \text{ (ft)}}{cmil}$$

Solution: The value for K can be found in Table 6-6, where the resistance and temperature coefficient of several types of materials are listed. The table indicates a value of 10.4 ohms per cmil foot for a copper conductor. The length (L) was given at 550 feet, and the area of 6 AWG wire is listed at 26,240 cmil as shown in Table 6-7.

$$R = \frac{10.4 \times 550}{26,240} = 0.218 \text{ ohm}$$

PROBLEM 2: An aluminum wire 2250 feet long cannot have a resistance greater than 0.2 ohm. What is the minimum size wire that may be used?

Solution: To find the size of wire, use

$$cmil = \frac{K \text{ (ohms per mil ft)} \times L \text{ (ft)}}{R \text{ (ohms)}} = \frac{17 \times 2250}{0.2}$$
$$= 191,250$$

The standard size conductor for this installation can be found in Table 6-7. Because the resistance cannot be greater than 0.2 ohm, the conductor cannot be smaller than 191,250 circular mils. The smallest acceptable standard conductor size is 4/0 AWG.

Good examples of when it becomes necessary to calculate the wire size for a particular installation can be seen in the following problems.

TABLE 6-7

Table 8 Conductor Properties

Size (AWG or kcmil)	Area mm²	Area Circular mils	Stranding Quantity	Stranding Diameter mm	Stranding Diameter in.	Overall Diameter mm	Overall Diameter in.	Overall Area mm²	Overall Area in.²	Copper Uncoated ohm/km	Copper Uncoated ohm/kFT	Copper Coated ohm/km	Copper Coated ohm/kFT	Aluminum ohm/km	Aluminum ohm/kFT
18	0.823	1620	1	—	—	1.02	0.040	0.823	0.001	25.5	7.77	26.5	8.08	42.0	12.8
18	0.823	1620	7	0.39	0.015	1.16	0.046	1.06	0.002	26.1	7.95	27.7	8.45	42.8	13.1
16	1.31	2580	1	—	—	1.29	0.051	1.31	0.002	16.0	4.89	16.7	5.08	26.4	8.05
16	1.31	2580	7	0.49	0.019	1.46	0.058	1.68	0.003	16.4	4.99	17.3	5.29	26.9	8.21
14	2.08	4110	1	—	—	1.63	0.064	2.08	0.003	10.1	3.07	10.4	3.19	16.6	5.06
14	2.08	4110	7	0.62	0.024	1.85	0.073	2.68	0.004	10.3	3.14	10.7	3.26	16.9	5.17
12	3.31	6530	1	—	—	2.05	0.081	3.31	0.005	6.34	1.93	6.57	2.01	10.45	3.18
12	3.31	6530	7	0.78	0.030	2.32	0.092	4.25	0.006	6.50	1.98	6.73	2.05	10.69	3.25
10	5.261	10380	1	—	—	2.588	0.102	5.26	0.008	3.984	1.21	4.148	1.26	6.561	2.00
10	5.261	10380	7	0.98	0.038	2.95	0.116	6.76	0.011	4.070	1.24	4.226	1.29	6.679	2.04
8	8.367	16510	1	—	—	3.264	0.128	8.37	0.013	2.506	0.764	2.579	0.786	4.125	1.26
8	8.367	16510	7	1.23	0.049	3.71	0.146	10.76	0.017	2.551	0.778	2.653	0.809	4.204	1.28
6	13.30	26240	7	1.56	0.061	4.67	0.184	17.09	0.027	1.608	0.491	1.671	0.510	2.652	0.808
4	21.15	41740	7	1.96	0.077	5.89	0.232	27.19	0.042	1.010	0.308	1.053	0.321	1.666	0.508
3	26.67	52620	7	2.20	0.087	6.60	0.260	34.28	0.053	0.802	0.245	0.833	0.254	1.320	0.403
2	33.62	66360	7	2.47	0.097	7.42	0.292	43.23	0.067	0.634	0.194	0.661	0.201	1.045	0.319
1	42.41	83690	19	1.69	0.066	8.43	0.332	55.80	0.087	0.505	0.154	0.524	0.160	0.829	0.253
1/0	53.49	105600	19	1.89	0.074	9.45	0.372	70.41	0.109	0.399	0.122	0.415	0.127	0.660	0.201
2/0	67.43	133100	19	2.13	0.084	10.62	0.418	88.74	0.137	0.3170	0.0967	0.329	0.101	0.523	0.159
3/0	85.01	167800	19	2.39	0.094	11.94	0.470	111.9	0.173	0.2512	0.0766	0.2610	0.0797	0.413	0.126
4/0	107.2	211600	19	2.68	0.106	13.41	0.528	141.1	0.219	0.1996	0.0608	0.2050	0.0626	0.328	0.100
250	127	—	37	2.09	0.082	14.61	0.575	168	0.260	0.1687	0.0515	0.1753	0.0535	0.2778	0.0847
300	152	—	37	2.29	0.090	16.00	0.630	201	0.312	0.1409	0.0429	0.1463	0.0446	0.2318	0.0707
350	177	—	37	2.47	0.097	17.30	0.681	235	0.364	0.1205	0.0367	0.1252	0.0382	0.1984	0.0605
400	203	—	37	2.64	0.104	18.49	0.728	268	0.416	0.1053	0.0321	0.1084	0.0331	0.1737	0.0529
500	253	—	37	2.95	0.116	20.65	0.813	336	0.519	0.0845	0.0258	0.0869	0.0265	0.1391	0.0424
600	304	—	61	2.52	0.099	22.68	0.893	404	0.626	0.0704	0.0214	0.0732	0.0223	0.1159	0.0353
700	355	—	61	2.72	0.107	24.49	0.964	471	0.730	0.0603	0.0184	0.0622	0.0189	0.0994	0.0303
750	380	—	61	2.82	0.111	25.35	0.998	505	0.782	0.0563	0.0171	0.0579	0.0176	0.0927	0.0282
800	405	—	61	2.91	0.114	26.16	1.030	538	0.834	0.0528	0.0161	0.0544	0.0166	0.0868	0.0265
900	456	—	61	3.09	0.122	27.79	1.094	606	0.940	0.0470	0.0143	0.0481	0.0147	0.0770	0.0235
1000	507	—	61	3.25	0.128	29.26	1.152	673	1.042	0.0423	0.0129	0.0434	0.0132	0.0695	0.0212
1250	633	—	91	2.98	0.117	32.74	1.289	842	1.305	0.0338	0.0103	0.0347	0.0106	0.0554	0.0169
1500	760	—	91	3.26	0.128	35.86	1.412	1011	1.566	0.02814	0.00858	0.02814	0.00883	0.0464	0.0141
1750	887	—	127	2.98	0.117	38.76	1.526	1180	1.829	0.02410	0.00735	0.02410	0.00756	0.0397	0.0121
2000	1013	—	127	3.19	0.126	41.45	1.632	1349	2.092	0.02109	0.00643	0.02109	0.00662	0.0348	0.0106

Notes:

1. These resistance values are valid **only** for the parameters as given. Using conductors having coated strands, different stranding type, and, especially, other temperatures changes the resistance.

2. Equation for temperature change: $R_2 = R_1 [1 + \alpha (T_2 - 75)]$ where $\alpha_{cu} = 0.00323$, $\alpha_{AL} = 0.00330$ at 75°C.

3. Conductors with compact and compressed stranding have about 9 percent and 3 percent, respectively, smaller bare conductor diameters than those shown. See Table 5A for actual compact cable dimensions.

4. The IACS conductivities used: bare copper = 100%, aluminum = 61%.

5. Class B stranding is listed as well as solid for some sizes. Its overall diameter and area is that of its circumscribing circle.

Reprinted with permission from NFPA 70-2011.

PROBLEM 3: A workshop is to be installed in a facility separate from the main building. The workshop is to contain a small arc welder, air compressor, various power tools, lights, and receptacles. It is determined that a 100-ampere, 120/240-volt, single-phase panelboard will be needed for this installation. The distance between the buildings is 206 ft (62.79 m). An extra 10 ft (3.05 m) of cable is to be added for connections, making a total length of 216 ft (65.84 m). The maximum current will be 100 amperes. The voltage drop, at full load, is to be kept to a maximum of 3 percent, as recommended by *210.19(A), Fine Print Note (FPN) 4.* An ambient temperature of 20°C is assumed. What size copper conductors should be used for this installation?

Solution: The first step is to determine the maximum amount of resistance the conductors can have without producing a voltage drop greater than 3 percent of the applied voltage.

The maximum voltage drop can be determined by multiplying the applied voltage by the decimal equivalent of 3 percent.

$$240 \times 0.03 = 7.2 \text{ volts}$$

Ohm's law can now be used to determine the resistance that will permit a voltage drop of 7.2 volts at 100 amperes.

$$R = \frac{E}{I} = \frac{7.2 \text{ volts}}{100 \text{ ampere}} = 0.072 \text{ ohm}$$

The length of cable between the main building and the workshop is 216 ft (66 m). Because current exists in two conductors at the same time, it is the same as having the conductors connected in series, which effectively doubles the length of the conductor. Therefore, the conductor length will be 432 ft (132 m).

$$\text{cmil} = \frac{K \text{ (ohms per mil ft)} \times L \text{ (ft)}}{R \text{ (ohms)}}$$

$$= \frac{10.4 \times 432}{0.072} = 62,400$$

A 2 AWG copper conductor may be used.

PROBLEM 4: This problem concerns conductors used in a 3-phase system. It is to be assumed that a motor is located 2500 ft (762 m) from its power source and operates on 560 volts. When the motor starts, the current will be 168 amperes. The voltage drop at the motor terminals shall not be greater than 5 percent of the source voltage during starting. What size aluminum conductors should be used for this installation?

Solution: First, find the maximum voltage drop that can be permitted at the load by multiplying the source voltage by 5 percent.

$$E = 560 \times 0.05 = 28 \text{ volts}$$

The second step is to determine the maximum amount of resistance of the conductors. To calculate this value, the maximum voltage drop will be divided by the starting current of the motor.

$$R = \frac{E}{I} = \frac{28}{168} = 0.166 \text{ ohm}$$

The third step is to calculate the length of the conductors. In the previous example, the length of the two conductors was added together to find the total amount of wire resistance. In a single-phase system, each conductor must carry the same amount of current. During any period of time, one conductor is supplying current from the source to the load, and the other conductor completes the circuit by permitting the same amount of current to flow from the load to the source.

In a balanced 3-phase circuit, there are three currents that are 120° out of phase with each other, Figure 6-3. These three conductors share the flow of current between source and load. In Figure 6-3, two lines labeled A and B have been drawn through the three current waveforms. Notice that at position A, the current in line 1 is maximum and in a positive direction. The currents in lines 2 and 3 are less than maximum and in a negative direction. This condition corresponds to the example shown in Figure 6-4. Notice that the maximum current exists in only one conductor.

Observe the line marking position B in Figure 6-3. The current in line 1 is zero, and the currents in lines 2 and 3 are in opposite directions and less than maximum. This condition is illustrated in Figure 6-5. Notice that only two of the three phase lines are conducting current, and that the current in each line is less than maximum.

Because the phase currents in a 3-phase system are never maximum at the same time, and at other times the current is divided between two phases, the

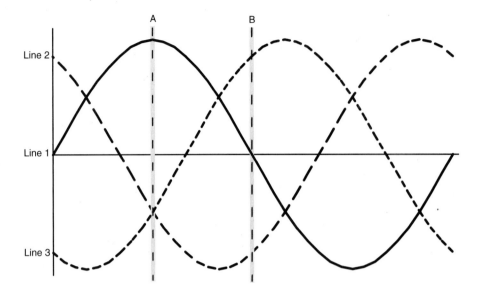

FIGURE 6-3   Currents of a 3-phase system are 120° out of phase with each other. (*Delmar/Cengage Learning*)

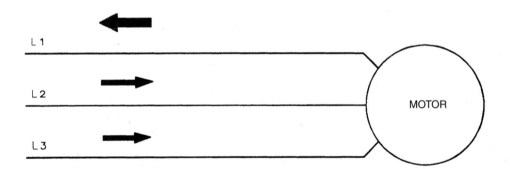

FIGURE 6-4   Current is maximum in one conductor and less than maximum in two conductors.
(*Delmar/Cengage Learning*)

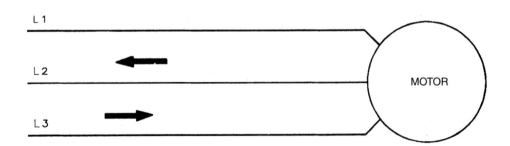

FIGURE 6-5   Currents in only two conductors. (*Delmar/Cengage Learning*)

total conductor resistance will not be the sum of two conductors. To calculate the resistance of conductors in a 3-phase system, a demand factor of 0.866 is used.

In this problem, the motor is located 2500 ft (762 m) from the source. The effective conductor length (Le) will be calculated by doubling the length of one conductor and then multiplying by 0.866.

$$Le = 2500 \text{ ft} \times 2 \times 0.866 = 4330 \text{ ft}$$
$$(= 762 \text{ m} \times 2 \times 0.866 = 1320 \text{ m})$$

Now that all the factors are known, the size of the conductor can be calculated using the formula:

where K = 17 (ohms per mil foot for aluminum)

$$\text{cmil} = \frac{K \times L}{R} = \frac{17 \times 4330}{0.166} = 443,434$$

Three 500-kcmil conductors will be used.

PARALLEL CONDUCTORS

Under certain conditions, it may become necessary or advantageous to connect conductors in parallel. One example of this condition is when conductor size is very large, as is the case in Problem 4 in the previous section. In that problem, it was calculated that the conductors supplying a motor 2500 ft from its source would have to be 500 kcmil. A 500-kcmil conductor is very large and difficult to handle. For this reason, it may be preferable to use parallel conductors for this installation. The *NEC* lists five conditions that must be met when conductors are connected in parallel (*310.10(H)*):

1. The conductors must be of the same length.

2. The conductors must be made of the same material. It is not permissible to use copper for one conductor and aluminum for the others.

3. The conductors must have the same circular mil area.

4. The conductors must use the same type of insulation.

5. The conductors must be terminated and connected in the same manner.

In the previous problem, the actual conductor size needed was calculated to be 443,434 circular mils. This circular mil area could be obtained by

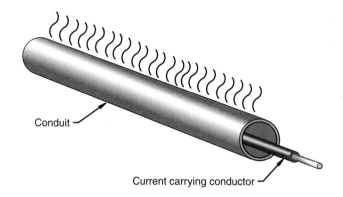

FIGURE 6-6   The current in the conductor induces eddy currents in the conduit causing the conduit to become hot. (*Delmar/Cengage Learning*)

connecting two 250-kcmil conductors in parallel for each phase, or three 3/0 AWG conductors in parallel for each phase.

> **NOTE:** Each 3/0 AWG conductor has an area of 167,800 circular mils. This is a total of 503,400 circular mils.

Another example of when it may be necessary to connect wires in parallel is when conductors of a large size must be run in conduit. The conductors of a single phase are not permitted to be run in metallic conduit, as shown in Figure 6-6 (*300.5(I)* and *300.20*). The reason for this is that when current exists in a conductor, a magnetic field is produced around the conductor. In an alternating current circuit, the current continuously changes direction and magnitude, which causes the magnetic field to cut through the wall of the metal conduit. This cutting action of the magnetic field induces a current called an eddy current into the metal of the conduit. Eddy currents can produce enough heat in high-current circuits to melt the insulation surrounding the conductors. All metal conduits can have eddy current induction, but conduits made of magnetic materials such as steel have an added problem with hysteresis loss. Hysteresis loss is caused by molecular friction. As the direction of the magnetic field reverses, the molecules of the metal are magnetized with the opposite polarity and swing to realign themselves. This continuous aligning and realigning of the

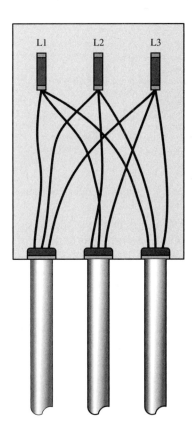

FIGURE 6-7 Each conduit contains a conductor from each phase. (*Delmar/Cengage Learning*)

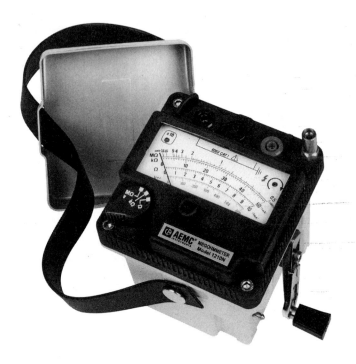

FIGURE 6-8 Hand-cranked meghometer. (*Courtesy AEMC® Instruments*)

molecules produces heat due to friction. Hysteresis losses become greater with an increase in frequency.

To correct this problem, a conductor of each phase must be run in each conduit, Figure 6-7. When all three phases are contained in a single conduit, the magnetic fields of the separate conductors cancel each other, resulting in no current being induced in the walls of the conduit.

TESTING WIRE INSTALLATIONS

After the conductors have been installed in conduits or raceways, it is accepted practice to test the installation for grounds and shorts. This test requires an ohmmeter, which not only can measure resistance in millions of ohms but also can provide a high enough voltage to ensure that the insulation will not break down when rated line voltage is applied to the conductors. Most ohmmeters operate with a maximum voltage that ranges from 1.5 to about 9 volts, depending on the type of ohmmeter and the setting of the

range scale. To test wire insulation, a megohmmeter is used with a voltage from about 250 to 5000 volts, depending on the model of the meter and the range setting. One model of a megohmmeter is shown in Figure 6-8. This instrument contains a hand crank that is connected to the rotor of a brushless dc generator. The advantage of this particular instrument is that it does not require the use of batteries. A range selector switch permits the meter to be used as a standard ohmmeter or as a megohmmeter. When it is used as a megohmmeter, the selector switch permits the test voltage to be selected. Test voltages of 100, 250, 500, and 1000 volts can be obtained.

A megohmmeter can also be obtained in battery-operated models, as shown in Figure 6-9. These models are small, lightweight, and particularly useful when it becomes necessary to test the dielectric of a capacitor.

Wire installations are generally tested for two conditions: shorts and grounds. Shorts are current paths that exist between conductors. To test an installation for shorts, the megohmmeter is connected across two conductors at a time, as shown in Figure 6-10. The circuit is tested at rated voltage or slightly higher. The megohmmeter indicates the resistance between the two conductors.

FIGURE 6-9 Battery-operated megohmeter. (*Courtesy AEMC® Instruments*)

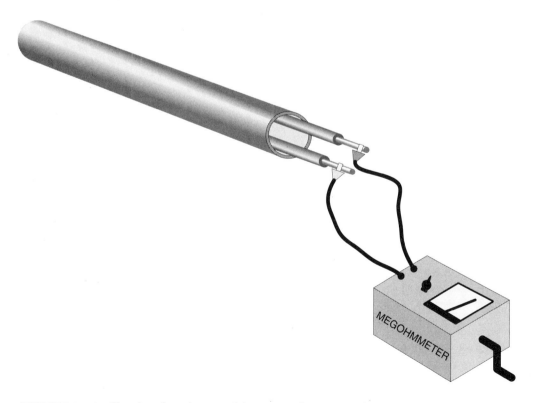

FIGURE 6-10 Testing for shorts with a megohmmeter. (*Delmar/Cengage Learning*)

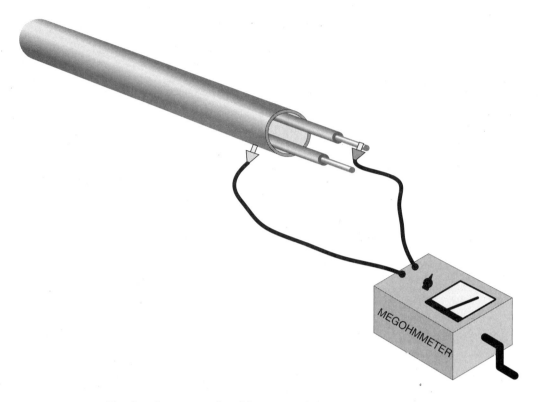

FIGURE 6-11   Testing for grounds with a megohmmeter. (*Delmar/Cengage Learning*)

Because both conductors are insulated, the resistance between them should be extremely high. Each conductor should be tested against every other conductor in the installation.

To test the installation for grounds, one lead of the megohmmeter is connected to the metallic raceway, as shown in Figure 6-11. The other meter lead is connected to one of the conductors. The conductor should be tested at rated voltage or slightly higher. Each conductor should be tested.

THE AMERICAN WIRE GAUGE (AWG)

The American Wire Gauge was standardized in 1857 and is used mainly in the United States for the diameters of round, solid, nonferrous electrical wire. The gauge size is important for determining the current carrying capacity of a conductor. Gauge sizes are determined by the number of draws necessary to produce a given diameter or wire. Electrical wire is produced by drawing it through a succession of dies, Figure 6-12. Each time a wire passes through a die, it is wrapped around a draw block several times.

The draw block provides the pulling force necessary to draw the wire through the die. A 24 AWG wire would be drawn through 24 dies, each having a smaller diameter. In the field, wire size can be

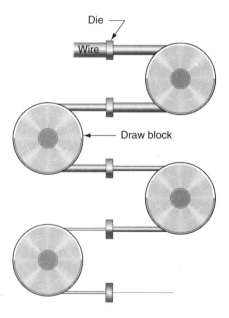

FIGURE 6-12   Electrical wire is made by drawing it though a succession of dies. (*Delmar/Cengage Learning*)

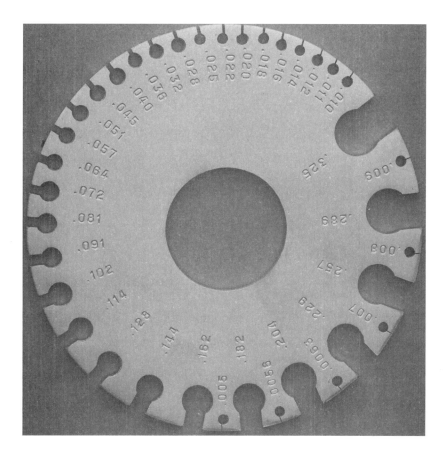

FIGURE 6-13   Typical wire gauge used to determine wire size. (*Delmar/Cengage Learning*)

determined with a wire gauge, Figure 6-13. One side of the wire gauge lists the AWG size of the wire, Figure 6-14. The opposite side of the wire gauge indicates the diameter of the wire in thousandths of an inch, Figure 6-15. When determining wire size, first remove the insulation from around the conductor. The slots in the wire gauge, not the holes behind the slots, are used to determine the size, Figure 6-16.

FIGURE 6-14   One side of the wire gauge lists the AWG size of wire. (*Delmar/Cengage Learning*)

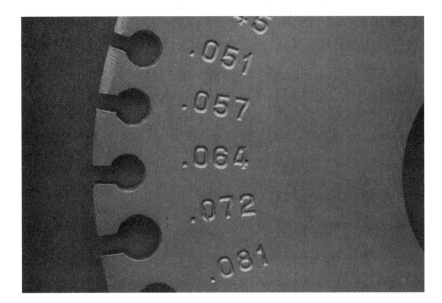

FIGURE 6-15 The opposite side of the wire gauge lists the diameter of the wire in thousandths of an inch. (*Delmar/Cengage Learning*)

FIGURE 6-16 The slot, not the hole behind the slot, determines the wire size. (*Delmar/Cengage Learning*)

REVIEW QUESTIONS

All answers should be written in complete sentences, calculations should be shown in detail, and *Code* references should be cited when appropriate.

Unless specified otherwise, the ambient temperature is 86°F (30°C), the location is dry, the termination has a temperature rating equal to or greater than that of the conductor, and the wire is copper. Where the phrase *allowable ampacity* is used, it refers to a value taken from one of the tables. Where the phrase *ampacity* is used, it refers to the allowable ampacity as corrected and adjusted and in compliance with *110.14(C).*

1. What is the temperature rating of a type XHHW conductor where used in a wet location?

_____

_____

_____

2. What types of conductors are approved for direct burial? _____

_____

_____

_____

3. What types of conductors are approved for underground use? _____

_____

_____

_____

4. Three 10 AWG, type THW conductors are to be installed between poles on individual insulators. What will the conductor ampacity be? _____

_____

_____

_____

5. Motor feeders consisting of six 1/0 AWG type THHN aluminum conductors are to be installed in a rigid metal conduit in an ambient temperature of 100°F (38°C). What will be the conductor ampacity? What will be the circuit ampacity? _____

_____

_____

_____

6. Explain what it means to install "conductors in parallel" and give five conditions that must be satisfied if this is done. _____

_____

_____

_____

_____

_____

_____

7. What is the size of the largest solid (not stranded) conductor approved for installation in raceways? _____

_____

_____

_____

8. How is a 4 AWG grounded conductor in a flat multiconductor cable identified?

_____

_____

9. What insulation colors are reserved for special uses? _____

_____

_____

_____

10. A single-phase, 86-ampere load is located 2800 ft (853 m) from the 480-volt electrical power source. What size aluminum conductors should be installed if the voltage drop cannot exceed 3 percent? _____

_____

_____

_____

11. A 3-phase, 480-volt motor with a starting current of 235 amperes is located 1800 ft (550 m) from the power source. What size copper conductors should be used to ensure that the voltage drop will not exceed 6 percent during starting?_____

_____

_____

_____

_____

12. What is the maximum noncontinuous load that can be connected to a 2 AWG, type THHN conductor? _____

_____

_____

_____

13. What is the maximum continuous load that can be connected to a 2 AWG, type THHN conductor? _____

_____

_____

_____

14. Calculations made in accord with _NEC Article 220_ indicate 110 amperes of continuous load and 40 amperes of noncontinuous load on a single-phase, 240/120-volt feeder. What is the minimum conductor size?_____

_____

_____

_____

_____

Signaling Systems

OBJECTIVES

After studying this chapter, the student should be able to

- describe and install the master clock.
- describe and install the program system.
- describe and install the paging system.
- describe and install the fire alarm system.

A signaling circuit is defined in *NEC Article 100* as any electric circuit that energizes signaling equipment. A signaling system may include one or more signaling circuits. For example, in the industrial building, there are several electrical systems that give recognizable visual and audible signals and are classified as signaling systems:

- a master clock;
- a program system;
- a paging or locating system; and
- a fire alarm system.

THE MASTER CLOCK

The master clock is a clock designed to drive some number of units that display the time. The display units are not actually clocks themselves but depend for their operation on signals received from the master clock. A display unit is shown in Figure 7-1. This type of display unit uses light-emitting diodes (LEDs) to indicate the time, rather than an analog display using numbers and hands. These types of displays are generally designed to accommodate large-size LEDs that can be seen from a long distance. If a display

FIGURE 7-1 Clock system display unit. (*Courtesy ESE*)

with a number larger than about 0.55 in. (14 mm) is desired, it will generally use a planar gas discharge display rather than an LED display. A diagram showing a master clock with several display units attached is shown in Figure 7-2.

Several methods can be used to sense time in the master clock unit. One of the most common methods for many years was to use a single-phase synchronous motor. The speed of the synchronous motor is proportional to its number of poles and the line frequency. This method uses the 60-hertz line frequency to measure time. This is the same method often used to operate electric clocks in the home. Sensing the line frequency is relatively accurate. Clocks using this method to sense time

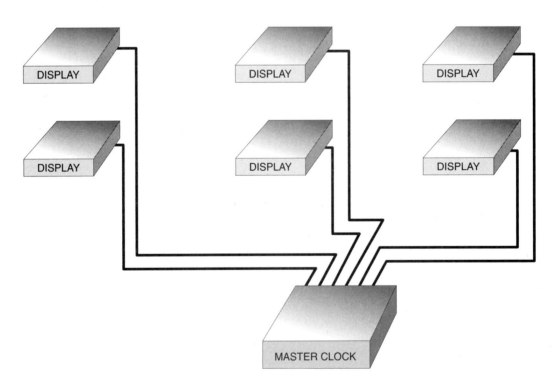

FIGURE 7-2 Master clock system. (*Delmar/Cengage Learning*)

FIGURE 7-3 Master clock. (*Courtesy ESE*)

are generally accurate to within a couple of minutes per month.

Another method that has become popular is the sensing of vibrations produced by a piece of quartz crystal. When an ac voltage is impressed across two faces of the crystal, it will resonate at some specific frequency. This resonant frequency is extremely constant and therefore can be used to accurately measure time. The frequency at which the quartz will resonate is inversely proportional to the size of the crystal. The smaller the crystal, the higher the resonant frequency will be. The shape of the crystal also plays a part in determining the resonant frequency. Quartz clocks are generally accurate to within 1 second per month.

The clock in this installation is shown in Figure 7-3. This master clock senses time by receiving a radio signal from WWV, a radio station that broadcasts time pulses. WWV is operated by the National Bureau of Standards and is used as a time standard throughout the United States. A cesium beam atomic clock is used to produce the pulses that are transmitted. WWV can be received on frequencies of 18, 20, and 60 kHz, and on frequencies of 2.5, 5, 10, 15, 20, and 25 MHz. At the beginning of each minute, a 1000-Hz signal is transmitted, except at the beginning of each hour, when a 1500-Hz signal is transmitted. The clock in this installation contains a radio receiver capable of receiving WWV pulses. The timing of the clock depends on the pulses received, and, in this way, the time clock is continually updated each minute.

The clock also contains a battery and battery charger. The battery is used to provide power to the clock in the event of a power failure. The battery can operate the clock for a period of at least 12 hours. During this time, the displays will be turned off to conserve battery power, but the clock continues to operate. The master clock also can be set to operate in a 12- or 24-hour mode.

THE PROGRAM SYSTEM

The program system is used to provide automatic signals for the operation of horns, bells, and buzzers. These devices are used in industry to signal the beginning and ending of shifts, lunch periods, and breaks. Different parts of the plant operate on different time schedules. Office workers, for example, begin and end work at different times than employees who work in the manufacturing area of the plant. Lunch and coffee break times also vary. For this reason, the program control system must be capable of providing different signals to different parts of the plant at the proper times.

The program controller used in this installation is shown in Figure 7-4. This controller is a microprocessor-based programmable timer. This unit has thirty-two separate output channels and can be programmed for up to 1000 events. Each channel contains a normally open reed relay. Each relay can be operated by momentary contact or latching, or the unit can be set so that there can be sixteen of each. A simple modification will permit sixteen double-pole relays to be used instead of thirty-two single-pole relays. The 1000 events can be entered into the unit randomly as opposed to entering them in chronological order. Cyclic events can be programmed to occur every minute, hour, day, or week, or in any combination desired. Any of the thirty-two output channels can be turned on at the same time.

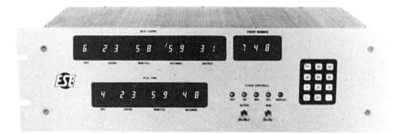

FIGURE 7-4 Programmable timer. (*Courtesy ESE*)

Programming is done with a twelve-button keyboard located on the front of the unit. The keyboard contains numbers 0 through 9, and CLEAR and ENTER buttons. Two toggle switches, also located on the front of the panelboard, are used to provide active/disable and run/enter functions. These switches permit programmed events to be viewed without interrupting the program. Once the timer has been programmed, it is possible to save the program on a cassette tape. This is done by connecting a cassette tape recorder to a jack provided on the rear of the timer. If for some reason the timer should have to be replaced, the same program can be loaded into the new unit from the cassette tape. This saves the time of having to reprogram the unit.

The program timer also contains a digital clock. The clock is operated by an internal crystal oscillator. A battery and battery charger are provided with the unit in case of power failure. With the addition of a serial time code generator, the program timer also can be used as a master clock. The time code generator gives the unit the capability of driving up to 100 display units.

THE PAGING SYSTEM

In many industrial installations, it is important to convey messages to all areas of the plant. When selecting a paging system, several factors should be taken into consideration, such as these:

1. What is the amount of area to be covered and the number of paging units needed?

2. The design of the system should permit expansion as the plant increases in size.

3. Should the paging system be voice, tone, or a combination of both?

4. Do areas of the plant require explosionproof or weatherproof equipment?

5. What is the ambient noise level of the plant?

A very important consideration when choosing the type of equipment to be used is the ambient or surrounding noise level. The chart shown in Figure 7-5 illustrates different levels of noise measured in decibels (dB). For a signal or voice to be heard, it should be at least 5 decibels louder than the surrounding noise level at the work station. Another important consideration is the distance of the speaker from the work station. As a general rule, sound decreases by 6 decibels each time the distance from the speaker is doubled.

PROBLEM: A speaker is rated to produce 110 decibels at a distance of 10 ft (3 m). The ambient noise level at the work station is measured at

Reaction	dB	Source Comparison
Uncomfortably loud (possible ear pain)	195	Circular saw at 2 ft (0.61 m)
	140	Jackhammer at 2 ft (0.61 m)
	120	Thunder (near)
Very loud	90 to 100	Industrial plant Wire mill boiler factory
Loud	80 to 90	Foundry factory Press room
Moderate	70 to 75	Normal conversation in office at 3 ft (0.91 m)
Quiet	40 to 55	Hospital room
Very quiet	30 to 35	Whisper at 2 ft (0.61 m)

FIGURE 7-5 Comparable sounds. (*Delmar/Cengage Learning*)

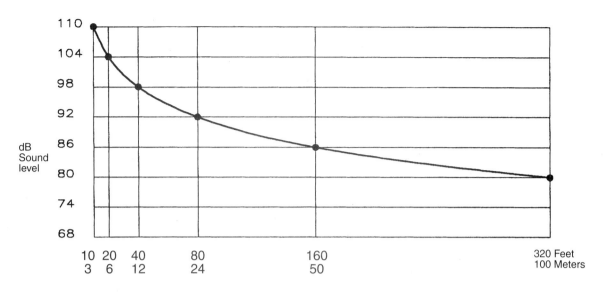

FIGURE 7-6 Effects of sound relating to distance. (*Delmar/Cengage Learning*)

80 decibels. If the speaker is mounted 160 ft (49 m) away from the work station, will the worker be able to clearly hear the messages?

Solution: Figure 7-6 illustrates the amount of sound decrease with distance. Notice that the chart starts with a value of 110 decibels at a distance of 10 ft (3 m), and decreases 6 decibels each time the distance is doubled. At a distance of 160 ft (49 m), the sound level should be 86 decibels. This is loud enough to permit the worker to hear the voice or tone.

The paging system chosen for this plant is manufactured by Audiosone Inc. It has the capability of producing both voice and tone signals. Two types of paging units will be used. The first type can be used to send voice messages only. The second type, shown in Figure 7-7, can send both voice and tone messages.

FIGURE 7-7 Unit used to send voice and tone messages. (*Courtesy Audiosone Inc.*)

FIGURE 7-8 Each unit contains an amplifier. (*Courtesy Audiosone Inc.*)

Each paging unit contains its own amplifier, Figure 7-8. This permits an almost unlimited number to be used when they are connected in parallel to a 4-conductor circuit. The system also can be expanded to use a voice evacuation alarm, shown in Figure 7-9. This unit permits taped messages to be used, which can instruct employees as to the nature of the emergency.

Four separate tones can be generated:

1. WAIL: conventional siren
2. HI-LO: alternating high and low (European siren)
3. WHOOP: ascending low to high, repeated
4. HORN: steady tone

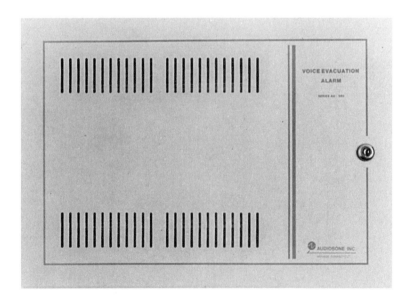

FIGURE 7-9 Voice evacuation alarm. (*Courtesy Audiosone Inc.*)

FIGURE 7-10   Speakers used with paging system. (*Courtesy Audiosone Inc.*)

The tones are to be used to announce different conditions. One is to be used as a fire signal and will be connected to the fire alarm system. The other three tones can be used to announce such conditions as plant evacuation, shift change, and so on. Two types of speakers will be used, Figure 7-10. These speakers will be located at strategic points throughout the plant.

THE FIRE ALARM SYSTEM

The fire alarm is part of a digital automation and control system manufactured by Jensen Electric Co. This system was chosen because of its flexibility and expansion capability. The central control unit (CCU) is a modern desktop computer. The processor, Figure 7-11, provides communication interface between the CCU and the system controllers. This system permits any number of processors (up to 100) to be connected to the CCU. The power equipment of the processor is 24 volts dc, which is provided by an uninterruptible power supply, Figure 7-12. The uninterruptible power supply produces a regulated 24 volts dc from a 120-volt ac 60-hertz supply line. Two 12-volt lead acid storage batteries are contained inside the

power supply. If the incoming ac power should fail, the lead acid batteries continue to provide power to the system. This ensures that control power to critical systems is maintained during a power failure.

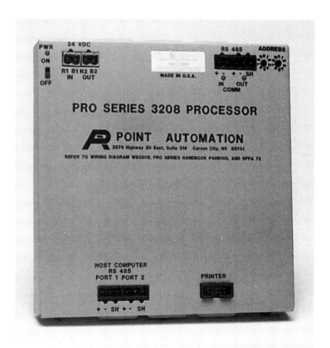

FIGURE 7-11   Processor unit. (*Courtesy Point Automation*)

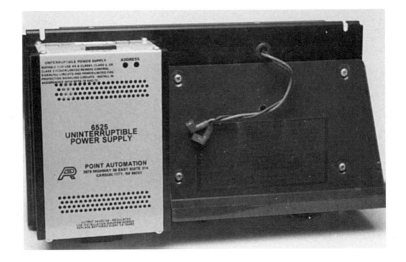

FIGURE 7-12 Power supply. (*Courtesy Point Automation*)

The controller, Figure 7-13, is an intelligent device that provides interface between the control modules and the processor. Each processor can handle up to 100 controllers. Each controller can contain up to four interface modules, Figure 7-14. This system can operate with only one input/output (I/O) point or with as many as 640,000.

The type of interface module used determines the system control function. Modules can be obtained that permit the system to be used for motor control, burglar alarm, ground-fault detection, or to interface with television cameras. Some modules permit analog input and output signals that operate from 0 to +5 volts dc, or 4 to 20 milliamps dc. The

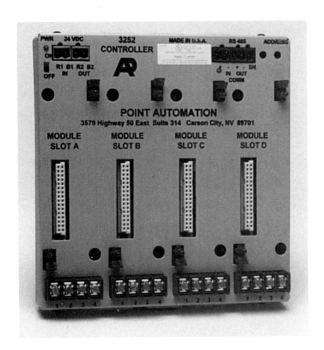

FIGURE 7-13 Programmable controller.
(*Courtesy Point Automation*)

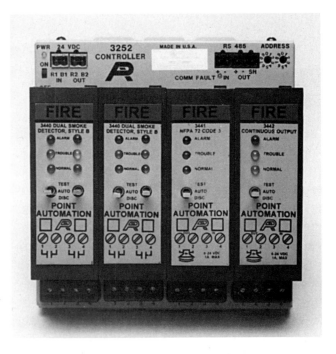

FIGURE 7-14 Controller with interface modules.
(*Courtesy Point Automation*)

FIGURE 7-15 Fire alarm modules.
(*Courtesy Point Automation*)

FIGURE 7-16 Smoke detector.
(*Courtesy Whelan Engineering Co.*)

fire alarm input and output modules are shown in Figure 7-15.

Each fire alarm module contains a momentary contact switch located on the front cover. This switch permits the alarm to be operated automatically or manually, or to be tested. Each module contains three LEDs. The green LED indicates normal circuit operation. A red LED indicates that a fire has been detected, and a yellow LED indicates trouble with the system. The conductor sizes and types for the alarm system are sized to meet the requirements of *725.1 (FPN)*, and conduit size is determined in accordance with *725.31(B)*.

The fire alarm system uses a combination of smoke detectors, Figure 7-16, and manual pull handles, Figure 7-17. Once the manual pull handles have been used, they must be reset with a key. The smoke detector, manufactured by Whelan Engineering Co., is powered by a separate 120-volt ac source and contains an internal horn that produces 86 decibels at 10 ft (3 m). This permits the smoke detector to be used in office areas without the addition of a separate audible alarm. An LED located on the front cover flashes every 4 seconds to indicate the detector is in working order. If smoke is detected, the LED will emit a steady glow.

When a fire condition is detected, two alarm devices are activated. The first is one of the tones produced by the paging system. The second is a high-intensity strobe light that produces 75 flashes per minute, Figure 7-18. The strobe light is powered directly by the smoke detector in

FIGURE 7-17 Fire alarm pull handles.
(*Courtesy Point Automation*)

FIGURE 7-18 High-intensity strobe light. (*Courtesy Whelan Engineering Co.*)

office areas. The fire alarm system is programmed so that when a fire is detected, all of the audible alarms produced by the paging system are activated. Not all of the strobe lights are activated, however. Only the strobe lights located in the area of the detected fire are permitted to flash. A basic line drawing of the fire alarm system is shown in Figure 7-19.

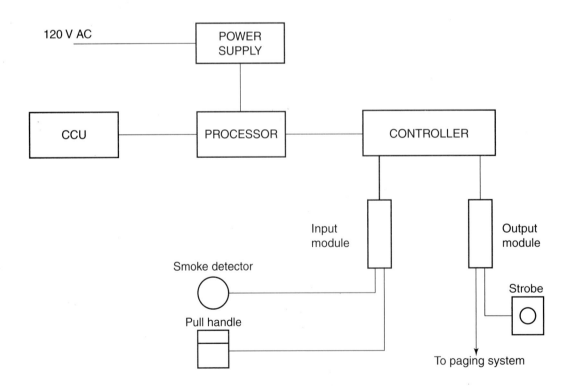

FIGURE 7-19 Basic fire alarm system. (*Delmar/Cengage Learning*)

All answers should be written in complete sentences, calculations should be shown in detail, and *Code* references should be cited when appropriate.

1. What is a signal circuit?_____

_____

_____

2. Where is the definition of a signal circuit found in the *NEC*? _____

_____

_____

3. Are the display units used in this installation actually clocks? _____

_____

_____

4. What is WWV? _____

_____

_____

5. What type of clock is used to provide the pulses broadcast by WWV?_____

_____

_____

6. What type of clock is used to operate the program timer?_____

_____

_____

7. How many separate events can be programmed in the program timer? _____

_____

_____

8. How many output channels are provided with the program timer?_____

_____

_____

9. What is the primary purpose of the paging system? _____

_____

_____

10. Name five factors that should be taken into consideration when selecting a paging system.

_____

_____

_____

11. Assume the surrounding noise level in a certain area of the plant is 80 decibels. If a message is to be clearly heard, what should be the minimum sound level of the message?

_____

_____

_____

12. How many tones can be generated by the paging system? _____

_____

_____

13. What is used as the central control unit for the fire alarm system? _____

_____

_____

14. What is the function of the processor?_____

_____

_____

15. What is the maximum number of processors that can be connected to the CCU? _____

_____

_____

16. What voltage is supplied to the processor? _____

_____

_____

17. What supplies power to the system if the incoming ac power should fail?_____

_____

_____

18. What is the function of the controller?_____

_____

_____

19. How many control modules can be connected to each controller? _____

_____

_____

20. What is the purpose of the switch located on the front of the fire alarm module? _____

_____

_____

21. What condition is indicated by each of the three LEDs located on the front of the fire alarm module?

Green _____

Red _____

Yellow _____

22. What two devices are used to indicate the presence of a fire?_____

_____

_____

23. Each smoke detector contains a separate internal horn. What is the sound level of this internal horn? _____

_____

_____

24. What two alarm devices are activated when a fire is detected?_____

_____

_____

25. When a fire is detected, do all the strobe lights throughout the plant flash? _____

_____

_____

Basic Motor Controls

OBJECTIVES

After studying this chapter, the student should be able to

- describe differences between contactors and motor starters.

- describe the different functions of fuses and overloads.

- list different types of overload relays and explain how they operate.

- connect basic control circuits using a schematic diagram.

- describe the differences between schematic and wiring diagrams.

- discuss the differences between schematic or ladder diagrams and wiring diagrams.

Anyone working as an electrician in industry should be able to connect and troubleshoot basic motor control circuits. Control circuits are used to start, stop, accelerate, decelerate, and protect motors. They may also consist of a number of sensing devices such as limit switches, float switches, pushbuttons, flow switches, pressure switches, temperature switches, and so on, that tell the circuit what action is to be performed. Motor control circuits can be divided into two major categories: 2 wire and 3 wire.

TWO-WIRE CONTROLS

Two-wire control circuits are the simplest. Two-wire controls basically consist of a switch used to connect or disconnect power to the motor, Figure 8-1. Many manual-type starters are designed as 2-wire controllers. Motor starters contain both a means to connect and disconnect the motor to and from the power source and also provide overload protection for the motor. Overload protection should not be confused with fuse or circuit protection. Fuses and circuit breakers protect the circuit from some types of high-current condition such as shorts and grounds. Overload protection is

designed to protect the motor from an overload condition. Assume, for example, that a motor has a full-load current rating of 10 amperes. Also assume that the motor is connected to a 20-ampere circuit. If the motor should become overloaded and the current increase to 15 amperes, the circuit breaker would never trip or the fuse never blow because the current draw is below the 20-ampere rating. The motor, however, will probably be damaged or destroyed because of the excessive current. Overloads are intended to open the circuit when the current exceeds the full-load current rating of the motor by 115 percent to 125 percent.

A single-phase manual starter with overload protection is shown in Figure 8-2. The starter resembles a single-pole switch with the addition of overload protection. A schematic diagram of the single-phase manual starter is shown in Figure 8-3.

A 3-phase manual starter, Figure 8-4, operates in a similar manner to the single-phase manual starter except that it provides three sets of contacts and three overloads. The starter is so designed that an overload on any phase of the 3-phase system will cause all three contacts to open. A schematic diagram of a 3-phase manual starter is shown in Figure 8-5.

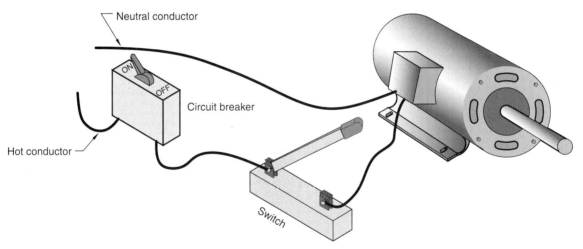

Single-phase motor controlled by a switch

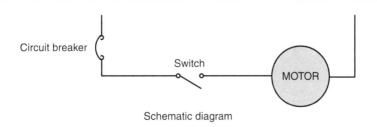

Schematic diagram

FIGURE 8-1   Pictorial and schematic diagram of a single-phase motor controlled by a switch.
(*Delmar/Cengage Learning*)

FIGURE 8-2 Single-phase manual motor starter with overload protection. (*Courtesy Square D*)

FIGURE 8-4 Three-phase manual starter. (*Courtesy Square D*)

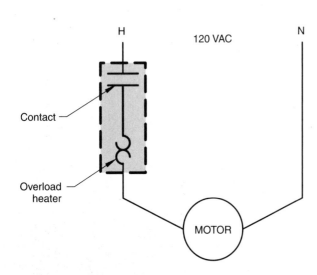

FIGURE 8-3 Schematic diagram of a single-phase manual starter. (*Delmar/Cengage Learning*)

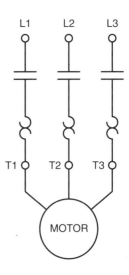

FIGURE 8-5 Schematic diagram of a 3-phase manual starter. (*Delmar/Cengage Learning*)

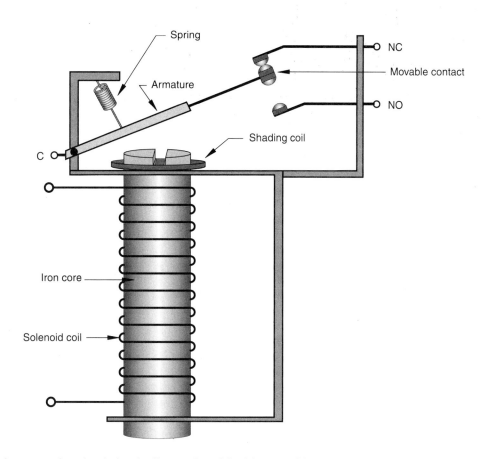

FIGURE 8-6   A magnetic relay is basically a solenoid with movable contacts attached. (*Delmar/Cengage Learning*)

THREE-WIRE CONTROLS

The designation *3-wire control* comes from the fact that three wires are run from a set of start-stop push buttons to a motor starter. Three-wire controls are characterized by the fact that they use magnetic contactors and starters. A magnetic contactor or relay is basically an electrical solenoid that closes a set of contacts when the coil is energized, Figure 8-6. A relay similar to the one illustrated in Figure 8-6 is shown in Figure 8-7. The terms used to describe motor control components can often be confusing. The term *relay* refers to a magnetically operated switch that contains small or auxiliary contacts. Auxiliary contacts are used as part of the control circuit and are not intended to connect a load to the line. *Contactors* contain large load contacts that are intended to handle large amounts of current. Load contacts are used to connect loads such as motors or other high-current devices to the power line. Contactors may or may

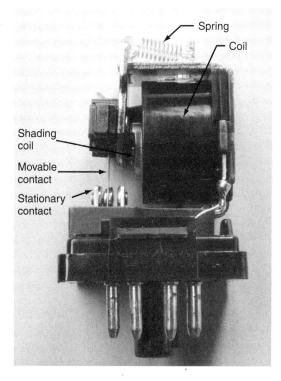

FIGURE 8-7   Eight-pin control relay. (*Delmar/Cengage Learning*)

FIGURE 8-8 Motor starters contain load contacts and overload protection for motors. (*Courtesy Square D*)

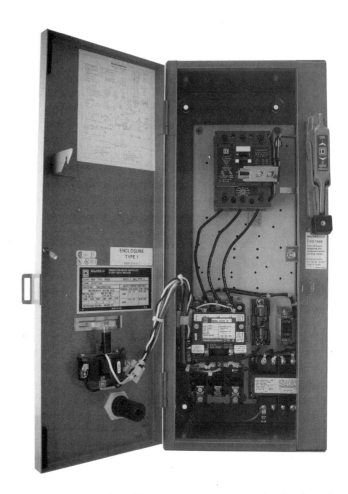

FIGURE 8-9 Combination starters contain the fused disconnect or circuit breaker, control transformer, and motor starter in one enclosure. (*Courtesy Square D*)

not also contain auxiliary contacts. *Motor starters* are contactors that are equipped with overload protection for a motor, Figure 8-8. *Combination starters* contain the fused disconnect or circuit breaker, control transformer, and motor starter in one enclosure, Figure 8-9.

SCHEMATIC SYMBOLS

Schematic and wiring diagrams are the written language of motor controls. To understand these drawings, it is helpful to understand some of the symbols employed when control diagrams are drawn. Although there is no set standard for drawing motor control symbols, the most commonly accepted are those used by the National Electrical Manufacturers Association (NEMA). When reading control schematics, all switch and contact symbols are drawn to indicate their position when the circuit is turned off or not operating. This is called the *normal* position for that contact or switch.

Switch contacts can be drawn as normally open (NO), normally closed (NC), normally open held closed (NOHC), or normally closed held open (NCHO). When a switch is to be shown as normally

open, it is drawn so that the movable contact is shown below and not touching the stationary contact, Figure 8-10A. If the contact is to be drawn normally closed, it is drawn so that the movable contact is above and touching the stationary contact, Figure 8-10B. There are some instances where a contact must be connected normally open, but when the circuit is not in operation, the contact is being held closed. A good example of this is the low-pressure switch on many air-conditioning circuits. The pressure in the system holds the contact closed. If the refrigerant should leak out, the reduced pressure will cause the switch to reopen and stop the operation of the compressor. A normally open held closed switch is drawn with the movable contact below but touching the stationary contact, Figure 8-10C. The switch is normally open because the movable contact is

NORMALLY OPEN SWITCH

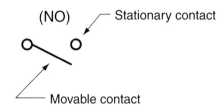

(NO) — Stationary contact

— Movable contact

The movable contact is drawn
below and not touching the
stationary contact

(A)

NORMALLY CLOSED SWITCH

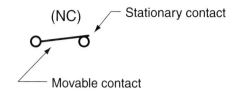

(NC) — Stationary contact

— Movable contact

The movable contact is drawn
above and touching the
stationary contact

(B)

NORMALLY OPEN HELD CLOSED SWITCH

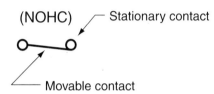

(NOHC) — Stationary contact

— Movable contact

Because the movable contact is drawn below
the stationary contact, the switch is normally
open. The symbol shows the movable
contact touching the stationary contact.
This indicates that the switch is being
held closed.

(C)

NORMALLY CLOSED HELD OPEN SWITCH

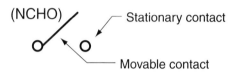

(NCHO) — Stationary contact

— Movable contact

Because the movable contact is drawn above
the stationary contact, the switch is normally
closed. The symbol shows the movable
contact not touching the stationary contact.
This indicates that the switch is being
held open.

(D)

FIGURE 8-10 Switches can be drawn as normally open, normally closed, normally open held closed, or normally closed held open. (*Delmar/Cengage Learning*)

drawn below the stationary contact. It is held closed because the movable contact is touching the stationary contact.

A normally closed held open switch can be drawn in a similar manner, Figure 8-10D. The switch is normally closed because the movable contact is shown above the stationary contact. It is being held open because the movable contact is not touching the stationary contact.

Other symbols are added to switch symbols to indicate a particular type of switch. A limit switch, for example, is shown with a wedge drawn below the movable contact. The wedge represents the bumper arm of the limit switch. Limit switch symbols are shown in Figure 8-11. Other symbols are used to indicate different types of switches (Figure 8-12). A

float switch, for example, uses a circle drawn on the bottom of a line. The circle represents a ball float. A flow switch uses a flag drawn on a line to represent the paddle that detects air or liquid flow. A pressure switch is drawn by connecting a semicircle to a line. The flat portion of the symbol represents a diaphragm used to sense pressure. A temperature switch or thermostat is indicated by drawing a zigzag line that represents a bimetal helix that expands and contracts with a change of temperature.

Normally open contact symbols are shown as parallel lines with connecting wires, Figure 8-13. Normally closed contact symbols are the same, with the exception that a diagonal line is drawn through the two parallel lines. Contacts are always shown in the de-energized or off position.

NORMALLY CLOSED LIMIT SWITCH

NORMALLY CLOSED HELD OPEN LIMIT SWITCH

NORMALLY OPEN LIMIT SWITCH

NORMALLY OPEN HELD CLOSED LIMIT SWITCH

FIGURE 8-11 Limit switches are indicated by drawing a wedge shape below the movable contact of a switch symbol. (*Delmar/Cengage Learning*)

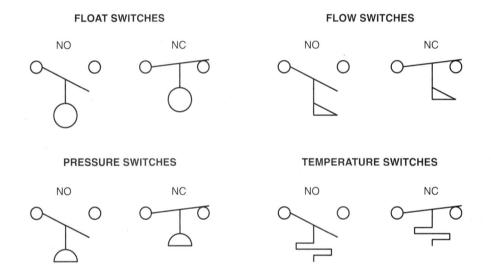

FLOAT SWITCHES   FLOW SWITCHES

NO   NC   NO   NC

PRESSURE SWITCHES   TEMPERATURE SWITCHES

NO   NC   NO   NC

FIGURE 8-12 Symbols used to represent different types of switches. (*Delmar/Cengage Learning*)

Another very common schematic symbol is the push button. Normally open push buttons are shown with the movable contact above and not touching the two stationary contacts, Figure 8-14. The symbol indicates that when finger pressure is applied to the movable contact, it travels downward and bridges the gap between the two stationary contacts. Normally closed push button symbols are drawn with the movable contact below and touching the stationary contacts. When pressure is applied to the movable contact, it travels downward and breaks the connection between the two stationary con-

tacts. Another very common push button symbol is the double-acting push button. Double-acting push buttons contain both normally open and normally closed contacts with one movable contact, Figure 8-15. A chart showing common control and electrical symbols is shown in Figure 8-16.

Relay, contactor, and starter coils are generally indicated by a circle. A number and/or letter is placed inside the circle to designate a particular coil. Coils with the letter M generally mean motor starter. The letters TR indicate a timer relay and the letters CR generally indicate a control relay.

NORMALLY
OPEN CONTACTS

NORMALLY
CLOSED CONTACTS

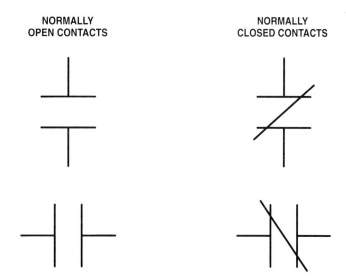

FIGURE 8-13   Normally open and normally closed contact symbols. (*Delmar/Cengage Learning*)

NORMALLY OPEN PUSH BUTTON

Movable contact

Stationary contacts

Normally open push buttons are drawn with
the movable contact above and not touching
the stationary contacts.

NORMALLY CLOSED PUSH BUTTON

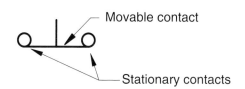

Movable contact

Stationary contacts

Normally closed push buttons are drawn
with the movable contact below and
touching the stationary contacts.

FIGURE 8-14   NEMA standard push-button symbols. (*Delmar/Cengage Learning*)

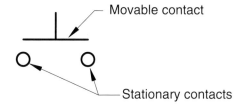

Normally closed contacts

Normally open contacts

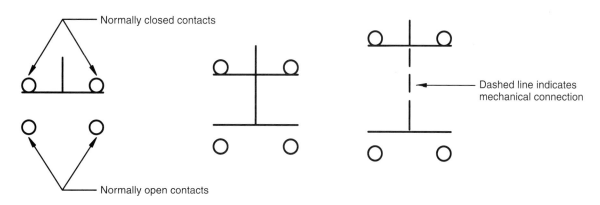

Dashed line indicates
mechanical connection

FIGURE 8-15   Different symbols used to represent double-acting push buttons. (*Delmar/Cengage Learning*)

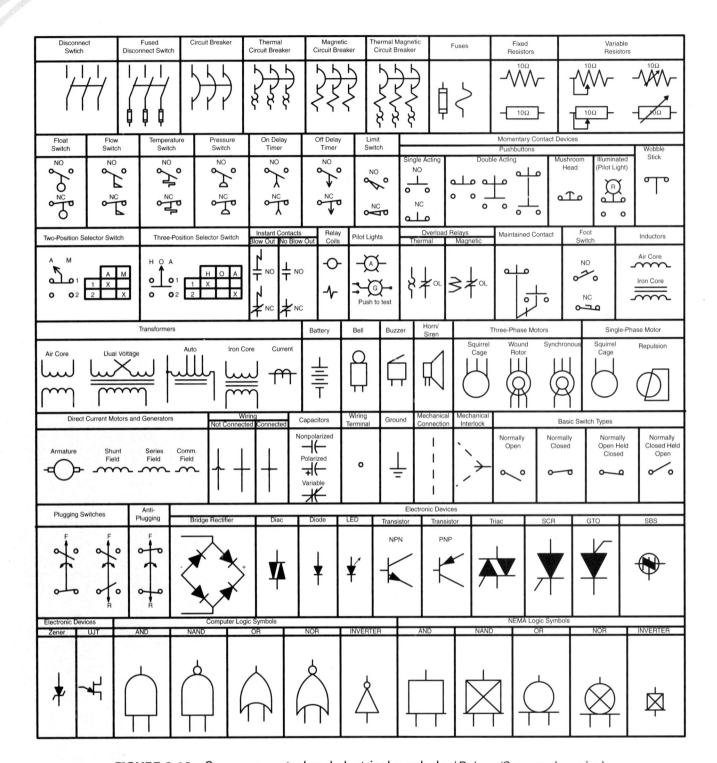

FIGURE 8-16 Common control and electrical symbols. (*Delmar/Cengage Learning*)

OVERLOAD RELAYS

Overload relays are designed to protect the motor from an overload condition. All overload relays contain two separate sections: the current-sensing section and the control contact section. The overload relay does not disconnect the motor from the power line when an over-load condition occurs. The overload relay contains a set of normally closed auxiliary contacts that are connected in series with the coil of the motor starter. If an overload condition occurs, the contacts open and disconnect power to the coil of the motor starter. This causes the load contacts on the starter to open and disconnect the motor from the power line.

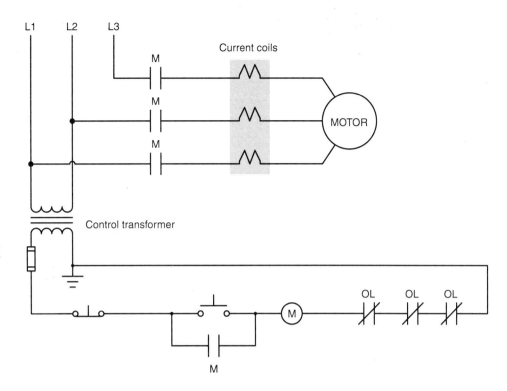

FIGURE 8-17 Magnetic overload relays sense motor current by connecting a current coil in series with the motor. (*Delmar/Cengage Learning*)

The current-sensing section of the overload relay can be magnetic, electronic, or thermal. Magnetic overload relays operate by connecting a current coil in series with the motor, Figure 8-17. If the motor current should become excessive, the magnetic field will become strong enough to cause the normally closed auxiliary contacts to open and de-energize the coil of the motor starter.

Electronic overload relays sense motor current by placing a current-carrying wire through a toroid transformer, Figure 8-18. The transformer measures the magnetic field strength of the conductor in

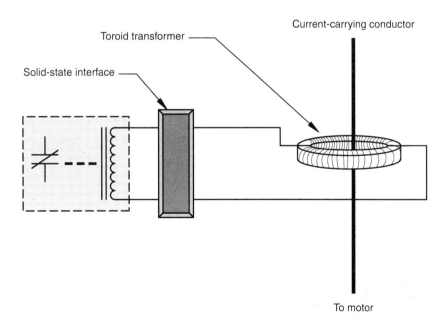

FIGURE 8-18 Electronic overload relays sense motor current by measuring the magnetic field strength around the conductor supplying power to the motor. (*Delmar/Cengage Learning*)

One-piece thermal unit

Solder pot (heat sensitive element) is an integral part of the thermal unit. It provides accurate response to overload current, yet prevents nuisance tripping.

Heating winding (heat producing element) is permanently joined to the solder pot, ensuring proper heat transfer and preventing misalignment in the field.

FIGURE 8-19 Thermal overload relays operate by connecting a heating element in series with the motor. (*Courtesy Square D*)

much the same way that a clamp-on-type ammeter measures the current in a conductor. If the current becomes excessive, the normally closed auxiliary contacts open and disconnect power to the motor starter coil.

Thermal overload relays are by far the most common. Thermal overload relays operate by connecting a heater element in series with the motor, Figure 8-19. The temperature of the heater is dependent on motor current and the ambient temperature. There are two types of thermal overload relays, the solder pot type and the bimetal strip type.

The solder pot–type overload relay works by placing a brass shaft inside a brass tube. A serrated wheel is attached to the brass shaft, Figure 8-20. Solder is used to bond the brass shaft to the brass tube. A lever arm between the contacts and serrated wheel holds the contacts in place, Figure 8-21. If the motor current becomes excessive, the heater will melt the solder and permit the serrated wheel to turn, causing the normally closed auxiliary contacts to open.

The bimetal strip-type overload uses a bimetal strip to open the normally closed auxiliary contacts if the motor current becomes excessive, Figure 8-22. There are other differences between the solder pot–type and bimetal strip–type overload relay. Bimetal strip–type overload relays generally permit the trip current to be adjusted to between 85 and 115 percent of the heater rating, Figure 8-23. Another difference is that bimetal strip–type overloads can generally be set to permit the contacts to reset automatically or manually after the bimetal strip has cooled sufficiently, Figure 8-24. Generally, the relay is adjusted for manual reset. Automatic reset should be used only if the sudden starting of a motor will not cause danger to personnel or damage equipment.

Regardless of the type of thermal overload relay used, the trip current of the relay is set by the size of the heater used with the relay. Manufacturers make different-sized heaters that are designed to open the contacts at different current levels. A chart provided by the manufacturer of the relay can be used to select the proper heater for a particular application.

When three single-phase overload relays are employed to protect a 3-phase motor, the three

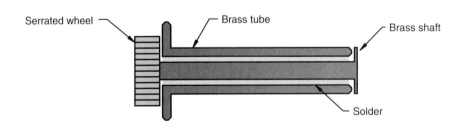

Serrated wheel —

Brass tube —

Brass shaft

Solder

FIGURE 8-20 Construction of a typical solder pot overload. (*Delmar/Cengage Learning*)

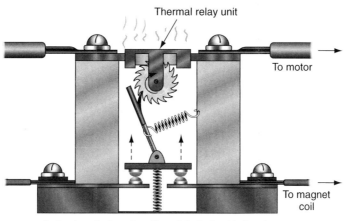

As heat melts the alloy, the ratchet wheel is free to
turn, and the spring pushes the contacts open.

FIGURE 8-21   Basic solder pot overload relay. (*Courtesy Square D*)

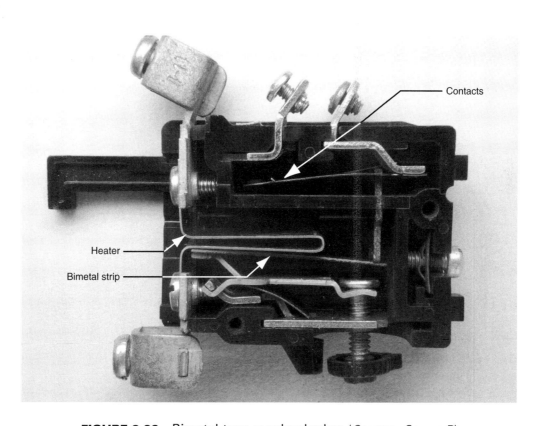

FIGURE 8-22   Bimetal-type overload relay. (*Courtesy Square D*)

auxiliary overload contacts are connected in series, as
shown in Figure 8-25. With this type of connection,
if one relay should trip, the motor starter coil is disconnected from the power line. Three-phase overload
relays, Figure 8-26, contain three separate heaters
but only one set of overload contacts. If an overload
occurs on any phase, it will open the normally closed
contacts. When a 3-phase overload relay is used, only
one set of normally closed contacts is connected in
series with the starter coil, Figure 8-27.

Although all overload relays contain a set
of normally closed contacts, some manufacturers

FIGURE 8-23 Bimetal strip–type overloads generally permit adjustment of the trip current. (*Delmar/Cengage Learning*)

FIGURE 8-24 Bimetal strip overload relays can generally be set to permit automatic or manual reset of the auxiliary contacts. (*Delmar/Cengage Learning*)

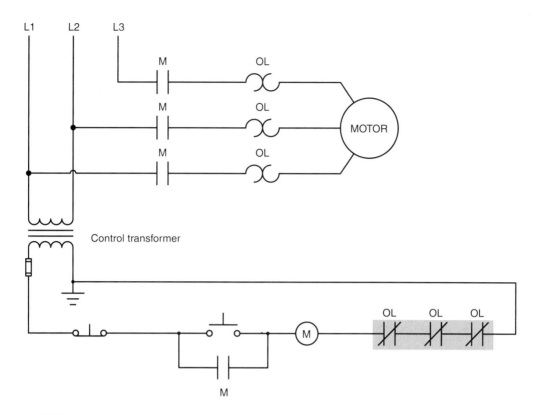

FIGURE 8-25 When three single overload relays are employed to protect a 3-phase motor, all normally closed overload contacts are connected in series. (*Delmar/Cengage Learning*)

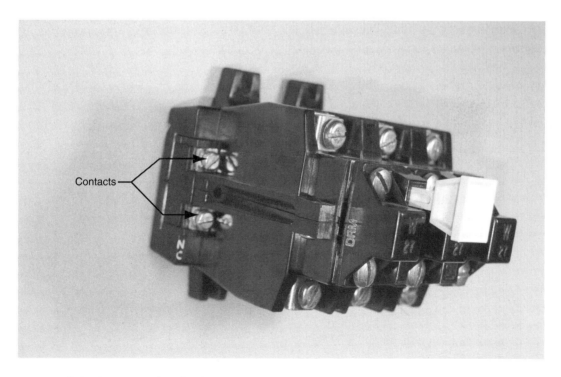

FIGURE 8-26 A 3-phase overload relay contains three separate heaters but only one set of normally closed contacts. (*Delmar/Cengage Learning*)

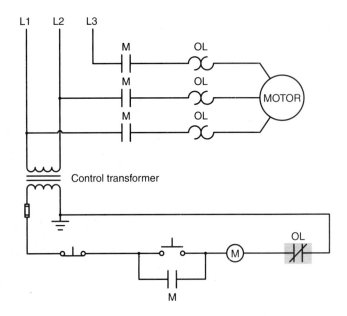

FIGURE 8-27 A 3-phase overload relay contains three heaters but only one set of normally closed contacts. (*Delmar/Cengage Learning*)

include a set of normally open contacts as well. There are two arrangements for normally open overload relay contacts. One arrangement contains two separate contacts, one normally open and the other normally closed. The second is basically a single-pole double-throw switch. The contacts contain a common terminal, a normally closed terminal, and a normally open terminal. The normally open terminal is sometimes labeled in some manner other than normally open. The overload relay shown in Figure 8-28 contains three terminals labeled COM. (common), OL. (overload), and ALAR. (alarm). The common terminal is connected to one side of the power supply for the

FIGURE 8-28 Overload relay with a common terminal, a normally closed terminal, and a normally open terminal. (*Delmar/Cengage Learning*)

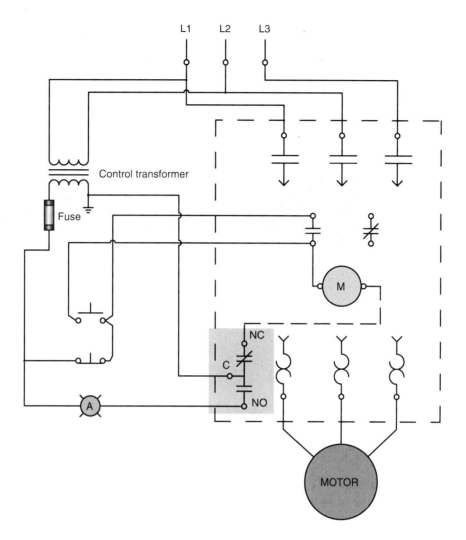

FIGURE 8-29   The normally open contact supplies power to a pilot warning lamp.
(*Delmar/Cengage Learning*)

control circuit. The overload terminal is connected in series with the coil of the motor starter, and the alarm terminal is connected to a pilot lamp, Figure 8-29. The pilot lamp indicates that the overload relay has tripped.

Another common method used for the normally open contacts is to supply power to the coil of a small control relay, Figure 8-30. The contact of the control relay can provide power to the input of a programmable logic controller. If the overload relay should trip, a signal is provided to the PLC to inform the circuit that the motor has tripped on overload. Interposing relays are used to prevent more than one power source from entering the control system either at the motor starter or at the PLC.

SCHEMATICS AND WIRING DIAGRAMS

Schematic or ladder diagrams and wiring diagrams both show electrical connections, but they differ greatly in appearance. The drawing in Figure 8-31 shows a schematic or ladder diagram and a wiring diagram of a start-stop push-button control. Although both of these circuits are identical electrically, they are different in appearance. Schematic diagrams show components in their electrical sequence without regard for physical location. Wiring diagrams show a pictorial representation of the components with connecting wires. Schematic diagrams are by far the more widely used in industry.

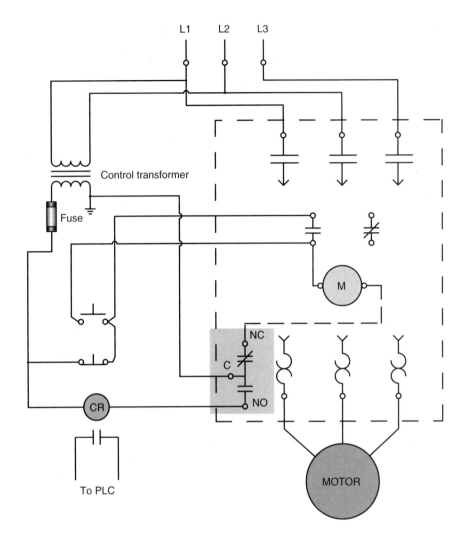

FIGURE 8-30 The normally open contact supplies power to the coil of a control relay.
(*Delmar/Cengage Learning*)

There are several rules that should be followed when reading a schematic diagram:

- All electrical components are shown in their off or de-energized position.

- Schematics should be read like a book, from top to bottom and from left to right.

- Contacts that contain the same letter or number as a coil are controlled by that coil regardless of where they are located in the drawing.

- When a circuit is completed to a coil, that coil will energize and all contacts controlled by that coil will change position. All normally open contacts will close and all normally closed contacts will open.

START-STOP PUSH-BUTTON CONTROL CIRCUIT

The start-stop push-button control circuit is often the beginning of more complex circuits. This circuit is often referred to as the basic circuit. To understand the operation of the circuit, refer to the schematic shown in Figure 8-32. Note that there are four normally open contacts labeled with the letter M. Also note that the coil of the motor starter is labeled with the letter M. This indicates that all the contacts that are labeled with an M are controlled by the coil labeled M. When the start button is pressed, Figure 8-33, a circuit is completed to M coil. When coil M energizes, all M contacts change from open to closed, Figure 8-34. The small auxiliary

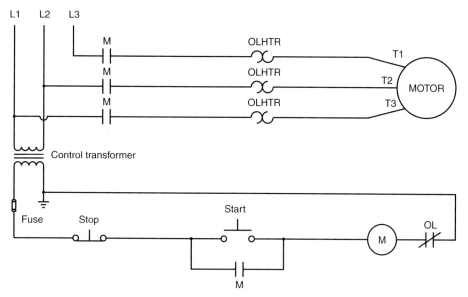

SCHEMATIC DIAGRAM OF A START-STOP PUSH BUTTON CONTROL

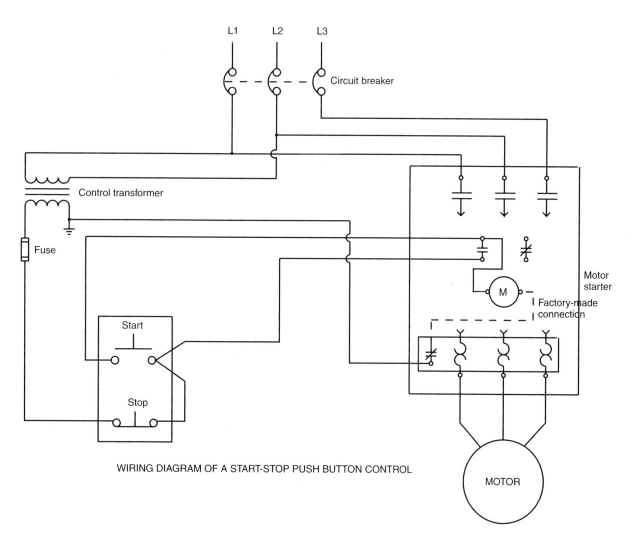

WIRING DIAGRAM OF A START-STOP PUSH BUTTON CONTROL

FIGURE 8-31   Schematic and wiring diagrams of a start-stop push-button control. (*Delmar/Cengage Learning*)

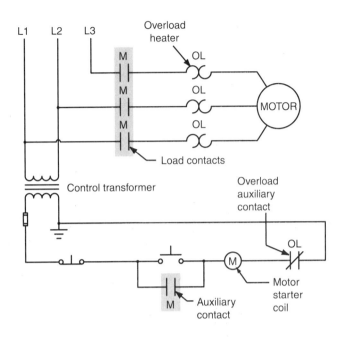

FIGURE 8-32 Basic start-stop push-button control circuit. (*Delmar/Cengage Learning*)

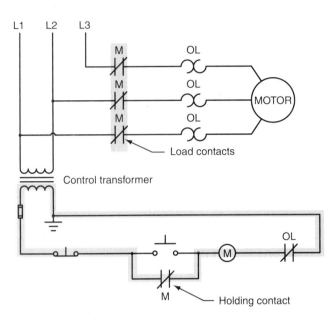

FIGURE 8-34 All contacts labeled M change position when the coil is energized. (*Delmar/Cengage Learning*)

M contact connected in parallel with the start push button closes to maintain the circuit to the coil when the push button is released. This contact is generally referred to as a hold, sealing, or maintaining contact because it holds the circuit closed after the start button is released. The three load contacts labeled M close and connect the motor to the power line. The

circuit will remain in operation until the stop button is pressed or an overload should occur causing the normally closed overload contacts to open, breaking the circuit to M coil, Figure 8-35. When coil M de-energizes, all M contacts return to their open position and the circuit is back to its original de-energized state.

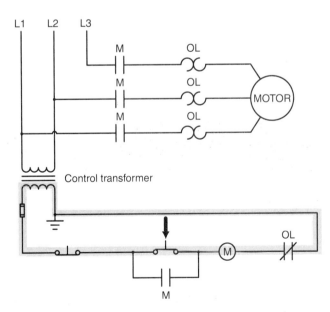

FIGURE 8-33 When the start button is pressed, a circuit is completed to coil M of the motor starter. (*Delmar/Cengage Learning*)

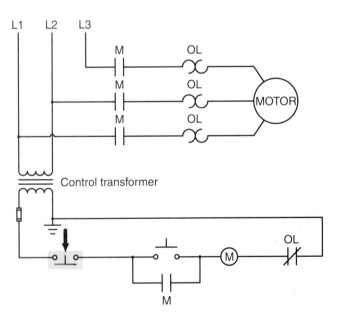

FIGURE 8-35 Pressing the stop button breaks the connection to the coil of M starter. (*Delmar/Cengage Learning*)

FORWARD-REVERSE CONTROL

Another very common control circuit found throughout industry is the forward-reverse control. Three-phase motors can be reversed by changing any two stator leads. Forward-reverse controls also employ interlocking to prevent both the forward and reverse coils from being energized at the same time. A typical forward-reverse control is shown in Figure 8-36. The dashed lines drawn from the F and R coils to a single line indicate mechanical interlocking. Mechanical interlocks are used to prevent both forward and reverse contactors from being energized at the same time. When one contactor is energized, a mechanism prevents the other from being able to close its contacts even if the coil should be energized. Electrical interlocking is accomplished by using the two normally closed auxiliary contacts connected in series with F and R coils, Figure 8-37. Note that the normally closed F contact is connected in series with the R contactor coil and the normally closed R contact is connected in series with the F contactor coil.

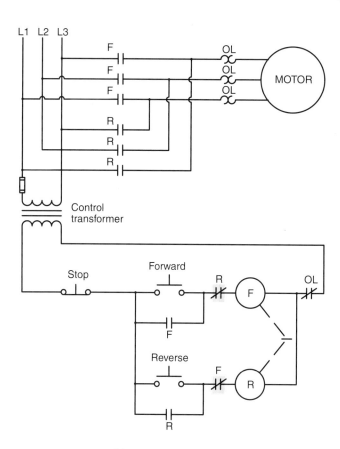

FIGURE 8-37 Normally closed auxiliary contacts are used to provide electrical interlock for the circuit. (*Delmar/Cengage Learning*)

When the forward push button is pressed, a circuit is completed through the normally closed R contact to the F coil. When F coil energizes, all F contacts change position, Figure 8-38. The three F load contacts close and connect the motor to the power line, causing the motor to run in what is considered the forward direction. The normally open F auxiliary contact connected in parallel with the forward push button closes to maintain the circuit when F push button is released. The normally closed F contact connected in series with R coil opens. This would prevent R coil from energizing if the reverse push button were to be pressed.

Before the motor can be operated in reverse, the stop push button must be pressed to break the circuit to F coil. When F coil de-energizes, all F contacts return to their normal positions, as shown in Figure 8-36. When the reverse push button is pressed, a circuit is completed through the now closed F auxiliary contact to R coil. When R coil energizes, all R contacts change position, Figure 8-39. The three R load contacts close and connect the motor

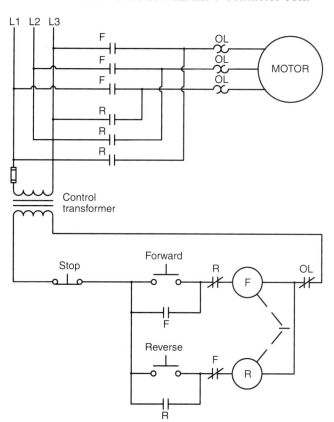

FIGURE 8-36 Basic forward-reverse control circuit. (*Delmar/Cengage Learning*)

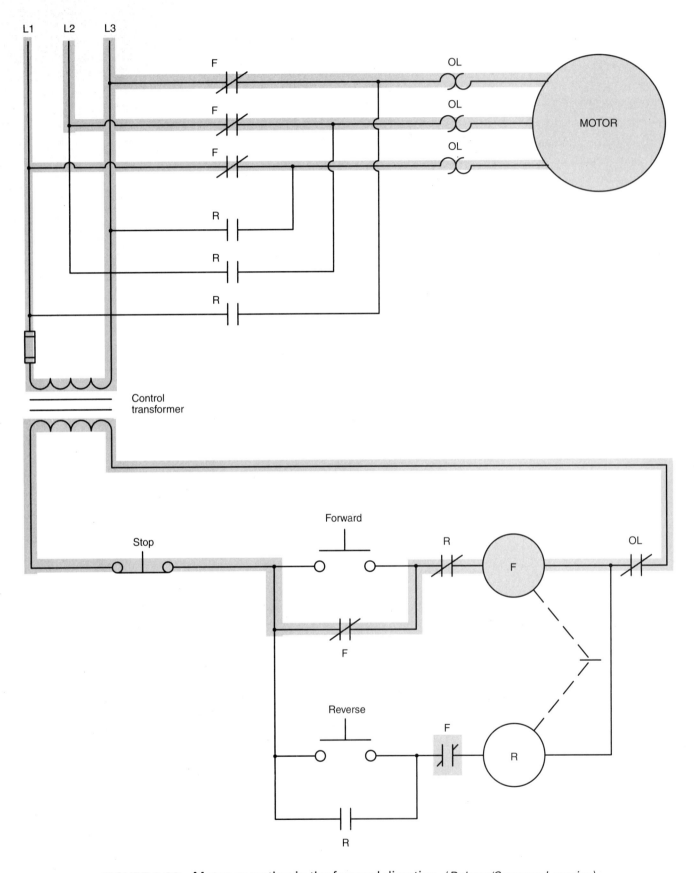

FIGURE 8-38 Motor operating in the forward direction. (*Delmar/Cengage Learning*)

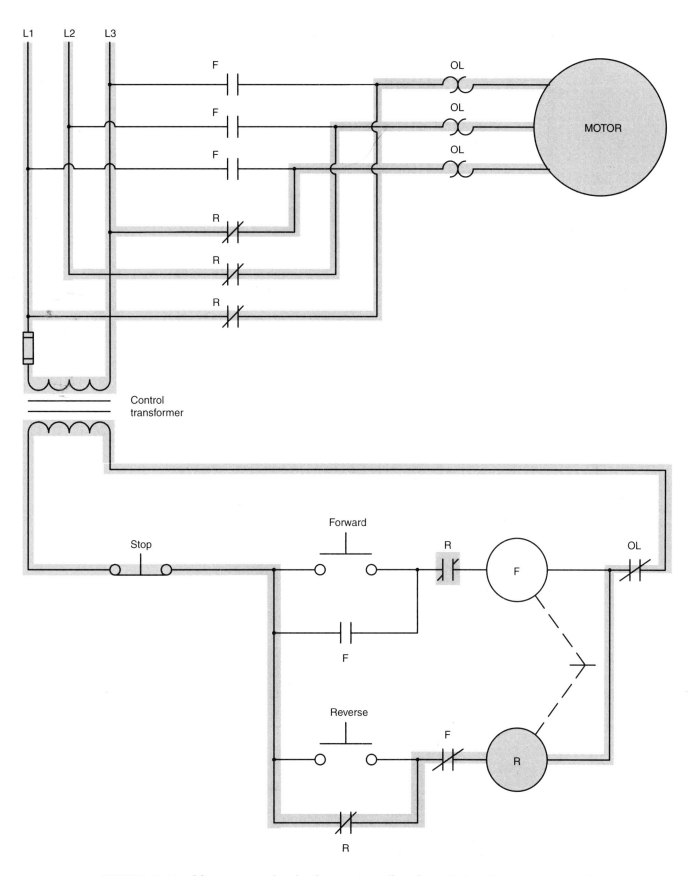

FIGURE 8-39 Motor operating in the reverse direction. (*Delmar/Cengage Learning*)

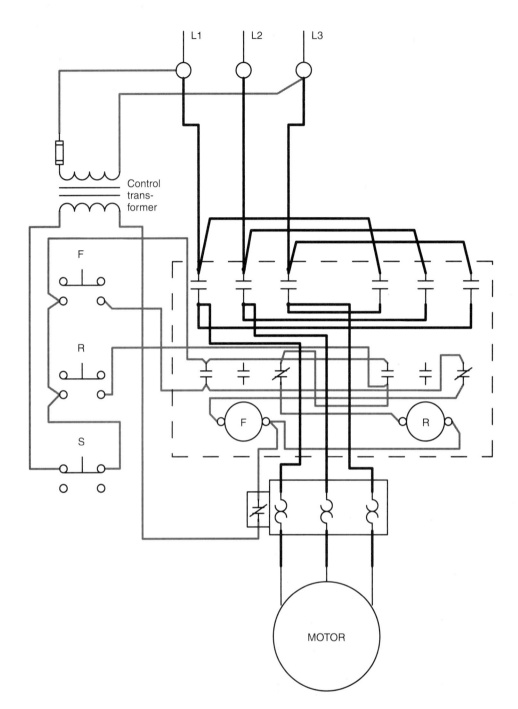

FIGURE 8-40   Wiring diagram for a forward-reverse control circuit with electrical interlock.
(*Delmar/Cengage Learning*)

to the power line. Note that the connections for L1 and L3 that go to the motor have been reversed. This causes the motor to operate in the reverse direction. Also note that the normally closed R auxiliary contact connected in series with F coil is now open, preventing a circuit from being established to F coil if the forward push button should be pressed. The wiring diagram of a forward-reverse control circuit with electrical interlocks is shown in Figure 8-40.

BASIC AIR-CONDITIONING CIRCUIT

A basic circuit for a central air-conditioning system is shown in Figure 8-41. The circuit ensures that the condenser fan is in operation before the compressor is permitted to start. A flow switch is used to sense the airflow caused by the condenser fan. The compressor is also protected from low pressure and

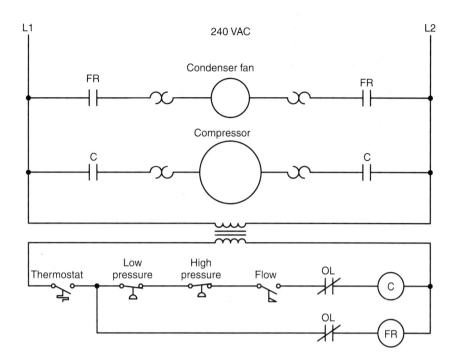

FIGURE 8-41   Basic control circuit for a central air conditioner. (*Delmar/Cengage Learning*)

high pressure by pressure switches. Both the condenser fan and compressor are also protected from overload by overload relays. A thermostat is used to control the operation of the circuit. A transformer is used to step the 240 volts of line voltage down to 24 volts for operation of the control circuit.

REVIEW QUESTIONS

All answers should be written in complete sentences, calculations should be shown in detail, and *Code* references should be cited when appropriate.

1. What are the two basic types of control circuits? _____

   _____

2. Describe the differences between relays, contactors, and motor starters. _____

   _____

   _____

3. The most commonly accepted set of schematic symbols are used by what organization?

   _____

4. When contact and switch symbols are drawn on a schematic, are they drawn in such a manner as to represent their state when the circuit is in operation or turned off? _____

   _____

5. True or false: A normally open push button symbol is drawn so that the movable contact is above and touching the stationary contacts. _____

6. True or false: A normally closed push button symbol is drawn so that the movable contact is below and touching the stationary contacts. _____

7. A limit switch is illustrated by drawing a wedge shape on the movable contact of a switch symbol. What does the wedge represent? _____

8. What are the two types of thermal overload relays? _____

9. Name two characteristics bimetal strip–type overload relays generally possess that solder pot types do not. _____
_____

10. If three single-phase overload relays are to be used to protect a 3-phase motor, how are the normally closed overload contacts arranged? _____
_____

11. When overload contacts are connected into a control circuit, are they connected in series with the starter coil or in parallel with the starter coil? _____

12. Explain how a schematic or ladder diagram differs from a wiring diagram in how it shows an electrical circuit. _____
_____
_____

13. In a start-stop push-button control circuit, what is the function of the normally open auxiliary contact connected in parallel with the start push button? _____
_____

14. In a forward-reverse control circuit, what is the function of the normally closed F and R auxiliary contacts? _____

15. Refer to the circuit shown in Figure 8-41. Describe the type of switch used for the following (normally open, normally closed, normally open held closed, or normally closed held open).

Thermostat   _____
Low-pressure switch   _____
High-pressure switch   _____
Flow switch   _____

Motors and Controllers

OBJECTIVES

After studying this chapter, the student should be able to

- describe the machine layout in the industrial building.
- describe the various types of motors used in the industrial building.
- explain the operation of the types of motor controllers used.
- describe how the motor branch circuits are installed.

Chapters 3 and 5 of this text detailed the method of distributing power to the various machines in the manufacturing area of the industrial building. Recall that plug-in busway is installed throughout the plant, and bus plugs are installed at selected points. With the use of rubber cord drops to each machine, power is supplied to the motor branch circuit that operates each machine.

THE MACHINES AND THEIR MOTORS

Sheet E-2 of the industrial building plans shows the layout of the 111 machines in the manufacturing area. The various types of machines are identified by a code number, as shown in the following list.

MA	Engine Lathes
MB	Turret Lathes
MC	Vertical Drills
MD	Multispindle Drills
ME	Milling Machines
MF	Shapers
MG	Vertical Boring Mills
MH	Planers
MI	Power Hacksaws
MJ	Band Saws
MK	Surface Grinders
ML	Cylindrical Grinders
MN	Punch Presses
MO	Special Machines

Each of these machines has a 3-phase motor rated at 460 volts. The current required by each motor is based on the horsepower rating of the motor. The current can be determined by the following equation:

$$\text{Amperes} = \frac{\text{hp} \times 746}{\text{volts} \times 1.73 \times \text{eff.} \times \text{PF}}$$

where
- hp = horsepower
- 1.73 = the square root of 3
- eff. = the assumed efficiency
- PF = the power factor (estimated)
- 746 = watts per horsepower

By applying this equation to the engine lathes (MA), the current required can be determined.

$$\text{Amperes} = \frac{5 \times 746}{460 \times 1.73 \times 0.82 \times 0.86} = 6.64$$

The efficiency of 82 percent and the power factor of 86 percent were taken from Table 9-1. Note that larger motors may have slightly higher efficiencies, whereas smaller motors usually have lower power factors and lower efficiencies. The values used in the equation are the assumed values at full load. When the motor is less than fully loaded, the values are much lower.

When the conductors to a motor are being selected, *430.6(A)* requires that the values given in *NEC Tables 430.247, 430.248, 430.249,* and *430.250* (reproduced in this book as Tables 9-1 through 9-4) be used in place of the actual full-load current of the motor, as determined by the equation given previously. Thus, for a 5-horsepower, 460-volt, 3-phase motor (look ahead to Table 10-4), a full-load current of 7.6 amperes is used to determine conductor sizes rather than the value of 6.64 amperes as calculated.

MOTOR TYPES

Several different types of motors having entirely different characteristics or patterns of performance are required for the various machine tools. One of the most commonly used motors is the squirrel-cage type. Refer to Figure 9-1 for a listing of motor control symbols.

SINGLE-SPEED SQUIRREL-CAGE INDUCTION MOTOR

The squirrel-cage type of induction motor does not have a conventional rotor winding. Instead, the laminated steel rotor has copper or aluminum bars that run axially around the periphery of the rotor. These bars are short-circuited by copper or aluminum end rings. When aluminum is used for the assembly, the bars and end rings are usually cast in one piece.

Three-phase squirrel-cage motors have a good starting torque, and their performance characteristics make them an ideal motor for general use. Figure 9-2 is a cutaway view of a 3-phase squirrel-cage motor.

An induction motor is much like a transformer, except that the secondary winding and core are mounted on a shaft set in bearings. This

TABLE 9-1

Motor efficiencies and power factors.

	Average Efficiencies and Power Factors for Polyphase Squirrel-Cage Induction Motors					
	Efficiencies			Power Factor		
Hp	One-half Load	Three-fourths Load	Full Load	One-half Load	Three-fourths Load	Full Load
¼	60.0	67.0	69.0	45	56	65
½	64.0	68.0	69.0	48	58	65
1	75.0	77.0	76.0	57	69	76
1 ½	75.0	77.0	78.0	64	76	81
2	77.0	80.0	81.0	68	79	84
3	80.0	82.0	81.0	70	80	84
5	80.0	82.0	82.0	76	83	86
7 ½	83.0	85.0	85.0	77	84	87
10	83.0	85.0	85.0	77	86	88
15	84.0	86.0	88.0	81	85	87
20	87.0	88.0	87.0	82	86	87
25	87.0	88.0	87.5	82	86	87
30	87.5	88.5	88.0	83	86.5	87
40	87.5	89.0	89.5	84	87	88
50	87.5	89.0	89.5	84	87	88
60	88.0	89.5	89.0	84	87	88
75	88.5	89.5	89.5	84	87	88
100	89.0	90.0	90.5	84	88	88
125	90.0	90.5	91.0	84	88	89
150	90.0	91.5	92.0	84	88	89
200	90.0	91.5	92.0	85	89	90
250	91.0	92.5	93.0	84	89	90
300	92.0	93.5	94.0	84	89	90

arrangement permits the secondary winding to rotate (hence the name rotor). An induction motor consists of two electrical circuits (the stator and the rotor) linked by a common magnetic circuit. Electric current applied to the stator winding induces a secondary current in the rotor winding. This winding is a closed circuit, either a short-circuit or nearly so. The induced current in the secondary always flows in a direction opposite to that of the applied current. In addition, the induced current lags 90° or one-quarter cycle behind the applied current. Magnetic fields are set up in the stator and rotor in a manner that gives rise to attracting and repelling forces. Because these forces are in the same direction (either clockwise or counterclockwise), a torque is produced and rotation results.

For example, Figure 9-3 shows that the north and south poles of the induction motor stator rotate at synchronous speed. That is, the poles of the stator and rotor are always in the position shown, with

respect to each other. Because unlike poles attract and like poles repel, forces are set up that produce rotation. The force acting at the rim of the rotor multiplied by the radius from the center of the rotor is called the torque. Torque can be determined by the following equation:

$$T = \frac{hp \times 5252}{rpm}$$

where
T = torque (in lb.-ft)
hp = motor horsepower
5252 = constant (33,000/2π)
rpm = rotor speed (in revolutions per minute)

Figure 9-3 also shows that the magnetic poles of the rotor are always midway between the magnetic poles of the stator so that the attracting and repelling forces work together. The frequency of the current in this case is 60 hertz (supplied by the power company). Sixty-hertz current is applied to the stator winding, but the frequency in the rotor

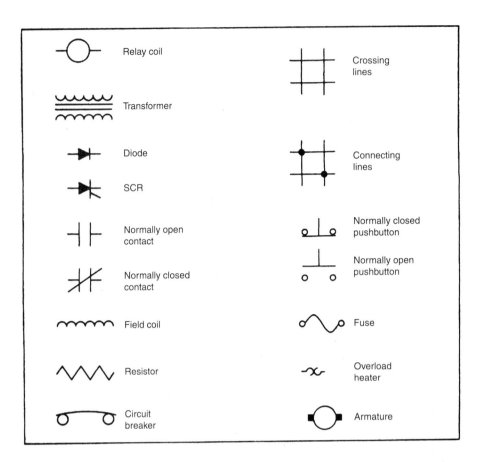

FIGURE 9-1 Electrical symbols. (*Delmar/Cengage Learning*)

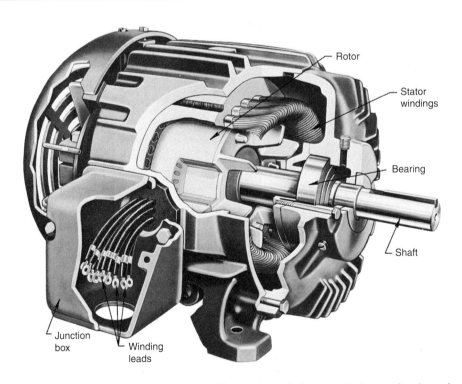

FIGURE 9-2 Cutaway view of 5-horsepower, totally enclosed, fan-cooled standard squirrel-cage motor.
(*Delmar/Cengage Learning*)

FIGURE 9-3 Diagram of 4-pole induction motor. (*Delmar/Cengage Learning*)

TABLE 9-2

Single-speed squirrel-cage induction motors.

Code Number	Number of Machines	Kind of Machines	Number of Motors	Hp of Motors
MA	20	Engine Lathes	20	5
MB	10	Turret Lathes	10	7.5
MC	12	Vertical Drills	12	1
ME	6	Milling Machines	18	10, 1, 1
MF	6	Shapers	6	7.5
MG	5	Boring Mills	15	3, 3, 3
MI	6	Power Hacksaws	6	3
MJ	4	Band Saws	4	5
MK	6	Surface Grinders	6	10
ML	10	Cylindrical Grinders	10	7.5
	5	Special Machines	5	5

is very low at operating speed and varies with the slip. The slip is the difference between the synchronous speed of the motor and its actual speed under full load.

The synchronous speed of an ac motor is obtained from the formula given as follows:

$$\text{Synchronous speed} = \frac{120 \times \text{frequency}}{\text{number of poles per phase}} = \text{rpm}$$

where the 120 is used to convert from seconds to minutes and to adjust for pairs of poles and the frequency is in cycles per second.

Thus, for a 4-pole, 60-hertz motor, the synchronous speed is

$$\text{Synchronous speed} = \frac{120 \times 60}{4} = 1800 \text{ rpm}$$

If the load causes the rotor to slip 75 rpm below the value of the synchronous speed, then the actual speed under full load is 1800 minus 75, or 1725 revolutions per minute.

Similarly, the synchronous speed of a 6-pole, 60-hertz motor is

$$\text{Synchronous speed} = \frac{120 \times 60}{6} = 1200 \text{ rpm}$$

Thus, with a full-load slip of 60 rpm, the actual full-load speed of the motor is 1200 minus 60, or 1140 rpm.

Induction Motors in the Industrial Plant

Many of the machines in the industrial building are driven by 3-phase, squirrel-cage, single-speed induction motors, Table 9-2. In addition to the motors listed in Table 9-2, other single-speed squirrel-cage motors are used as listed in Table 9-3.

TABLE 9-3

List of single-speed squirrel-cage motors.

Number of Motors	Descriptions of Motors
6	3-hp motors driving the six ventilating blowers on the roof
6	12.5-hp motors (two to a unit) installed in the three liquid chillers used in conjunction with the air-conditioning equipment
10	2-hp motors for the fan coil units
23	¼-hp motors used at the twenty-three machines equipped with oil fog precipitation units

The smaller sizes of squirrel-cage motors use controllers known as across-the-line starters, Figure 9-4. The controller is a magnetic switch or contactor including overload relays that provide running protection for the motor. A pushbutton station can provide a means for starting, stopping, reversing, or jogging the motor, Figure 9-5 and Figure 9-6.

Four-Speed Squirrel-Cage Motors

The manufacturing area of the industrial building contains eight multispindle drills (MD) that are equipped with 4-speed, two-winding, squirrel-cage induction motors. For this type of motor, each of the two windings can produce two different speeds for a total of four speeds. The windings are not used at the same time as each winding provides two speeds, one of which is half the value of the other. For example, the first winding can supply 600 rpm and 1200 rpm and the second winding can supply 900 rpm and 1800 rpm.

The four speeds are obtained by changing the number of poles in the motor. That is, certain leads of the motor are connected to the power supply for each speed desired, Table 9-4. This type of motor has fourteen terminal leads. The controller is usually arranged so that the sequence of speeds is Reverse–Off–600–900–1200–1800 rpm, Figure 9-7.

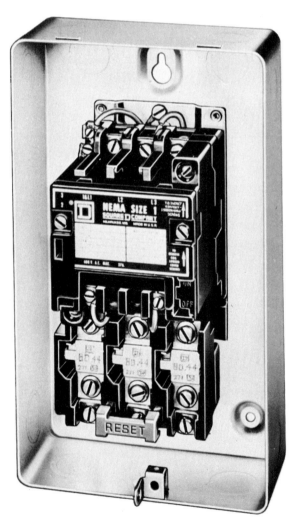

FIGURE 9-4   Across-the-line magnetic motor starter. (*Delmar/Cengage Learning*)

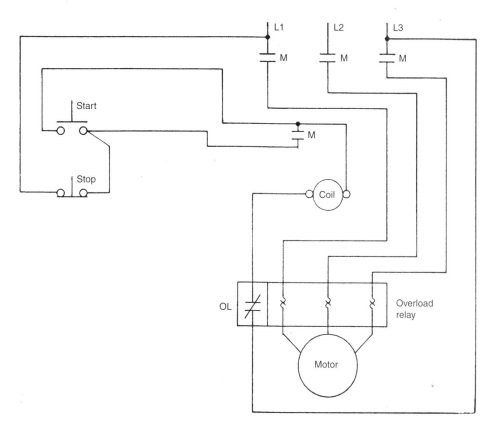

FIGURE 9-5 Wiring diagram of an across-the-line starter. (*Delmar/Cengage Learning*)

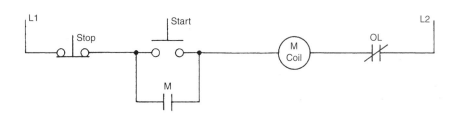

FIGURE 9-6 Schematic diagram of an across-the-line starter. (*Delmar/Cengage Learning*)

TABLE 9-4

Connection of leads for various desired speeds.

Speed	L1	L2	L3	Together
R 600	T2	T1	T3, T7	—
F 600	T1	T2	T3, T7	—
F 900	T11	T12	T13–T17	—
F 1200	T6	T4	T5	T1, T2, T3, T7
F 1800	T16	T14	T15	T11, T12, T13, T17

Note: All other terminals are left open.

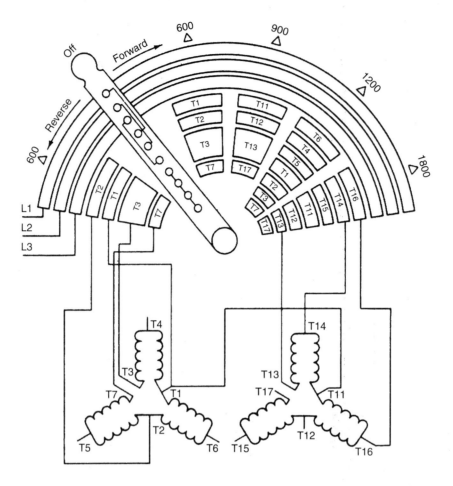

This multispeed squirrel-cage induction motor is shown partially wired. The motor has fourteen terminal leads. Controllers are made in various forms such as a drum type, a cam type, and an automatic push-button type. Numbered terminals on the motor are connected to the corresponding numbers on the controller.

FIGURE 9-7 Four-speed, two-winding induction motor and controller. (*Delmar/Cengage Learning*)

Primary Resistance Starters

The milling machines are equipped with primary resistance starters. For this type of starter or controller, the heavy starting current results in a voltage drop while it passes through the primary resistors; thus, there is a lower voltage value at the motor terminals. The motor accelerates gently with less torque than is the case when line starters are used. When the motor has almost reached its normal speed, a time-delay relay (set for about 5 seconds) closes a second contactor to short out the primary resistors. At this point, the motor receives the full line voltage and accelerates to its normal speed, Figure 9-8.

Reduced Voltage Starters

The motors used on the surface grinders have another type of controller called a reduced voltage starter, Figure 9-9. This type of controller uses an autotransformer to obtain a reduced voltage. When the starting push-button is depressed, a magnetic 5-pole contactor (S) connects the autotransformer to the line. Taps are made from the autotransformer at a value of about 70 percent of the line voltage at the start portion of the cycle. Several seconds after the motor begins to rotate, a timing relay opens the first contactor (S) and closes a second 3-pole contactor (R). This action disconnects the autotransformer from the line

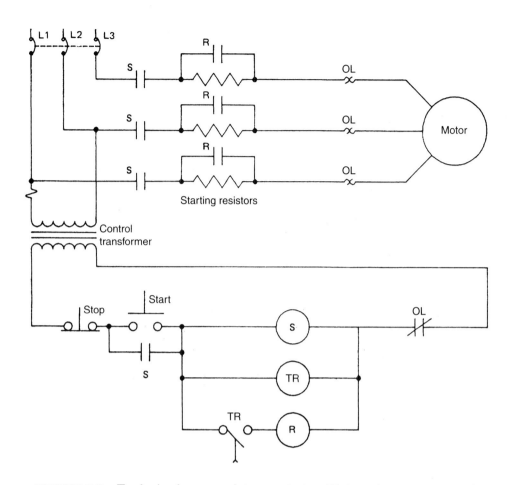

FIGURE 9-8   Typical primary resistance starter. (*Delmar/Cengage Learning*)

and connects the motor directly across the line. As a result, the motor is accelerated to its normal speed, Figure 9-10.

The controllers covered in the chapter to this point are used with squirrel-cage motors. There are several other types of controllers used on other motors of various types in the industrial building.

THE WOUND-ROTOR INDUCTION MOTOR

The punch presses (MN) are equipped with wound-rotor induction motors. These motors operate on the same rotating magnetic field principle as a squirrel-cage induction motor. The difference between the two motors is the construction of the rotor. The rotor of the squirrel-cage induction motor contains bars connected together at each end by shorting rings.

The rotor of the wound-rotor induction motor contains a 3-phase winding very similar to the stator winding. The stator winding leads are marked L1, L2, and L3. The rotor winding leads are marked M1, M2, and M3. One end of each rotor winding is connected together with the others to form a wye connection. The other end of the winding is connected to slip rings (collector rings) located on the rotor shaft, Figure 9-11. Low-resistance carbon brushes in contact with the slip rings provide connection to external resistors, Figure 9-12. The ability to control the amount of resistance connected to the rotor circuit causes the wound-rotor induction motor to exhibit several desirable characteristics over other types of 3-phase motors:

1. The amount of starting current can be controlled by controlling the amount of resistance connected to the rotor. Induction motors are

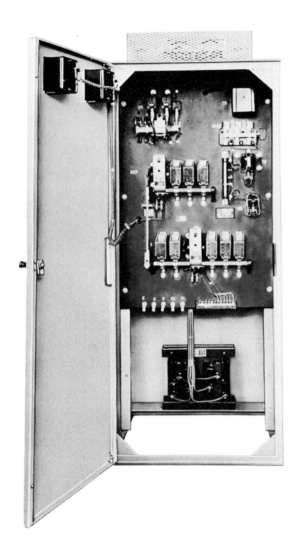

FIGURE 9-9 Reduced voltage starter.
(*Delmar/Cengage Learning*)

very similar to transformers. The stator winding is the primary and the rotor is the secondary. Limiting the amount of rotor (secondary) current limits the stator (primary) current also.

2. The wound-rotor induction motor exhibits the highest amount of starting torque per ampere of starting current of any 3-phase motor. There are three factors that determine the amount of torque developed by an induction motor:

   a. The magnetic field strength of the stator.

   b. The magnetic field strength of the rotor.

   c. The phase angle difference between rotor and stator current.

The torque reaches maximum when the rotor and stator currents are in phase with each other. Adding resistance to the rotor circuit causes the rotor current to be more in phase with the stator current, thus producing a greater amount of starting torque.

3. The speed of the wound-rotor induction motor can be controlled by the amount of resistance connected in the rotor circuit. Because the amount of resistance in the rotor circuit controls the current in both the rotor and stator windings, it controls the strength of the magnetic fields in the rotor and stator windings. Controlling the magnetic field strength controls the amount of torque produced by the motor. Inserting resistance in the rotor circuit results in a reduction of torque, which causes a greater amount of slip between the speed of the rotor and the speed of the rotating magnetic field. When resistance is decreased, magnetic field strength increases, causing an increase in torque and a corresponding increase in speed. Full motor speed is obtained when all the resistance has been shorted out of the rotor circuit.

Secondary Resistance Controller

The punch press motors can be operated in any of three speeds. The control circuit is shown in Figure 9-13. Motor speed is controlled by two sets of three resistors each connected in the rotor circuit. Three push buttons permit the motor to be operated in any of the three speeds. The speed can be changed at any time by pressing the appropriate button. However, there is a time delay of three seconds between any increases in speed. If the third speed button is pressed, for example, the motor will start in the first or lowest speed. After a three-second delay, the motor will increase to the second speed. After another three-second delay, the motor will increase to third or full speed. Motor speed can be decreased by pressing either of the other two push buttons. There is no time delay if the motor speed is decreased.

The wound-rotor induction has a larger than normal shaft diameter because of its ability to develop high torque. It is possible for the motor to

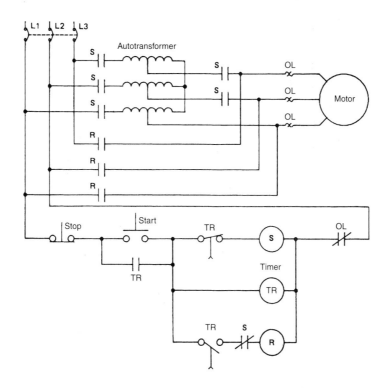

FIGURE 9-10 Schematic of an autotransformer-type reduced voltage starter. (*Delmar/Cengage Learning*)

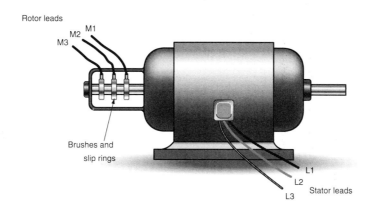

FIGURE 9-11 Wound-rotor induction motor. (*Delmar/Cengage Learning*)

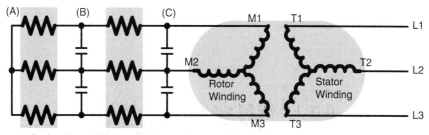

Connection at (A) results in first or low speed operation.
Connection at (B) results in second speed operation.
Connection at (C) results in third or high speed operation.

FIGURE 9-12 Wound-rotor motor connections. (*Delmar/Cengage Learning*)

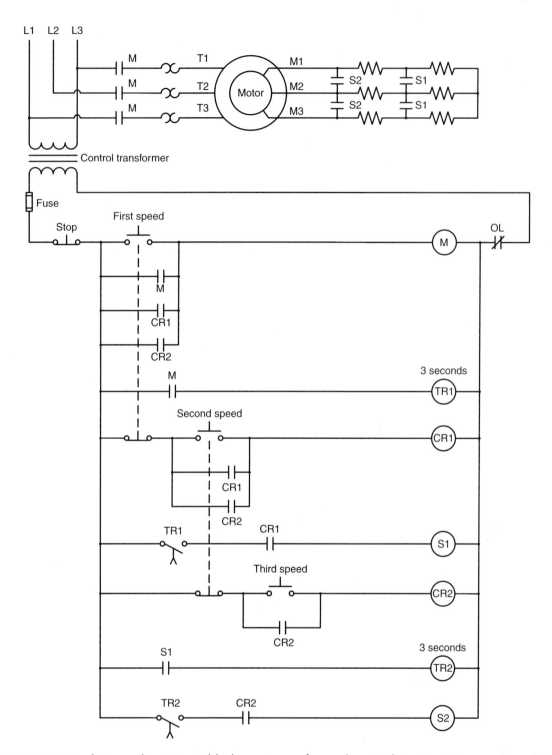

FIGURE 9-13   Automatic starter with three steps of speed control. (*Delmar/Cengage Learning*)

develop torque that is 300 percent above normal running torque. This can create a great amount of stress on the shaft.

The wound-rotor motor is used for extra heavy-duty starting. Typical applications include the use of this type of motor with pumps having an extremely high back pressure, or with machines having a very high static inertia. The secondary resistance controller is used to bring the motor up to speed smoothly. In addition, this controller is used in normal running operations to adjust the torque and speed to any desired values.

⊶∿⊷ DETERMINING DIRECTION OF ROTATION FOR 3-PHASE MOTORS

On many types of machinery, the direction of rotation of the motor is critical. The direction of rotation of any 3-phase motor can be changed by reversing two of its stator leads. This causes the direction of the rotating magnetic field to reverse. When a motor is connected to a machine that will not be damaged when its direction of rotation is reversed, power can be momentarily applied to the motor to observe its direction of rotation. If the rotation is incorrect, any two line leads can be interchanged to reverse the motor's rotation.

When a motor is to be connected to a machine that can be damaged by incorrect rotation, however, the direction of rotation must be determined before the motor is connected to its load. This can be accomplished in two basic ways. One way is to make electrical connection to the motor before it is mechanically connected to the load. The direction of rotation can then be tested by momentarily applying power to the motor before it is coupled to the load.

There may be occasions when this is not practical or convenient. It is possible to determine the direction of rotation of a motor before power is connected to it with the use of a phase rotation meter, as shown in Figure 9-14. The phase rotation meter is used to compare the phase rotation of two different 3-phase connections. The meter contains six terminal leads. Three of the leads are connected to one side of the meter and labeled "Motor." Each of these three motor leads is labeled A, B, or C. The *line* leads are located on the other side of the meter, and each of these leads is labeled A, B, or C.

To determine the direction of rotation of the motor, first zero the meter by following the instructions provided by the manufacturer. Then set the meter selector switch to motor, and connect the three *motor* leads of the meter to the "T" leads of the motor, as shown in Figure 9-15. The phase rotation meter contains a zero-center voltmeter. One side of the voltmeter is labeled "INCORRECT," and the other side is labeled "CORRECT." While observing the zero-center voltmeter, turn the motor shaft in the direction of desired rotation. The zero-center voltmeter will immediately swing in the CORRECT or INCORRECT direction. When the motor shaft stops turning, the needle may swing in the opposite direction. It is the *first* indication of the voltmeter that is to be used.

If the voltmeter needle indicated CORRECT, label the motor "T" leads A, B, or C to correspond with the *motor* leads from the phase rotation meter. If the voltmeter needle indicated INCORRECT, change any two of the *motor* leads from the phase rotation meter and again turn the motor shaft. The voltmeter needle should now indicate CORRECT. The motor "T" leads can now be labeled to correspond with the *motor* leads from the phase rotation meter.

After the motor "T" leads have been labeled A, B, or C to correspond with the leads of the phase rotation meter, the rotation of the line supplying power to the motor must be determined. Set the selector switch on the phase rotation meter to the *line* position. After making certain the power has been turned off, connect the three *line* leads of the phase rotation meter to the motor supply line, Figure 9-16. Turn on the power and observe the zero-center voltmeter. If the meter is pointing in the CORRECT direction, turn off the power and label the line leads A, B, or C to correspond with the *line* leads of the phase rotation meter.

If the voltmeter is pointing in the INCORRECT direction, turn off the power and change any two of the leads from the phase rotation meter. When the

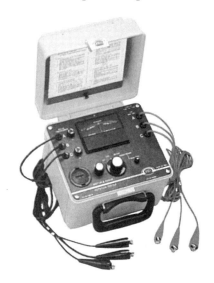

FIGURE 9-14 Phase rotation meter. (*Courtesy of Megger®*)

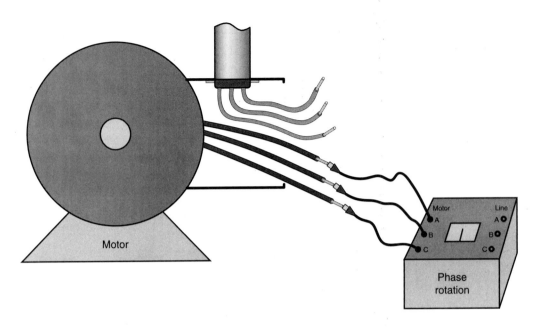

FIGURE 9-15 Connecting the phase rotation meter to the motor. (*Delmar/Cengage Learning*)

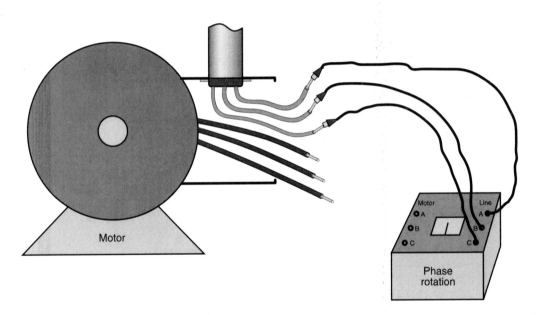

FIGURE 9-16 Connecting the phase rotation meter to the line. (*Delmar/Cengage Learning*)

power is turned on, the voltmeter should point in the CORRECT direction. Turn off the power and label the line leads A, B, or C to correspond with the leads from the phase rotation meter.

Now that the motor "T" leads and the incoming power leads have been labeled, connect the line lead labeled A to the "T" lead labeled A; the line lead labeled B to the "T" lead labeled B; and the line lead labeled C to the "T" lead labeled C. When

power is connected to the motor, it will operate in the proper direction.

Notice that the phase rotation meter can be used to determine the *phase rotation* of two different connections. It cannot determine which of the three phase lines is A, B, or C, or which line lead is L1, L2, or L3. The phase rotation meter can be used to determine the rotation of two separate 3-phase systems. For example, assume all the short-circuit protective devices

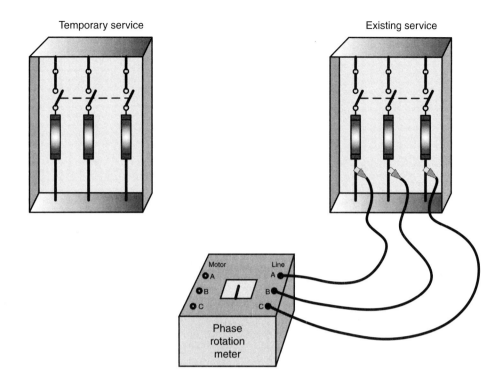

Temporary service

Existing service

FIGURE 9-17 Testing the phase rotation of the existing service. (*Delmar/Cengage Learning*)

and switch gear for an existing 3-phase system must be replaced. To minimize downtime, a temporary 3-phase service will be connected to supply power while the existing switch gear is being replaced. It is critical that the phase rotation of the temporary service be the same as the existing service when power is applied. The phase rotation meter can be used to ensure the connection is correct.

The first step is to connect the *line* leads of the phase rotation meter to the existing power, Figure 9-17. If the zero-center voltmeter indicates CORRECT, label the load side of the service A, B, and C to correspond with the leads of the phase rotation meter. If the voltmeter indicates INCORRECT, change two of the meter leads. This should cause the phase rotation meter to indicate CORRECT. Label the load side of the service to correspond with A, B, or C of the phase rotation meter leads.

Before connecting the temporary service to the load side of the circuit, connect the phase rotation meter to the line side of the temporary service, Figure 9-18. Obtain a CORRECT reading on the phase rotation meter by changing two of the meter leads if necessary. After the correct reading has been obtained, label the service leads A, B, and C to correspond with the leads of the phase rotation meter. If the marked temporary service leads are connected to their like-marked load leads, the phase rotation of the temporary service will be the same as the existing service.

CONNECTING DUAL-VOLTAGE 3-PHASE MOTORS

Many of the 3-phase motors used in industry are designed to be operated on two voltages, such as 240 to 480 volts. Motors of this type contain two sets of windings per phase. Most dual-voltage motors bring out nine "T" leads at the terminal box. There is a standard method used to number these leads, as shown in Figure 9-19. Starting with terminal #1, the leads are numbered in a decreasing spiral as shown. Another method of determining the proper lead numbers is to add three to each terminal. For example, starting with lead #1, add three to one. Three plus one equals four. The phase winding that begins with #1 ends with #4. Now add three to four.

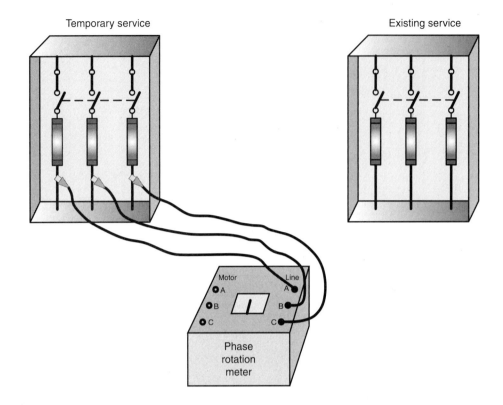

FIGURE 9-18 Testing the phase rotation of the temporary service. (*Delmar/Cengage Learning*)

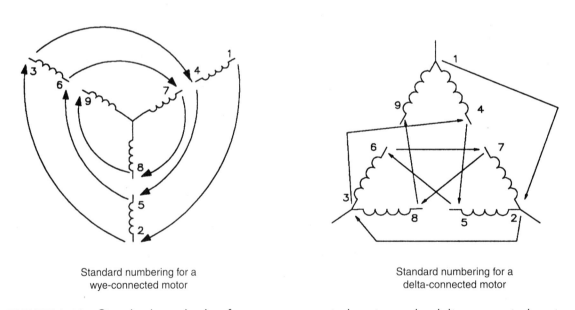

Standard numbering for a wye-connected motor

Standard numbering for a delta-connected motor

FIGURE 9-19 Standard numbering for a wye-connected motor and a delta-connected motor. (*Delmar/Cengage Learning*)

Three plus four equals seven. The beginning of the second winding for phase one is seven. This method will work for the windings of all phases. If in doubt, draw a diagram of the phase windings and number them in a spiral.

Three-phase motors can be constructed to operate in either wye or delta. If a motor is to be connected to high voltage, the phase windings will be connected in series. In Figure 9-20, a schematic diagram and terminal connection chart for high

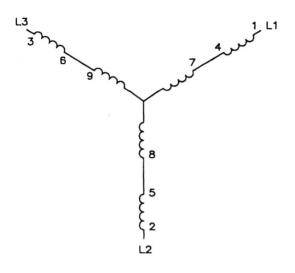

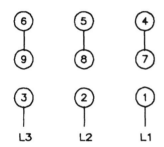

FIGURE 9-20   High-voltage wye connection. (*Delmar/Cengage Learning*)

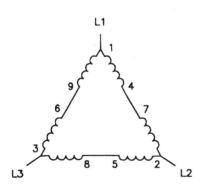

FIGURE 9-21   High-voltage delta connection. (*Delmar/Cengage Learning*)

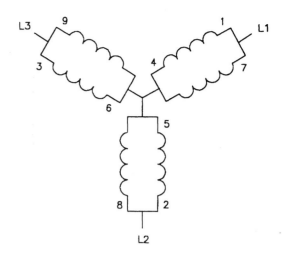

FIGURE 9-22   Stator windings connected in parallel. (*Delmar/Cengage Learning*)

voltage are shown for a wye-connected motor. In Figure 9-21, a schematic diagram and terminal connection chart for high voltage are shown for a delta-connected motor.

When a motor is to be connected for low-voltage operation, the phase windings must be connected in parallel. Figure 9-22 shows the basic schematic diagram for a wye-connected motor with parallel phase windings. In actual practice, however, it is not possible to make this exact connection with a nine-lead motor. The schematic shows that terminal #4 connects to the other end of the phase winding that starts with terminal #7. Terminal #5 connects to the other end of winding #8, and terminal #6 connects to the other end of winding #9. In actual motor construction, the opposite ends of windings 7, 8, and 9 are connected together inside the motor and are not brought outside the motor case. The problem is solved, however, by forming a second wye connection by connecting terminals 4, 5, and 6, as shown in Figure 9-23.

The phase winding of a delta-connected motor must also be connected in parallel for use on low voltage. A schematic for this connection is shown in Figure 9-24. A connection diagram and terminal connection chart for this hookup are shown in Figure 9-25.

Some dual-voltage motors will contain twelve "T" leads instead of nine. In this instance, the opposite ends of terminals 7, 8, and 9 are brought

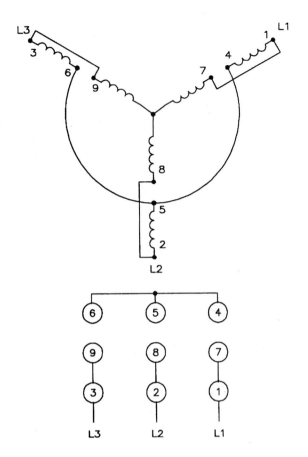

FIGURE 9-23 Low-voltage wye connection.
(*Delmar/Cengage Learning*)

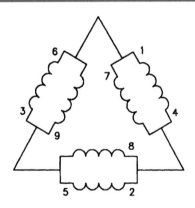

FIGURE 9-24 Parallel delta connection.
(*Delmar/Cengage Learning*)

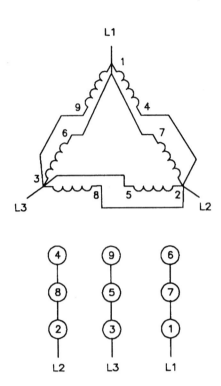

FIGURE 9-25 Low-voltage delta connection.
(*Delmar/Cengage Learning*)

DUAL-VOLTAGE SINGLE-PHASE MOTORS

Many single-phase motors are designed to be connected to either 120 or 240 volts. Most dual-voltage single-phase motors will be of the split-phase type, which contains both run and starting windings. Figure 9-27 shows the schematic diagram of a split-phase motor designed for dual-voltage operation. This particular motor contains two run windings and two start windings. The lead numbers for single-phase motors are also numbered in a standard manner. One of the run windings has lead numbers of T1 and T2. The other run winding has its leads numbered T3 and T4. This particular motor uses two different sets of start winding leads. One set is labeled T5 and T6, and the other set is labeled T7 and T8.

If the motor is to be connected for high-voltage operation, the run windings and start windings will be connected in series, as shown in Figure 9-28. The start windings are then connected in parallel with the run windings. It should be noted that if the opposite direction of rotation is desired, T5 and T8 will be changed.

out for connection. Figure 9-26 shows the standard numbering for both delta- and wye-connected motors. Twelve leads are brought out if the motor is intended to be used for wye-delta starting. When this is the case, the motor must be designed for normal operation with its windings connected in delta. If the windings are connected in wye during starting, the starting current of the motor is greatly reduced.

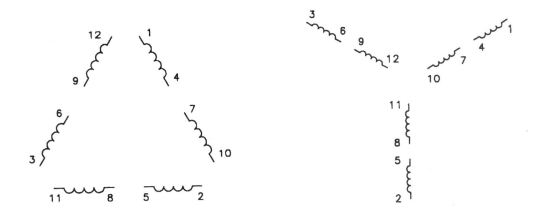

FIGURE 9-26 Twelve-lead motor. (*Delmar/Cengage Learning*)

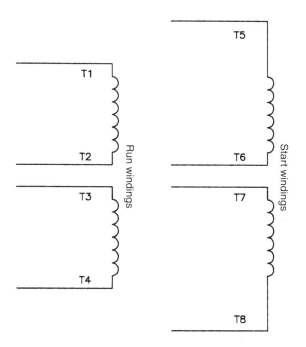

FIGURE 9-27 Single-phase dual-voltage motor. (*Delmar/Cengage Learning*)

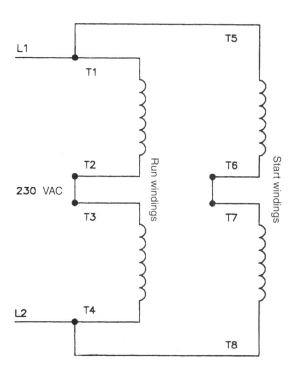

FIGURE 9-28 High-voltage connection for single-phase motor with two run windings and two start windings. (*Delmar/Cengage Learning*)

For low-voltage operation, the windings must be connected in parallel, as shown in Figure 9-29. This connection is made by first connecting the run windings in parallel by hooking T1 and T3 together, and T2 and T4 together. The start windings are paralleled by connecting T5 and T7 together, and T6 and T8 together. The start windings are then connected in parallel with the run windings. If the opposite direction of rotation is desired, T5 and T6, and T7 and T8 should be reversed.

Not all dual-voltage single-phase motors contain two sets of start windings. Figure 9-30 shows the schematic diagram of a motor that contains two sets of run windings and only one start winding. In this illustration, the start winding is labeled T5 and T6. It should be noted, however, that some motors identify the start winding by labeling it T5 and T8, as shown in Figure 9-31.

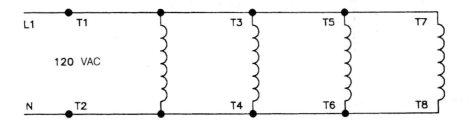

FIGURE 9-29 Low-voltage connection for single-phase motor with two start windings. (*Delmar/Cengage Learning*)

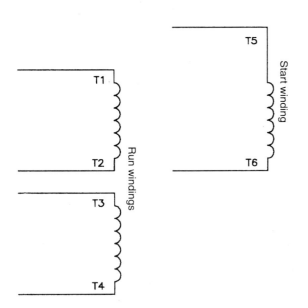

FIGURE 9-30 Dual-voltage motor with one start winding labeled T5 and T6. (*Delmar/Cengage Learning*)

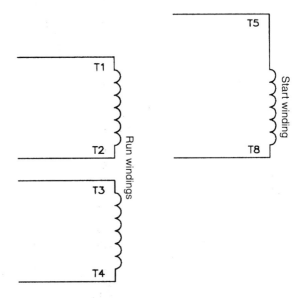

FIGURE 9-31 Dual-voltage motor with one start winding labeled T5 and T8. (*Delmar/Cengage Learning*)

Regardless of which method is used to label the terminal leads of the start winding, the connection will be the same. If the motor is to be connected for high-voltage operation, the run windings will be connected in series, and the start winding will be connected in parallel with one of the run windings, as shown in Figure 9-32. In this type of motor, each winding is rated at 120 volts. If the run windings are connected in series across 240 volts, each winding will have a voltage drop of 120 volts. By connecting the start winding in parallel across only one run winding, it will receive only 120 volts when power is applied to the motor. If the opposite direction of rotation is desired, T5 and T8 should be changed.

If the motor is to be operated on low voltage, the windings are connected in parallel, as shown in

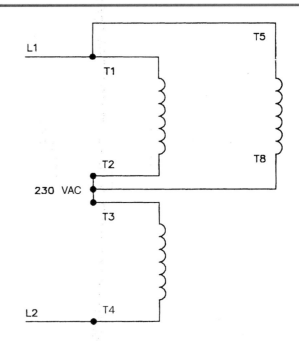

FIGURE 9-32 High-voltage connection with one start winding. (*Delmar/Cengage Learning*)

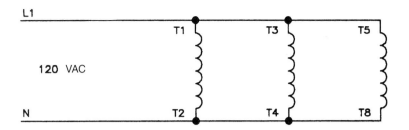

FIGURE 9-33 Low-voltage connection for a single-phase motor with one start winding. (*Delmar/Cengage Learning*)

Figure 9-33. Because all windings are connected in parallel, each will receive 120 volts when power is applied to the motor.

DETERMINING DIRECTION OF ROTATION FOR SINGLE-PHASE MOTORS

The direction of rotation of a single-phase motor can generally be determined when the motor is connected. The direction of rotation is determined by facing the back or rear of the motor. Figure 9-34 shows a connection diagram for rotation. If clockwise rotation is desired, T5 should be connected to T1. If counterclockwise rotation is desired, T8 (or T6) should be connected to T1. It should be noted that this connection diagram assumes the motor

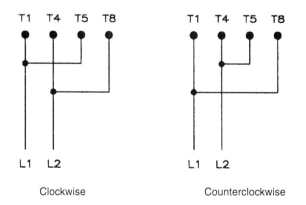

FIGURE 9-34 Determining direction of rotation for single-phase motors. (*Delmar/Cengage Learning*)

contains two sets of run and two sets of start windings. The type of motor used will determine the actual connection. For example, Figure 9-32 shows the connection of a motor with two run windings and only one start winding. If this motor were to be connected for clockwise rotation, terminal T5 should be connected to T1 and terminal T8 should be connected to T2 and T3. If counterclockwise rotation were desired, terminal T8 would be connected to T1, and terminal T5 would be connected to T2 and T3.

Direct-Current Motors

Two types of machine tools in the industrial building require dc motors. Five vertical boring mills (MG) require one dc motor each, and three planers (MH) require the same type of motors and controllers.

The motors used are standard compound wound dc motors and are all rated at 25 horsepower. The motors are not operated from regular dc sources, but rather are operated from 480-volt ac lines through electronic controllers that rectify the current (change it from ac to dc).

Direct-current compound motors contain a rotating armature and a stationary field. The field also serves as the frame or housing of the motor. End bells or end brackets support the shaft bearings. The armature has a winding that is connected to a commutator. Brush holders and carbon brushes mounted on the front end bell contact the commutator, which rotates when the motor is running.

A Exciting and commutating field coils

B Coils encased in a heat-resistant polyester compound

C Brush rigging

D Fan blades

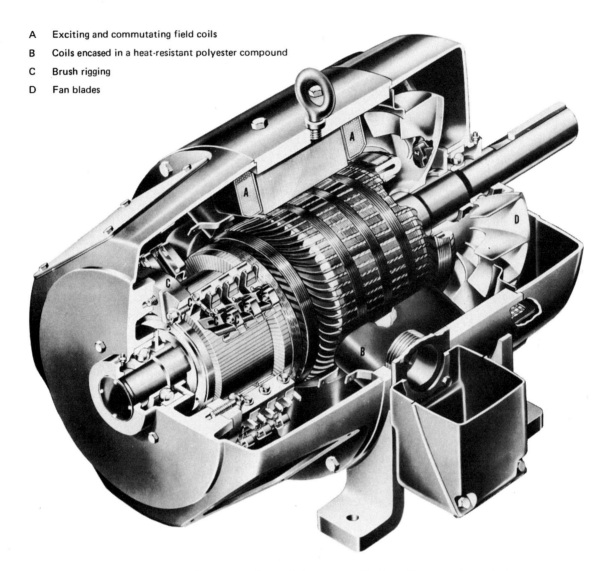

FIGURE 9-35 Cutaway view of dc motor. (*Delmar/Cengage Learning*)

A typical dc motor is shown in Figure 9-35.

The compound wound field consists of two separate field windings. The shunt field is wound with relatively small wire and has thousands of turns. The series field is wound with large wire and has only a few turns. The field windings or coils are placed on pole pieces attached to the frame or yoke.

Compound wound motors have an even number of poles, with the smaller motor sizes usually having two or four poles, and the larger motor sizes having a larger number of poles. The field frame of a dc motor is shown in Figure 9-36.

A part of the shunt field and a part of the series field are wound on each pole piece. The windings on each alternate pole piece are made in opposite directions, clockwise and counterclockwise. In this manner, each pole piece is alternately magnetized north and south. The ends of the shunt winding (two ends) and the series field winding (two ends) are brought out to the motor terminal box.

Commutating poles or interpoles are also provided. These very small pole pieces are placed midway between the main pole pieces. The interpoles are wound with a few turns of heavy wire. As with the main pole pieces, the interpoles are also wound in an alternate clockwise and counterclockwise manner. The pole pieces are connected permanently in series with the armature brush holders and are considered to be a part of the armature circuit. Interpoles counteract the

FIGURE 9-36 Field frame of a dc motor. (*Delmar/Cengage Learning*)

distortion of the field magnetism caused by the rotation of the heavily magnetized armature in the field flux. As a result, sparking or arcing at the brushes is reduced.

A dc compound motor can be connected in several ways. When the shunt field spans only the armature, it is known as a *short shunt connection*, Figure 9-37. If, on the other hand, the shunt field spans both the armature and the series field, it is called a *long shunt connection*. When the motor is connected short shunt, the shunt field current is added to the series field current. This generally causes a slight overcompounding of the motor, which permits it to exhibit stronger torque characteristics. When the motor is connected long shunt, it exhibits better speed regulation.

If the motor terminal connections are made so that the series field magnetism aids or strengthens

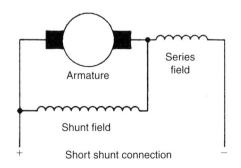
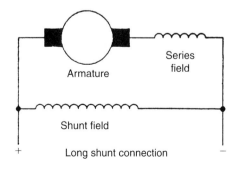

FIGURE 9-37 Direct-current motor connection. (*Delmar/Cengage Learning*)

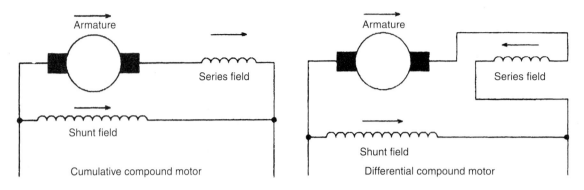

FIGURE 9-38 Connections for cumulative and differential compound connection motors. (*Delmar/Cengage Learning*)

the magnetism produced by the shunt field, then the motor is said to be a cumulative compound motor, Figure 9-38.

If the motor terminal connections are reversed so that the magnetism of the series field opposes or weakens the magnetism of the shunt field, the motor is called a differential compound motor, Figure 9-38.

Although the differential compound motor gives a more constant speed at all loads, the motor is somewhat unstable. For this reason, this type of motor is not used in as many applications as the cumulative compound motor.

The strength of the shunt field is constant. However, because the series field is connected in series with the armature, the strength of the series field varies with the load on the motor. When the motor is running at idle speed (no output), the series field contributes almost no magnetism to that of the shunt field. When the motor is loaded, the series field increases the magnetism of the shunt field to

produce more torque and cause a slight drop in the motor speed.

The armature/commutator also has poles. This component of the motor is a wrought copper cylinder with segments or bars. These segments are insulated from one another and serve as a mounting to which the armature winding is connected. A 2-pole armature has a coil span equal to the diameter of the armature, less a few slots. A 4-pole armature has a coil span equal to one-quarter of the circumference of the armature, less a few slots. A 6-pole armature has a coil span equal to one-sixth of the circumference of the armature, less one or two slots Figure 9-39. The coil span arrangement depends entirely on the number of poles present. A 4-pole armature cannot be used with a 2-pole field. The two units, the armature and the field, must be wound with the same number of poles.

It was shown that for ac motors, the torque and horsepower are proportional to the square

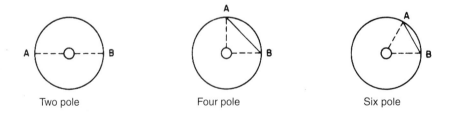

The coil span is the line A–B. With a 2-pole motor, coil span is 180. A 4-pole motor has a coil span of 90, and a 6-pole motor has a coil span of 60. In actual practice, coil span is chorded and is slightly less than the full span.

FIGURE 9-39 Relation of armature coil span to number of poles. (*Delmar/Cengage Learning*)

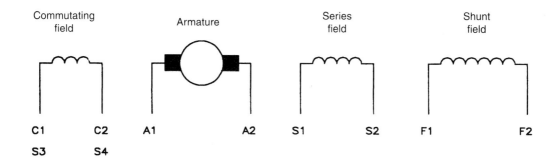

FIGURE 9-40 Lead identification for dc machines. (*Delmar/Cengage Learning*)

of the voltage applied, and the speed of rotation depends upon the frequency and the number of poles in the motor. However, the performance of a dc motor depends upon entirely different factors. The speed of a dc motor increases when the voltage increases and decreases if the field strength is increased or if there is an increase in the number of poles or turns of wire wound on the armature.

For all motors, the horsepower output is

$$hp = \frac{2\,\pi\,F \times R \times S}{33,000} \ \text{ or } \ \frac{T \times S}{5250}$$

where   F = force, in pounds
R = radius, in feet
S = speed, in rpm
T = torque, in foot-pounds

TERMINAL IDENTIFICATION FOR DIRECT-CURRENT MOTORS

The terminal leads of dc machines are labeled so they can be identified when they are brought outside the motor housing to the terminal box. Figure 9-40 illustrates this standard identification. Terminals A1 and A2 are connected to the armature through the brushes. The ends of the series field are identified with S1 and S2, and the ends of the shunt field are marked F1 and F2. Some dc machines will provide access to another set of windings called the commutating field or interpoles. The ends of this winding will be labeled C1 and C2, or S3 and S4. It is common practice to provide access to the interpole

winding on machines designed to be used as motors or generators.

DETERMINING THE DIRECTION OF ROTATION OF A DIRECT-CURRENT MOTOR

The direction of rotation of a dc motor is determined by facing the commutator end of the motor. This is generally the back or rear of the motor. If the windings have been labeled in a standard manner, it is possible to determine the direction of rotation when the motor is connected. Figure 9-41

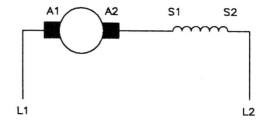

Counterclockwise rotation

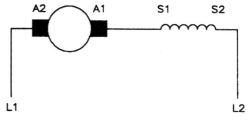

Clockwise rotation

FIGURE 9-41 Series motor. (*Delmar/Cengage Learning*)

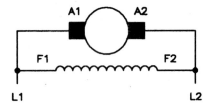

Counterclockwise rotation

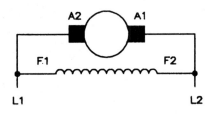

Clockwise rotation

FIGURE 9-42   Shunt motor. (*Delmar/Cengage Learning*)

illustrates the standard connections for a series motor. The standard connections for a shunt motor are illustrated in Figure 9-42, and the standard connections for a compound motor are shown in Figure 9-43.

The direction of rotation of a dc motor can be reversed by changing the connections of the arma-

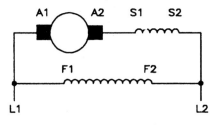

Counterclockwise rotation

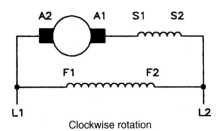

Clockwise rotation

FIGURE 9-43   Compound motor. (*Delmar/Cengage Learning*)

ture leads or the field leads. It is common practice to change the connection of the armature leads. This is done to prevent changing a cumulative compound motor into a differential compound motor.

DIRECT-CURRENT POWER SUPPLIES

The use of direct-current motors in industry creates a need for a supply of dc power. Because most of industry operates on ac power, the dc power needed is generally produced within the industrial plant. The most common method to convert ac voltage to dc voltage is by the use of solid-state components.

A simple half-wave rectifier is shown in Figure 9-44. The diode is used to convert the ac voltage to dc voltage. The diode operates like an electric check valve; it permits the current to flow through it in only one direction. When the voltage applied to the cathode end of the diode is more negative than the voltage applied to the anode end, the diode becomes *forward biased*. This permits current to flow through the load resistor and then through the diode to complete the circuit. When the voltage applied to the cathode end of the diode becomes more positive than the voltage applied to the anode end, the diode becomes *reverse biased* and turns off. When the diode is reverse biased, no current flows in the circuit. The waveforms in Figure 9-44 illustrate this condition. The negative half of the ac input wave has been cut off to produce the dc output wave. This type of rectifier is called a half-wave because only one-half of the ac waveform is used. The output voltage is pulsating. It turns on and off, but the direction of current flow never reverses. Because the output voltage never reverses direction, it is direct current.

Single-Phase, Full-Wave Rectifiers

Full-wave rectification of single-phase ac can be obtained by using either of two circuits. Figure 9-45 shows these two types of full-wave rectifiers: the two-diode type and the bridge type. The *two-diode* rectifier requires the use of a center-tapped transformer. It is the more efficient of the two because there is a voltage drop across only two

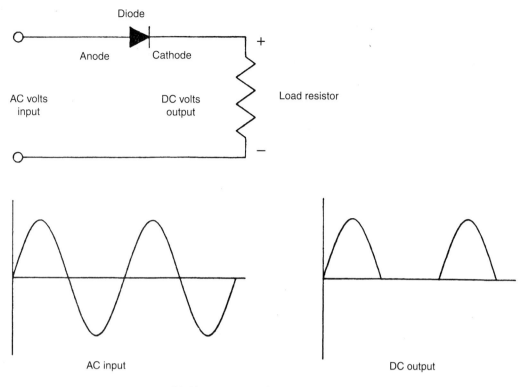

FIGURE 9-44 Half-wave rectifier. (*Delmar/Cengage Learning*)

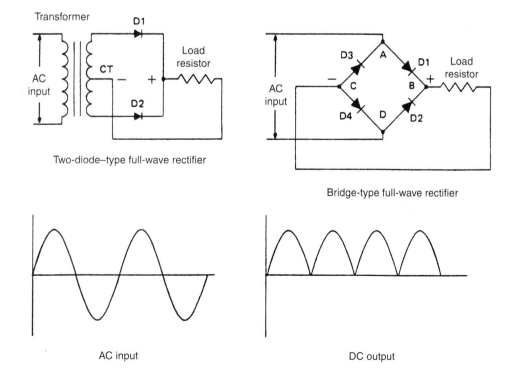

FIGURE 9-45 Single-phase, full-wave rectifiers. (*Delmar/Cengage Learning*)

diodes instead of four. To understand the operation of this rectifier, assume the voltage applied to the cathode of diode D1 to be negative, and the voltage applied to the cathode of diode D2 to be positive. Because diode D1 has a negative voltage applied to its cathode, it is forward biased and current can flow through it. Diode D2, however, is reverse biased and no current can flow through it. The current must flow from the center tap of the transformer, through the load resistor, and complete the circuit through diode D1 back to the transformer.

During the next half cycle of ac voltage, a negative voltage is applied to the cathode of diode D2 and a positive voltage is applied to the cathode of diode D1. Diode D2 is now forward biased and diode D1 is reverse biased. Current can flow from the center tap of the transformer, through the load resistor, and then complete the circuit through diode D2 back to the transformer. Notice in this rectifier that current flowed through the load resistor during both half cycles of ac voltage. Because both cycles of ac voltage were changed into dc, it is full-wave rectification.

The *bridge-type* rectifier requires the use of four diodes, but it does not require the use of a center-tapped transformer. To understand the operation of this type of rectifier, assume the voltage applied to point A of the rectifier to be positive, and the voltage applied to point D to be negative. Current can flow through diode D4 to point C of the rectifier. Because diode D3 is reverse biased, the current must flow through the load resistor to point B of the rectifier. The current then flows through diode D1 to point A and back to complete the circuit. During the next half cycle, the voltage applied to point A is negative, and the voltage applied to point D is positive. Current can now flow through diode D3 to point C of the rectifier. Because diode D4 is reverse biased, the current must flow through the load resistor to point B of the rectifier. At this point, the current flows through diode D2 to complete the circuit. Notice the current flowed through the load resistor during both half cycles of ac voltage.

Average Value of Voltage

When ac voltage is changed into dc voltage, the output dc value of voltage will not be the same as the ac input voltage. To determine the output dc voltage, the average value must be found. The average value of dc voltage for a full-wave rectifier can be found by multiplying the peak ac value by 0.637. For example, assume an ac voltage has a root mean square (RMS) value of 120 volts. To determine the value of dc voltage after rectification, change this RMS value into a peak value by multiplying by 1.414.

$$120 \times 1.414 = 169.68 \text{ volts}$$

Now change the peak value into the average value by multiplying by 0.637.

$$169.68 \times 0.637 = 108 \text{ volts dc}$$

If a half-wave rectifier is used, the answer is divided by 2.

Three-Phase Rectifiers

Most of industry operates on 3-phase power instead of single phase. When it is necessary to change ac voltage to dc voltage, it is generally done with 3-phase rectifiers. There are two basic types of 3-phase rectifiers: the half-wave rectifier and the full-wave rectifier. A 3-phase, half-wave rectifier is shown in Figure 9-46. The 3-phase, half-wave rectifier requires the use of a wye-connected transformer with a center tap to complete the circuit. Notice that only three diodes are used to make this connection. The average dc output voltage for a 3-phase, half-wave rectifier is 0.827 of peak. If the peak voltage of this rectifier is 169.68 volts, the output dc voltage will be

$$169.68 \times 0.827 = 140.32 \text{ volts dc}$$

A 3-phase, full-wave type of rectifier is shown in Figure 9-47. This rectifier does not require the use of a wye-connected transformer with center tap, so it can be used with a wye- or delta-connected system. This rectifier does, however, require the use of six diodes. The average dc output voltage for this rectifier is 0.955 of peak. If the ac voltage applied to this rectifier has a peak value of 169.68 volts, the output dc voltage will be

$$169.68 \times 0.955 = 162 \text{ volts dc}$$

Silicon Controlled Rectifiers

When the amount of dc output voltage must be varied, silicon controlled rectifiers (SCRs) are generally used instead of diodes. The reason for this is that the SCR can be turned on at different

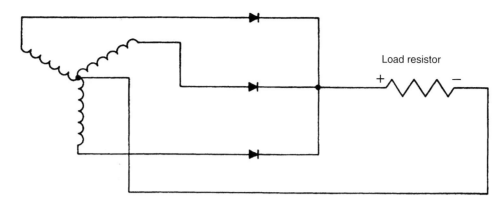

Three-phase, half-wave rectifier

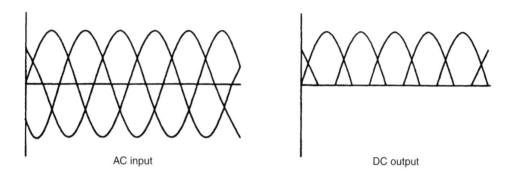

AC input

DC output

FIGURE 9-46 Three-phase, half-wave rectifier. (*Delmar/Cengage Learning*)

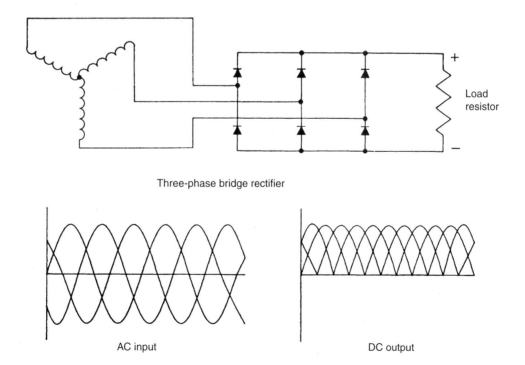

Three-phase bridge rectifier

AC input

DC output

FIGURE 9-47 Three-phase, full-wave rectifier. (*Delmar/Cengage Learning*)

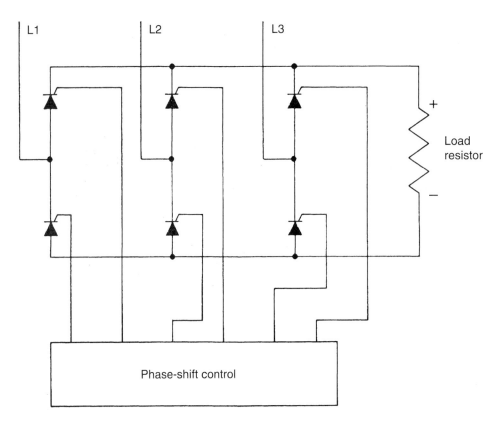

FIGURE 9-48 Three-phase rectifier using SCRs. (*Delmar/Cengage Learning*)

points during the ac waveform applied to it. This permits the output voltage to be varied from 0 volts to the full output voltage of the power supply. The heart of the SCR controller is the phase-shift control. The *phase-shift control* determines when the SCRs turn on during the ac voltage cycle applied to it. A 3-phase bridge rectifier using SCRs and a phase-shift control is shown in Figure 9-48. A basic diagram of an SCR control system for a dc motor is shown in Figure 9-49. Notice that all sensor controls are connected to the phase-shift control. The operator control permits the operator to determine the amount of output voltage that is to be applied to the armature of the motor. This, in turn, determines the speed of the motor.

The field-failure control senses the current flow through the shunt field of the motor. If the shunt-field current should drop below a predetermined level, a signal is sent to the phase-shift control, which turns off the SCRs.

The current-limit control senses the input current to the controller. If a predetermined amount of

current should be sensed, the phase-shift control will not permit the SCRs to turn on more and produce more current flow. This is designed to prevent damage to the motor and controller if the motor should become shorted or stalled.

The electrotachometer is connected to the shaft of the motor. Its function is to sense the speed of the motor. If the motor speed should decrease, a signal is sent to the phase-shift control, and the SCRs turn on more to provide more voltage to the armature of the motor. If the motor speed should increase, the phase-shift control will cause the SCRs to turn on less of the time; thus, the voltage applied to the armature will decrease.

VARIABLE-FREQUENCY DRIVES

Although dc motors are still used in many industries, they are being replaced by variable-frequency drives controlling squirrel-cage induction motors.

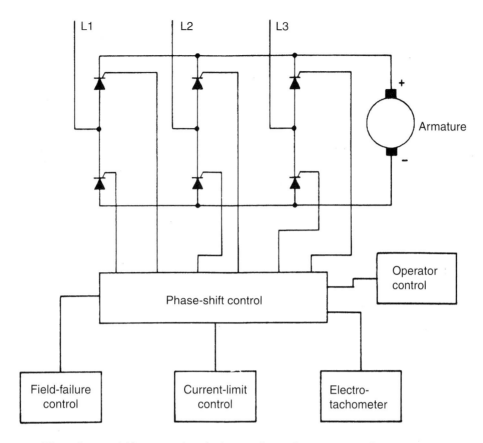

FIGURE 9-49 The phase-shift control unit determines the output voltage. (*Delmar/Cengage Learning*)

The advantage of a dc motor compared to an ac motor is the fact that the speed of the dc motor can be controlled. Although the wound-rotor induction motor does permit some degree of speed control, it does not have the torque characteristics of a dc motor. Direct-current motors can develop maximum torque at 0 rpm. Variable-frequency drives can give these same speed control and torque characteristics to squirrel-cage induction motors. A variable-frequency drive and ac squirrel-cage motor are less expensive to purchase than a comparable dc drive and dc motor. Variable-frequency drives and squirrel-cage motors have less downtime and maintenance problems than dc drives and dc motors.

Variable-Frequency Drive Operating Principles

The operating principle of all polyphase ac motors is the rotating magnetic field. The speed of the rotating field is called *synchronous* speed and is controlled by two factors: frequency of the applied voltage and the number of stator poles. Table 9-5 lists the synchronous speed for different numbers of poles at different frequencies. Variable-frequency drives control motor speed by controlling the frequency of the power supplied to the motor.

TABLE 9-5

Motor revolutions per minute for different poles and frequencies.

Poles	Frequency in Hertz					
	60	50	40	30	20	10
2	3600	3000	2400	1800	1200	600
4	1800	1500	1200	900	600	300
6	1200	1000	800	600	400	200
8	900	750	600	450	300	150

SCRs control output voltage

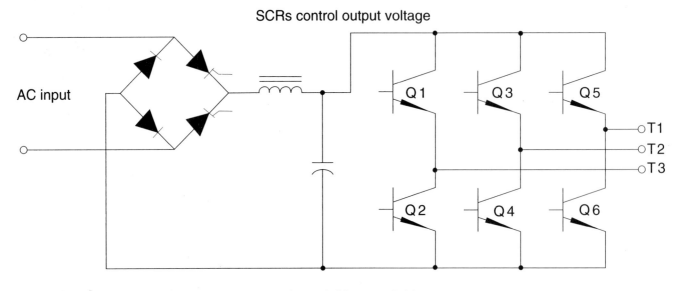

FIGURE 9-50 Basic schematic of a variable-speed drive. (*Delmar/Cengage Learning*)

Voltage and Current Considerations

A critical factor that must be considered when the frequency supplying a motor load is reduced is overcurrent. Motor current is limited by the inductive reactance of the stator winding, and inductive reactance is proportional to the frequency applied to the inductor.

$$X_L = 2\pi fL$$

If the frequency is reduced, the inductive reactance will be reduced also. To overcome this problem, the voltage must be reduced in proportion to the frequency.

If a motor operates on 480 volts at 60 hertz, the voltage should be reduced to 400 volts when the frequency is decreased to 50 hertz, 320 volts at a frequency of 40 hertz, 240 volts at a frequency of 30 hertz, and so on.

Basic Construction of a Variable-Frequency Drive

Most variable-frequency drives operate by first changing the ac voltage into dc and then changing it back to ac at the desired frequency. There are several methods used to change the dc voltage back into ac. The method employed is generally determined by the manufacturer, age of the equipment, and the size motor the drive must control. Variable-frequency drives intended to control the speed of motors up to 500 horsepower generally use transistors. In the

circuit shown in Figure 9-50, a single-phase bridge changes the alternating current into direct current.

The bridge rectifier uses two SCRs and two diodes. The SCRs permit the output voltage of the rectifier to be controlled. As the frequency decreases, the SCRs fire later in the cycle and lower the output voltage to the transistors. A choke coil and capacitor bank are used to filter the output voltage before transistors Q_1 through Q_6 change the dc voltage back into ac. An electronic control unit is connected to the bases of transistor Q_1 through Q_6. The control unit converts the dc voltage back into 3-phase alternating current by turning transistors on or off at the proper time and in the proper sequence. Assume, for example, that transistors Q_1 and Q_4 are switched on at the same time. This permits stator winding T_3 to be connected to a positive voltage and T_2 to be connected to a negative voltage. Current can flow through Q_4 to T_2, through the motor stator winding and through T_3 to Q_1.

Now assume that transistors Q_1 and Q_4 are switched off and transistors Q_3 and Q_6 are switched on. Current will now flow through Q_6 to stator winding T_1, through the motor to T_2, and through Q_3 to the positive of the power supply.

Because the transistors are turned completely on or completely off, the waveform produced is a square wave instead of a sine wave, Figure 9-51. Induction motors will operate on a square wave without a great problem. Some manufacturers design units

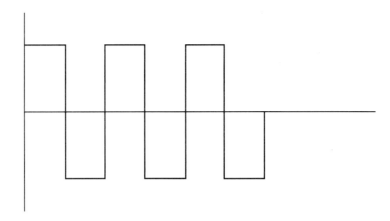

FIGURE 9-51 Square wave voltage waveform. (*Delmar/Cengage Learning*)

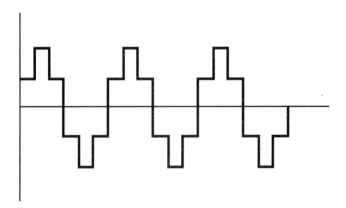

FIGURE 9-52 A stepped waveform approximates a sine wave. (*Delmar/Cengage Learning*)

that will produce a stepped waveform, as shown in Figure 9-52. The stepped waveform is used because it more closely approximates a sine wave.

Some Related Problems

The circuit illustrated in Figure 9-50 employs the use of SCRs in the power supply and junction transistors in the output stage. SCR power supplies control the output voltage by chopping the incoming waveform. This can cause harmonics on the line that cause overheating of transformers and motors, and can cause fuses to blow and circuit breakers to trip. When bipolar junction transistors are employed as switches, they are generally driven into saturation by supplying them with an excessive amount of base-emitter current. Saturating the transistor causes the collector-emitter voltage to drop to between 0.04 and 0.03 volt.

This small voltage drop allows the transistor to control large amounts of current without being destroyed. When a transistor is driven into saturation, however, it cannot recover or turn off as quickly as normal. This greatly limits the frequency response of the transistor.

IGBTs

Many transistor-controlled variable drives now employ a special type of transistor called an insulated gate bipolar transistor (IGBT). IGBTs have an insulated gate very similar to some types of field effect transistors (FETs). Because the gate is insulated, it has a very high impedance. The IGBT is a voltage-controlled device, not a current-controlled device. This gives it the ability to turn off very quickly. IGBTs can be driven into saturation to provide a very low voltage drop between emitter and collector, but they do not suffer from the slow recovery time of common junction transistors. The schematic symbol for an IGBT is shown in Figure 9-53.

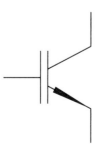

FIGURE 9-53 Schematic symbols for an IGBT. (*Delmar/Cengage Learning*)

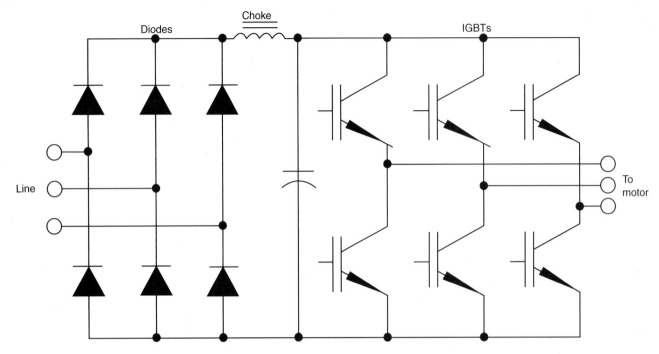

FIGURE 9-54 Variable-frequency drives using IGBTs with diodes in the rectifier instead of SCRs. (*Delmar/Cengage Learning*)

Drives using IGBTs generally use diodes to rectify the ac voltage into dc, not SCRs, Figure 9-54. The 3-phase rectifier supplies a constant dc voltage to the transistors. The output voltage to the motor is controlled by pulse-width modulation (PWM). PWM is accomplished by turning the transistor on and off several times during each half cycle, Figure 9-55. The output voltage is an average of the peak or maximum voltage and the amount of time the transistor is turned on or off. Assume that 480-volt, 3-phase ac

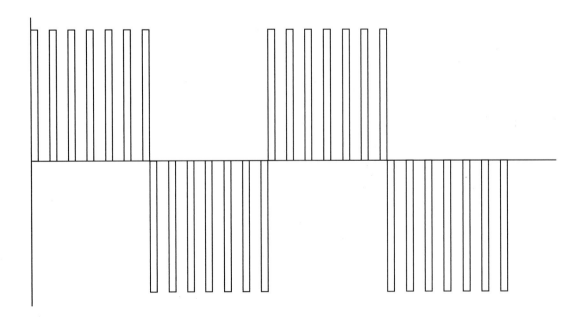

FIGURE 9-55 PWM is accomplished by turning the voltage on and off several times during each half cycle. (*Delmar/Cengage Learning*)

is rectified to dc and filtered. The dc voltage applied to the IGBTs is approximately 630 volts. The output voltage to the motor is controlled by the switching of the transistors. Assume that the transistor is on for 10 microseconds and off for 20 microseconds. In this example, the transistor is on for one-third of the time and off for two-thirds of the time. The voltage applied to the motor would be 210 volts (630/3).

Advantages and Disadvantages of IGBT Drives

A great advantage of drives using IGBTs is the fact that SCRs are generally not used in the power supply and this greatly reduces problems with line harmonics. The greatest disadvantage is that the fast switching rate of the transistors can cause voltage spikes in the range of 1600 volts to be applied to the motor. These voltage spikes can destroy some motors. Line length from the drive to the motor is of great concern with drives using IGBTs. The shorter the line length, the better.

Inverter-Rated Motors

Due to the problem of excessive voltage spikes caused by IGBT drives, some manufacturers produce a motor that is "inverter rated." These motors are specifically designed to be operated by variable-frequency drives. They differ from standard motors in several ways:

1. Many inverter-rated motors contain a separate blower to provide continuous cooling for the motor regardless of the speed. Many motors use a fan connected to the motor shaft to help draw air though the motor. When the motor speed is reduced, the fan cannot maintain sufficient air flow to cool the motor.

2. Inverter-rated motors generally have insulating paper between the windings and the stator core, Figure 9-56. The high-voltage spikes produce high currents that produce a high magnetic field. This increased magnetic field causes the motor windings to move. This movement can eventually cause the insulation to wear off the wire and produce a grounded motor winding.

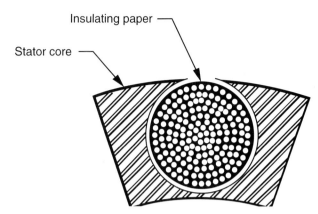

FIGURE 9-56 Insulated paper placed between the windings and the stator frame. (*Delmar/Cengage Learning*)

3. Inverter-rated motors generally have phase paper added to the terminal leads. Phase paper is insulating paper added to the terminal leads that exit the motor. The high-voltage spikes affect the beginning lead of a coil much more than the wire inside the coil. The coil is an inductor that naturally opposes a change of current. Most of the insulation stress caused by high-voltage spikes occurs at the beginning of a winding.

4. The magnet wire used in the construction of the motor windings has a higher rated insulation than other motors.

5. The case size is larger than most 3-phase motors. The case size is larger because of the added insulating paper between the windings and the stator core. Also, a larger case size helps cool the motor by providing a larger surface area for the dissipation of heat.

Variable-Frequency Drives Using SCRs and GTOs

Variable-frequency drives intended to control motors of more than 500 horsepower generally use SCRs or gate turnoff devices (GTOs). GTOs are similar to SCRs except that conduction through the GTO can be stopped by applying a negative voltage, negative with respect to the cathode,

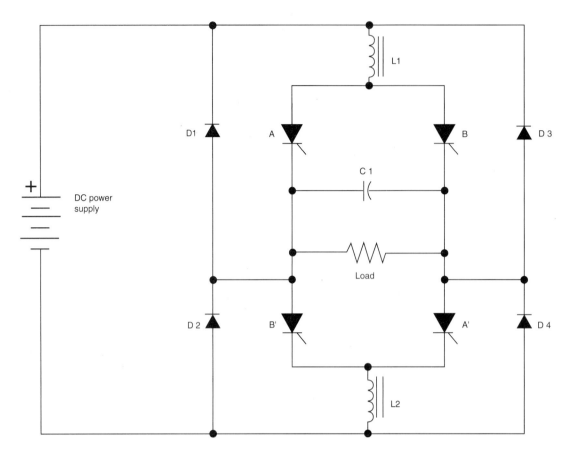

FIGURE 9-57 Changing dc to ac using SCRs. (*Delmar/Cengage Learning*)

to the gate. SCRs and GTOs are thyristors and have the ability to handle a greater amount of current than transistors. An example of a single-phase circuit used to convert dc voltage to ac voltage with SCRs is shown in Figure 9-57.

In this circuit, the SCRs are connected to a control unit that controls the sequence and rate at which the SCRs are gated on. The circuit is constructed so that SCRs A and A' are gated on at the same time and SCRs B and B' are gated on at the same time. Inductors L1 and L2 are used for filtering and wave shaping. Diodes D1 through D4 are clamping diodes and are used to prevent the output voltage from becoming excessive. Capacitor C1 is used to turn one set of SCRs off when the other set is gated on. This capacitor must be a true ac capacitor because it will be charged to the alternate polarity each half cycle. In a converter intended to handle large amounts of power, capacitor C1 will be a bank of capacitors.

To understand the operation of the circuit, assume that SCRs A and A' are gated on at the same time. Current will flow through the circuit as shown in Figure 9-58. Notice the direction of current flow through the load and that capacitor C1 has been charged to the polarity shown. When an SCR is gated on, it can be turned off only by permitting the current flow through the anode-cathode section to drop below a certain level called the holding current level. As long as the current continues to flow through the anode-cathode, the SCR will not turn off.

Now assume that SCRs B and B' are turned on. Because SCRs A and A' are still turned on, two current paths now exist through the circuit. The positive charge on capacitor C1, however, causes the negative electrons to see an easier path. The current will rush to charge the capacitor to the opposite polarity, stopping the current flowing through SCRs A and A' permitting them to turn off. The current

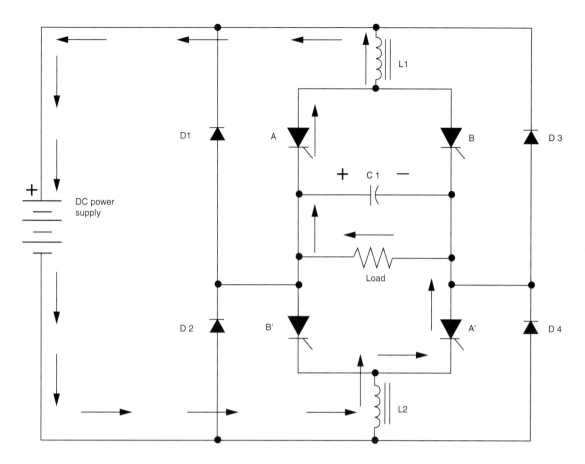

FIGURE 9-58   Current flows through SCRs A and A'. (*Delmar/Cengage Learning*)

now flows through SCRs B and B' and charges the capacitor to the opposite polarity, Figure 9-59. Notice that the current now flows through the load in the opposite direction, which produces alternating current across the load.

To produce the next half cycle of AC current, SCRs A and A' are gated on again. The positively charged side of the capacitor will now cause the current to stop flowing through SCRs B and B' permitting them to turn off. The current again flows through the load in the direction indicated in Figure 9-58. The frequency of the circuit is determined by the rate at which the SCRs are gated on.

Features of Variable-Frequency Control

Although the primary purpose of a variable-frequency drive is to provide speed control for an ac motor, most drives provide functions that other types of controls do not. Many variable-frequency drives can provide the low-speed torque that is so desirable in dc motors. It is this feature that permits ac squirrel-cage motors to replace dc motors for many applications. Many variable-frequency drives also provide current limit and automatic speed regulation for the motor. Current limit is generally accomplished by connecting current transformers to the input of the drive and sensing the increase in current as load is added. Speed regulation is accomplished by sensing the speed of the motor and feeding this information back to the drive.

Another feature of variable-frequency drives is acceleration and deceleration control, sometimes called *ramping*. Ramping is used to accelerate or decelerate a motor over some period of time. Ramping permits the motor to bring the load up to speed slowly as opposed to simply connecting the motor directly to the line. Even if the speed control is set in the maximum position when the start button

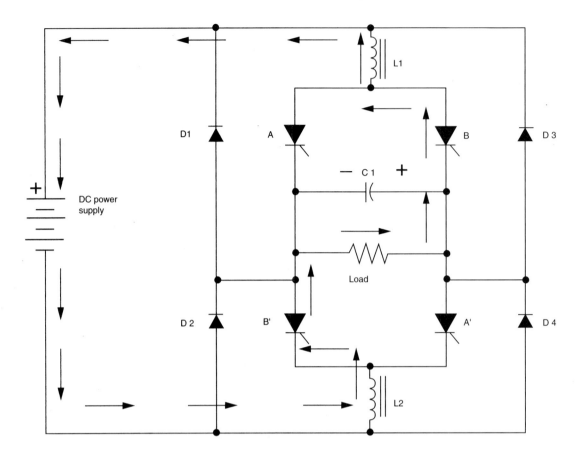

FIGURE 9-59 Current flows through SCRs B and B'. (*Delmar/Cengage Learning*)

is pressed, ramping permits the motor to accelerate the load from zero to its maximum rpm over several seconds. This feature can be a real advantage for some types of loads, especially gear drive loads. In some units the amount of acceleration and deceleration time can be adjusted by setting potentiometers on the main control board. Other units are completely digitally controlled and the acceleration and deceleration times are programmed into the computer memory.

Some other adjustments that can usually be set by changing potentiometers or programming the unit are as follows:

Current limit: This control sets the maximum amount of current the drive is permitted to deliver to the motor.

Volts per hertz: This sets the ratio by which the voltage changes as frequency increases or decreases.

Maximum hertz: This control sets the maximum speed of the motor. Most motors are intended to operate between 0 and 60 hertz, but some drives permit the output frequency to be set above 60 hertz, which would permit the motor to operate at higher-than-normal speed. The maximum hertz control can also be set to limit the output frequency to a value less than 60 hertz, which would limit the motor speed to a value less than normal.

Minimum hertz: This sets the minimum speed the motor is permitted to run.

Some variable-frequency drives permit adjustment of current limit, maximum and minimum speed, ramping time, and so on, by adjustment of trim resistors located on the main control board. Other drives employ a microprocessor as the controller. The values of current limit, speed, ramping time, and so on, for these drives are programmed into the unit and are much easier to make and are generally more accurate than adjusting trim resistors. Variable-frequency drives are shown in Figure 9-60.

FIGURE 9-60 Variable-frequency drives. (*Courtesy Reliance Electric*)

REVIEW QUESTIONS

All answers should be written in complete sentences, calculations should be shown in detail, and *Code* references should be cited when appropriate.

1. What is the source of electric power for the motor branch circuits? _____

2. What provides branch-circuit protection and where is it located? _____

3. What information is needed to select the correct size motor branch-circuit protection?

4. What would be the full-load current of a 15-horsepower, 3-phase, 480-volt motor that operates with an 85 percent efficiency and 75 percent power factor? _____

5. What current value would be used to size the branch-circuit overcurrent protection of the motor described in the previous question? _____

_____

_____

6. Describe the construction of the squirrel-cage motor. _____

_____

_____

7. Describe the construction of the wound rotor motor. _____

_____

_____

8. Describe the construction of the dc motor. _____

_____

_____

9. What will be the synchronous speed of a 50-hertz, 8-pole motor? _____

_____

_____

10. Describe the operation of a line starter. _____

_____

_____

_____

11. Describe the operation of the primary resistance starter. _____

_____

_____

_____

12. Describe the operation of the autotransformer starter. _____

_____

_____

_____

13. The control circuit for an autotransformer starter is shown in Figure 9-10. Describe what happens when the start button is pushed. _____

_____

_____

_____

14. A connection diagram for a bridge-type full-wave rectifier is shown in Figure 9-45. Describe the current path for both a positive and a negative pulse of ac input. _____

_____

_____

_____

15. What is the principle of operation of all polyphase motors? _____

_____

_____

16. What two factors determine synchronous field speed? _____

_____

_____

_____

_____

17. As the frequency applied to a motor is reduced, what must be done to prevent excessive current to the motor? _____

_____

_____

Motor Installation

OBJECTIVES

After studying this chapter, the student should be able to

- determine the ampere rating of various motors using the *NEC*.

- determine the conductor size for installing a single motor.

- determine the conductor size for a multimotor installation.

- determine the overload size for various motors.

- determine the fuse or circuit-breaker size for a single-motor installation.

- determine the fuse or circuit-breaker size for a multimotor installation.

MOTOR TABLES

There are different types of motors, such as dc, single-phase ac, 2-phase ac, and 3-phase ac. Different tables are used to list the running current for different types of motors.

- Table 10-1 (*NEC Table 430.247*) lists the currents for dc motors;

- Table 10-2 (*NEC Table 430.248*) lists the currents for single-phase ac motors;

- Table 10-3 (*NEC Table 430.249*) lists the full-load currents for 2-phase ac motors;

- Table 10-4 (*NEC Table 430.250*) lists the full-load currents for 3-phase motors.

The tables list the amount of current the motor is expected to require under full-load conditions. A motor will have a lower current requirement if it is not under a full load.

These tables list the ampere rating of the motors according to horsepower and connected voltage. It should be noted that according to *430.6(A)*, the currents listed in these tables are to be used in determining conductor size, fuse size, and ground-fault protection instead of the motor's nameplate rating. The motor overload size, however, is to be determined by using the current listed on the nameplate of the motor.

DIRECT-CURRENT MOTORS

Table 10-1 lists the full-load running current for dc motors. The horsepower rating of the motor is given in the far left column. Rated voltages are listed across the top of the table. The table shows that a 1-horsepower motor will have a full-load current of 12.2 amperes when connected to 90 volts dc.

TABLE 10-1

Table 430.247 Full-Load Current in Amperes, Direct-Current Motors
The following values of full-load currents[*] are for motors running at base speed.

| Horsepower | Armature Voltage Rating[*] | | | | | |
	90 Volts	120 Volts	180 Volts	240 Volts	500 Volts	550 Volts
¼	4.0	3.1	2.0	1.6	—	—
⅓	5.2	4.1	2.6	2.0	—	—
½	6.8	5.4	3.4	2.7	—	—
¾	9.6	7.6	4.8	3.8	—	—
1	12.2	9.5	6.1	4.7	—	—
1½	—	13.2	8.3	6.6	—	—
2	—	17	10.8	8.5	—	—
3	—	25	16	12.2	—	—
5	—	40	27	20	—	—
7½	—	58	—	29	13.6	12.2
10	—	76	—	38	18	16
15	—	—	—	55	27	24
20	—	—	—	72	34	31
25	—	—	—	89	43	38
30	—	—	—	106	51	46
40	—	—	—	140	67	61
50	—	—	—	173	83	75
60	—	—	—	206	99	90
75	—	—	—	255	123	111
100	—	—	—	341	164	148
125	—	—	—	425	205	185
150	—	—	—	506	246	222
200	—	—	—	675	330	294

[*]These are average dc quantities.

If the 1-horsepower motor is designed to be connected to 240 volts dc, it will have a full-load current of 4.7 amperes.

SINGLE-PHASE ALTERNATING-CURRENT MOTORS

The current ratings for single-phase ac motors are given in Table 10-2. These ratings are for motors that operate at normal speeds and torques. Motors especially designed for low speed and high torque, or multispeed motors, shall have their running current determined from the nameplate rating of the motor.

 EXAMPLE

A 3-horsepower single-phase motor is connected to a 208-volt supply. To properly size the conductors it is necessary to know the full-load current. The full-load current for single-phase motors is given in Table 10-2 (*NEC Table 430.248*). The full-load current for a 3-horsepower motor connected to a 208-volt single-phase ac supply is given as 18.7 amperes.

TABLE 10-2

Table 430.248 Full-Load Currents in Amperes, Single-Phase Alternating-Current Motors

The following values of full-load currents are for motors running at usual speeds and motors with normal torque characteristics. The voltages listed are rated motor voltages. The currents listed shall be permitted for system voltage ranges of 110 to 120 and 220 to 240 volts.

Horsepower	115 Volts	200 Volts	208 Volts	230 Volts
1/6	4.4	2.5	2.4	2.2
1/4	5.8	3.3	3.2	2.9
1/3	7.2	4.1	4.0	3.6
1/2	9.8	5.6	5.4	4.9
3/4	13.8	7.9	7.6	6.9
1	16	9.2	8.8	8.0
1 1/2	20	11.5	11.0	10
2	24	13.8	13.2	12
3	34	19.6	18.7	17
5	56	32.2	30.8	28
7 1/2	80	46.0	44.0	40
10	100	57.5	55.0	50

Reprinted with permission from NFPA 70-2011.

TABLE 10-3

Table 430.249 Full-Load Current, Two-Phase Alternating-Current Motors (4-Wire)

The following values of full-load current are for motors running at speeds usual for belted motors and motors with normal torque characteristics. Current in the common conductor of a 2-phase, 3-wire system will be 1.41 times the value given. The voltages listed are rated motor voltages. The currents listed shall be permitted for system voltage ranges of 110 to 120, 220 to 240, 440 to 480, and 550 to 600 volts.

Horsepower	Induction-Type Squirrel Cage and Wound Rotor (Amperes)				
	115 Volts	230 Volts	460 Volts	575 Volts	2300 Volts
1/2	4.0	2.0	1.0	0.8	—
3/4	4.8	2.4	1.2	1.0	—
1	6.4	3.2	1.6	1.3	—
1 1/2	9.0	4.5	2.3	1.8	—
2	11.8	5.9	3.0	2.4	—
3	—	8.3	4.2	3.3	—
5	—	13.2	6.6	5.3	—
7 1/2	—	19	9.0	8.0	—
10	—	24	12	10	—
15	—	36	18	14	—
20	—	47	23	19	—
25	—	59	29	24	—
30	—	69	35	28	—
40	—	90	45	36	—
50	—	113	56	45	—
60	—	133	67	53	14
75	—	166	83	66	18
100	—	218	109	87	23
125	—	270	135	108	28
150	—	312	156	125	32
200	—	416	208	167	43

Reprinted with permission from NFPA 70-2011.

TWO-PHASE MOTORS

Although 2-phase motors are seldom used, Table 10-3 lists rated full-load currents for these motors. Like single-phase motors, 2-phase motors that are especially designed for low speed-high torque applications, and multispeed motors use the nameplate rating instead of the value shown in the chart. When a 2-phase, 3-wire system is used, the size of the neutral conductor must be increased by the square root of 2, or 1.41. This is because the voltages of a 2-phase system are 90° out of phase with each other, as shown in Figure 10-1.

Line 1 ————————

Line 2 — — — — — —

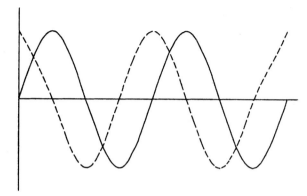

FIGURE 10-1 Voltages of a 2-phase system.
(*Delmar/Cengage Learning*)

EXAMPLE

A 60-horsepower, 460-volt motor is connected to a 2-phase supply. To determine the phase and neutral currents, refer to Table 10-3 (*NEC Table 430.249*). The phase current is given as 67 amperes. To determine the neutral current, multiply the phase current by 1.41 ($\sqrt{2}$). Neutral current = 67 × 1.41 = 94.5 amperes.

THREE-PHASE MOTORS

Table 10-4 (*NEC Table 430.250*) is used to determine the full-load current of a 3-phase motor. The full-load current of low speed–high torque and multispeed motors is to be determined from the nameplate rating. The table has a note that addresses synchronous motors. Notice that the right side of the table is devoted to the full-load currents of synchronous-type motors. The currents listed are for motors that are to be operated at unity or 100 percent power factor. Because synchronous motors are often made to have a leading power factor by overexcitation of the rotor current, the full-load current rating must be increased when this

is done. If the motor is operated at 90 percent power factor, the rated full-load current in the table is to be increased by 10 percent. If the motor is to be operated at 80 percent power factor, the full-load current is to be increased by 25 percent.

EXAMPLE

A 150-horsepower, 460-volt synchronous motor is to be operated at 80 percent power factor. What will be the full-load current rating of the motor?

151 × 1.25 = 188.75, or 189 amperes

DETERMINING CONDUCTOR SIZE FOR A SINGLE MOTOR

As stated by *430.6(A)*, the conductor size for a motor connection shall be based on the values from *NEC* tables that are included in this text as Table 10-1, Table 10-2, Table 10-3, and Table 10-4, instead of motor nameplate current. *NEC 430.22* states that the conductors supplying a single motor shall have an ampacity of not less than 125 percent of the motor full-load current. *NEC Article 310* is used to select the conductor size after the ampacity has been determined. The exact table employed will be determined by the wiring conditions. Probably the most frequently used table is *NEC Table 310.15(B)(16)*, which is provided in this text as Table 6-2.

Another factor that must be taken into consideration when determining the conductor size is the temperature rating of the devices and terminals as specified in *110.14(C)*. This section states that the conductor will be selected and coordinated so as to not exceed the lowest temperature rating of any connected termination, any connected conductor, or any connected device. This means that regardless of the temperature rating of the conductor, the allowable ampacity must be selected from a column that does not exceed the temperature rating of the terminations. The conductors listed in the first column of Table 6-2

TABLE 10-4

Table 430.250 Full-Load Current, Three-Phase Alternating-Current Motors

The following values of full-load currents are typical for motors running at speeds usual for belted motors and motors with normal torque characteristics.

The voltages listed are rated motor voltages. The currents listed shall be permitted for system voltage ranges of 110 to 120, 220 to 240, 440 to 480, and 550 to 600 volts.

Horsepower	Induction-Type Squirrel Cage and Wound Rotor (Amperes)							Synchronous-Type Unity Power Factor* (Amperes)			
	115 Volts	200 Volts	208 Volts	230 Volts	460 Volts	575 Volts	2300 Volts	230 Volts	460 Volts	575 Volts	2300 Volts
½	4.4	2.5	2.4	2.2	1.1	0.9	—	—	—	—	—
¾	6.4	3.7	3.5	3.2	1.6	1.3	—	—	—	—	—
1	8.4	4.8	4.6	4.2	2.1	1.7	—	—	—	—	—
1½	12.0	6.9	6.6	6.0	3.0	2.4	—	—	—	—	—
2	13.6	7.8	7.5	6.8	3.4	2.7	—	—	—	—	—
3	—	11.0	10.6	9.6	4.8	3.9	—	—	—	—	—
5	—	17.5	16.7	15.2	7.6	6.1	—	—	—	—	—
7½	—	25.3	24.2	22	11	9	—	—	—	—	—
10	—	32.2	30.8	28	14	11	—	—	—	—	—
15	—	48.3	46.2	42	21	17	—	—	—	—	—
20	—	62.1	59.4	54	27	22	—	—	—	—	—
25	—	78.2	74.8	68	34	27	—	53	26	21	—
30	—	92	88	80	40	32	—	63	32	26	—
40	—	120	114	104	52	41	—	83	41	33	—
50	—	150	143	130	65	52	—	104	52	42	—
60	—	177	169	154	77	62	16	123	61	49	12
75	—	221	211	192	96	77	20	155	78	62	15
100	—	285	273	248	124	99	26	202	101	81	20
125	—	359	343	312	156	125	31	253	126	101	25
150	—	414	396	360	180	144	37	302	151	121	30
200		552	528	480	240	192	49	400	201	161	40
250	—	—	—	—	302	242	60	—	—	—	—
300	—	—	—	—	361	289	72	—	—	—	—
350	—	—	—	—	414	336	83	—	—	—	—
400	—	—	—	—	477	382	95	—	—	—	—
450	—	—	—	—	515	412	103	—	—	—	—
500	—	—	—	—	590	472	118	—	—	—	—

*For 90 and 80 percent power factor, the figures shall be multiplied by 1.1 and 1.25, respectively.

Reprinted with permission from NFPA 70-2011.

P₅ 71

have a temperature rating of 60°C (140°F), the conductors in the second column have a rating of 75°C (167°F), and the conductors in the third column have a rating of 90°C (194°F). The temperature ratings of devices such as circuit breakers, fuses, and terminals are found in UL product directories. The temperature rating may be found on the piece of equipment. As a general rule, the temperature rating of devices will not exceed 75°C (167°F).

When the termination temperature rating is not listed or known, *110.14(C)* further states that for circuits rated at 100 amperes or less, or for 14 AWG through 1 AWG conductors, the ampacity of the wire, regardless of its temperature rating, will be selected from the 60°C (140°F) column. This does not mean that only those insulation types listed in the 60°C (140°F) column must be used, but that the ampacity must be selected from that column. For example, assume that a copper conductor with type XHHW insulation is to be connected to a 50-ampere circuit breaker that does not have a listed temperature rating. According to Table 6-2, an 8 AWG copper conductor with XHHW insulation is rated to carry 55 amperes as given in the 90°C (194°F) column, but the temperature rating of the circuit breaker is unknown.

Therefore, the wire size must be selected from the ampacity ratings in the 60°C (140°F) column. A 6 AWG conductor may be used.

NEC 110.14(C)(1)(a)(4) has a special provision for motors marked with NEMA design code B, C, or D. This section states that conductors rated at 75°C (167°F) or higher may be selected from the 75°C (167°F) column in the table even if the ampacity is 100 amperes or less. This code will not apply to motors that do not have a NEMA design code marking on their nameplate. Most motors manufactured before 1996 will not have a NEMA design code. The NEMA design code letter should not be confused with the code letter that indicates the type of squirrel-cage rotor used in the motor.

PROBLEM: A 30-horsepower, 3-phase squirrel-cage induction motor is connected to a 480-volt line. The conductors are run in conduit to the motor. The termination temperature rating of the devices is not known. Copper conductors with THWN insulation are to be used for motor connection. The motor nameplate is not marked with a NEMA design code. What size conductor must be used?

Solution: The first step is to determine the full-load current of the motor. This is determined from Table 10-4. The table indicates a current of 40 amperes for this motor. This current must be increased by 25 percent according to *430.22*.

$$40 \times 1.25 = 50 \text{ amperes}$$

Table 6-2 (*NEC Table 310.15(B)(16)*) is used to determine the conductor size. Locate the column that contains THWN insulation. THWN is located in the 75°C (167°F) column. Because this circuit is less than 101 amperes and the termination temperature is not known, the conduction size must be selected from the allowable ampacities listed in the 60°C (140°F) column. A size 6 AWG conductor with type THWN insulation may be used.

For circuits rated over 100 amperes, or for conductor sizes larger than 1 AWG, *110.14(C)* states that the ampacities listed in the 75°C (167°F) column may be used to select wire sizes unless a conductor with a 60°C (140°F) temperature rating has been selected for use. For example, types TW and UF insulation are listed in the 60°C (140°F) column. If one of these two

insulation types has been specified, the wire size must be chosen from the 60°C (140°F) column regardless of the ampere rating of the circuit.

OVERLOAD SIZE

In determining the overload size for a motor, the nameplate current rating of the motor is to be used instead of the current values listed in the tables (*430.6*). Other factors such as the service factor (SF) or temperature rating (°C) of the motor are also to be used to determine the overload size of a motor. The temperature rating of the motor is an indication of the type of insulation used in the motor windings and should not be confused with termination temperature discussed in *110.14(C)*.

NEC 430.32, which covers overload protection in continuous-duty motors, is reproduced here as Table 10-5; *430.32(A)(1)* is used to determine the overload size for motors of 1 horsepower or more. The overload size is based on a percentage of the full-load current of the motor.

PROBLEM: A 25-horsepower, 3-phase induction motor has a nameplate rating of 32 amperes. The nameplate also shows a temperature rise of 30°C. What is the overload size for this motor?

Solution: As indicated by *430.32(A)(1)* in Table 10-5, the overload size is to be 125 percent of the full-load current rating of the motor.

$$32 \times 1.25 = 40 \text{ amperes}$$

If for some reason this overload size does not permit the motor to start without tripping out, *430.32(C)* permits the overload size to be increased to a maximum of 140 percent for this motor. If this increase in overload size does not solve the starting problem, the overload may be shunted out of the circuit during the starting period in accordance with *430.35(A)* and *430.35(B)*.

DETERMINING LOCKED-ROTOR CURRENT

There are two basic methods for determining the locked-rotor current (starting current) of a squirrel-cage motor, depending on the information available. If the nameplate lists code letters that range from A

TABLE 10-5

430.32 Continuous-Duty Motors.

(A) More Than 1 Horsepower. Each motor used in a continuous duty application and rated more than 1 hp shall be protected against overload by one of the means in 430.32(A)(1) through (A)(4).

(1) Separate Overload Device. A separate overload device that is responsive to motor current. This device shall be selected to trip or shall be rated at no more than the following percent of the motor nameplate full-load current rating:

Motors with a marked service factor 1.15 or greater	125%
Motors with a marked temperature rise 40°C or less	125%
All other motors	115%

Modification of this value shall be permitted as provided in 430.32(C). For a multispeed motor, each winding connection shall be considered separately.

Where a separate motor overload device is connected so that it does not carry the total current designated on the motor nameplate, such as for wye-delta starting, the proper percentage of nameplate current applying to the selection or setting of the overload device shall be clearly designated on the equipment, or the manufacturer's selection table shall take this into account.

> Informational Note: Where power factor correction capacitors are installed on the load side of the motor overload device, see 460.9.

(2) Thermal Protector. A thermal protector integral with the motor, approved for use with the motor it protects on the basis that it will prevent dangerous overheating of the motor due to overload and failure to start. The ultimate trip current of a thermally protected motor shall not exceed the following percentage of motor full-load current given in Table 430.248, Table 430.249, and Table 430.250:

Motor full-load current 9 amperes or less	170%
Motor full-load current from 9.1 to, and including, 20 amperes	156%
Motor full-load current greater than 20 amperes	140%

If the motor current-interrupting device is separate from the motor and its control circuit is operated by a protective device integral with the motor, it shall be arranged so that the opening of the control circuit will result in interruption of current to the motor.

(3) Integral with Motor. A protective device integral with a motor that will protect the motor against damage due to failure to start shall be permitted if the motor is part of an approved assembly that does not normally subject the motor to overloads.

(4) Larger Than 1500 Horsepower. For motors larger than 1500 hp, a protective device having embedded temperature detectors that cause current to the motor to be interrupted when the motor attains a temperature rise greater than marked on the nameplate in an ambient temperature of 40°C.

(B) One Horsepower or Less, Automatically Started. Any motor of 1 hp or less that is started automatically shall be protected against overload by one of the following means.

(1) Separate Overload Device. By a separate overload device following the requirements of 430.32(A)(1).

For a multispeed motor, each winding connection shall be considered separately. Modification of this value shall be permitted as provided in 430.32(C).

(2) Thermal Protector. A thermal protector integral with the motor, approved for use with the motor that it protects on the basis that it will prevent dangerous overheating of the motor due to overload and failure to start. Where the motor current-interrupting device is separate from the motor and its control circuit is operated by a protective device integral with the motor, it shall be arranged so that the opening of the control circuit results in interruption of current to the motor.

(3) Integral with Motor. A protective device integral with a motor that protects the motor against damage due to failure to start shall be permitted (1) if the motor is part of an approved assembly that does not subject the motor to overloads, or (2) if the assembly is also equipped with other safety controls (such as the safety combustion controls on a domestic oil burner) that protect the motor against damage due to failure to start. Where the assembly has safety controls that protect the motor, it shall be so indicated on the nameplate of the assembly where it will be visible after installation.

(4) Impedance-Protected. If the impedance of the motor windings is sufficient to prevent overheating due to failure to start, the motor shall be permitted to be protected as specified in 430.32(D)(2)(a) for manually started motors if the motor is part of an approved assembly in which the motor will limit itself so that it will not be dangerously overheated.

> Informational Note: Many ac motors of less than 1/20 hp, such as clock motors, series motors, and so forth, and also some larger motors such as torque motors, come within this classification. It does not include split-phase motors having automatic switches that disconnect the starting windings.

(C) Selection of Overload Device. Where the sensing element or setting or sizing of the overload device selected in accordance with 430.32(A)(1) and 430.32(B)(1) is not sufficient to start the motor or to carry the load, higher size sensing elements or incremental settings or sizing shall be

TABLE 10-5 *continue*

permitted to be used, provided the trip current of the overload device does not exceed the following percentage of motor nameplate full-load current rating:

Motors with marked service factor 1.15 or greater	140%
Motors with a marked temperature rise 40°C or less	140%
All other motors	130%

If not shunted during the starting period of the motor as provided in 430.35, the overload device shall have sufficient time delay to permit the motor to start and accelerate its load.

> Informational Note: A Class 20 or Class 30 overload relay will provide a longer motor acceleration time than a Class 10 or Class 20, respectively. Use of a higher class overload relay may preclude the need for selection of a higher trip current.

(D) One Horsepower or Less, Nonautomatically Started.

(1) Permanently Installed. Overload protection shall be in accordance with 430.32(B).

(2) Not Permanently Installed.

(a) *Within Sight from Controller.* Overload protection shall be permitted to be furnished by the branch-circuit short-circuit and ground-fault protective device; such device, however, shall not be larger than that specified in Part IV of Article 430.

Exception: Any such motor shall be permitted on a nominal 120-volt branch circuit protected at not over 20 amperes.

(b) *Not Within Sight from Controller.* Overload protection shall be in accordance with 430.32(B).

(E) Wound-Rotor Secondaries. The secondary circuits of wound-rotor ac motors, including conductors, controllers, resistors, and so forth, shall be permitted to be protected against overload by the motor-overload device.

Reprinted with permission from NFPA 70-2011.

to V, they indicate the type of rotor bars used when the rotor was made. Different types of bars are used to make motors designed for different applications. The type of bar largely determines the locked-rotor current of the motor. The locked-rotor current of the motor is used to determine the maximum starting current. *NEC Table 430.7(B)*, reproduced here as Table 10-6, lists the different code letters and gives the locked-rotor kilovolt-amperes per horsepower. The starting current can be determined by multiplying the kVA rating by the horsepower rating and then dividing by the applied voltage.

PROBLEM: A 15-horsepower, 3-phase squirrel-cage motor with a code letter of K is connected to a 240-volt line. Determine the locked-rotor current.

Solution: The table lists 8.0 to 8.99 kVA per horsepower for a motor with a code letter of K. An average value of 8.5 will be used.

$$8.5 \times 15 = 127.5 \text{ kVA (or 127,500 VA)}$$

$$\frac{127,500}{240 \times \sqrt{3}} = 306.7 \text{ amperes}$$

TABLE 10-6

Table 430.7(B) Locked-Rotor Indicating Code Letters

Code Letter	Kilovolt-Amperes per Horsepower with Locked Rotor
A	0–3.14
B	3.15–3.54
C	3.55–3.99
D	4.0–4.49
E	4.5–4.99
F	5.0–5.59
G	5.6–6.29
H	6.3–7.09
J	7.1–7.99
K	8.0–8.99
L	9.0–9.99
M	10.0–11.19
N	11.2–12.49
P	12.5–13.99
R	14.0–15.99
S	16.0–17.99
T	18.0–19.99
U	20.0–22.39
V	22.4 and up

Reprinted with permission from NFPA 70-2011.

The locked-rotor current for motors with NEMA design codes is determined using *NEC Tables 430.251(A)* and *430.251(B)*. *NEC Table 430.251(A)* lists the locked-rotor currents for single-phase motors, and *Table 430.251(B)* lists the locked-rotor currents for polyphase motors.

SHORT-CIRCUIT PROTECTION

The ratings of the short-circuit protective devices are set forth in *NEC Table 430.52*, reproduced here as Table 10-7. The far left-hand column lists the type of motor that is to be protected. To the right of this are four columns that list different types of

TABLE 10-7

Table 430.52 Maximum Rating or Setting of Motor Branch-Circuit Short-Circuit and Ground-Fault Protective Devices

	Percentage of Full-Load Current			
Type of Motor	Nontime Delay Fuse[1]	Dual Element (Time-Delay) Fuse[1]	Instantaneous Trip Breaker	Inverse Time Breaker[2]
Single-phase motors	300	175	800	250
AC polyphase motors other than wound-rotor	300	175	800	250
Squirrel cage — other than Design B energy-efficient	300	175	800	250
Design B energy-efficient	300	175	1100	250
Synchronous[3]	300	175	800	250
Wound rotor	150	150	800	150
Direct current (constant voltage)	150	150	250	150

Note: For certain exceptions to the values specified, see 430.54.
[1]The values in the Nontime Delay Fuse column apply to Time-Delay Class CC fuses.
[2]The values given in the last column also cover the ratings of nonadjustable inverse time types of circuit breakers that may be modified as in 430.52(C)(1), Exception No. 1 and No. 2.
[3]Synchronous motors of the low-torque, low-speed type (usually 450 rpm or lower), such as are used to drive reciprocating compressors, pumps, and so forth, that start unloaded, do not require a fuse rating or circuit-breaker setting in excess of 200 percent of full-load current.

short-circuit protective devices: nontime delay fuse, dual-element time-delay fuse, instantaneous trip circuit breaker, and inverse-time circuit breaker. Although it is permissible to use nontime delay fuses and instantaneous trip circuit breakers, most motor circuits are protected by dual-element time-delay fuses or inverse-time circuit breakers.

Each of these columns lists a percentage of the motor current that is to be used in determining fuse size. The current listed in the appropriate motor table is to be used instead of nameplate current. According to *430.52(C)(1)*, the protective device is to have a rating or setting not exceeding the value calculated in accord with Table 10-7. Exception 1 of this section, however, states that if the calculated value does not correspond to a standard size or rating of a fuse or circuit breaker, it shall be permissible to use the next higher standard size. The standard sizes of fuses and circuit breakers are listed in *NEC 240.6*, reproduced here as Table 10-8.

Table 10-7 lists squirrel-cage motor types by design letter instead of code letter. Design letters are assigned by NEMA. As required by *430.7(A)(9)*, motor nameplates are marked with design letter B, C, or D. Motors manufactured before this requirement, however, do not list design letters on the nameplate. Most common squirrel-cage motors used in industry actually fall into the design B classification and, for the purposes of selecting fuse size, are considered to be design B unless otherwise listed.

PROBLEM 1: A 100-horsepower, 3-phase squirrel-cage induction motor is connected to a 240-volt line. A dual-element time-delay fuse is to

TABLE 10-8

240.6 Standard Ampere Ratings.

(A) Fuses and Fixed-Trip Circuit Breakers. The standard ampere ratings for fuses and inverse time circuit breakers shall be considered 15, 20, 25, 30, 35, 40, 45, 50, 60, 70, 80, 90, 100, 110, 125, 150, 175, 200, 225, 250, 300, 350, 400, 450, 500, 600, 700, 800, 1000, 1200, 1600, 2000, 2500, 3000, 4000, 5000, and 6000 amperes. Additional standard ampere ratings for fuses shall be 1, 3, 6, 10, and 601. The use of fuses and inverse time circuit breakers with nonstandard ampere ratings shall be permitted.

be used as the short-circuit protective device. What is the correct fuse rating?

Solution: Table 10-4 (*NEC Table 430.250*) lists a full-load current of 248 amperes for this motor. Table 10-7 (*NEC Table 430.52*) indicates that a dual-element time-delay fuse is to be set at 175 percent of the full-load current for an ac polyphase (more than one phase) squirrel-cage motor. Because this motor does not list a design letter on the nameplate, it will be assumed that the motor is design B.

$$248 \times 1.75 = 434 \text{ amperes}$$

The next standard fuse rating above this calculated value listed in Table 10-8 [*240(A)*] is 450 amperes. A 450-ampere fuse will be used to protect this motor circuit.

If for some reason this fuse will not permit the motor to start, *430.52(C)(1), Exception No. 2(b)*, states that the rating of a dual-element time-delay fuse may be increased to a maximum of 225 percent of the full-load motor current.

It should be noted that because this is a maximum value, the fuse size cannot be increased to the next higher value, but must be reduced to the closest standard value without exceeding the maximum value. For example, 248 amperes × 2.25 (225%) = 558 amperes. In this case, a 500-ampere fuse would be employed.

Determine the conductor size, overload size, and short-circuit protection for the following motors.

PROBLEM 2: A 40-horsepower, 240-volt dc motor has a nameplate current rating of 132 amperes. The conductors are to be copper with TW insulation. The short-circuit protective device is to be an instantaneous trip circuit breaker. Refer to Figure 10-2. The termination temperature rating of the connected devices is not known.

Solution: The conductor size must be determined from the current listed in Table 10-1. This current is to be increased by 25 percent.

$$140 \times 1.25 = 175 \text{ amperes}$$

Table 6-2 (*NEC Table 310.15(B)(16)*) is used to find the conductor size. Although *110.14(C)* states that for circuit currents of 100 amperes or greater,

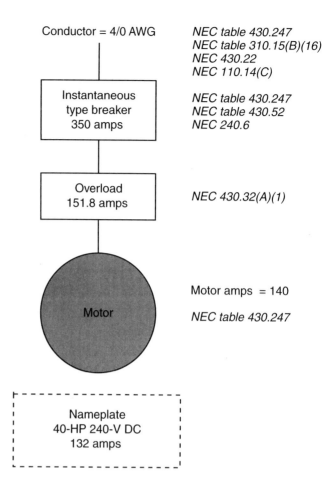

FIGURE 10-2 Values for Problem 2.
(*Delmar/Cengage Learning*)

the allowable ampacity of the conductor is to be determined from the 75°C (167°F) column, in this instance, the specified insulation type is located in the 60°C (140°F) column. Therefore, the conductor size must be determined using the 60°C (140°F) column. A 4/0 AWG conductor will be used.

The circuit-breaker rating is selected from Table 10-7. The current value from Table 10-1 is used instead of the motor nameplate rating. Under direct-current motors (constant voltage), the instantaneous trip circuit-breaker rating is given at 250 percent.

$$140 \times 2.50 = 350 \text{ amperes}$$

Because 350 amperes is one of the standard ratings of circuit breakers listed in Table 10-8, that rating breaker will be used as the short-circuit protective device.

The overload size is determined from *430.32(A)(1)*. Because there is no service factor or temperature rise listed for the motor, the heading ALL OTHER MOTORS will be used. The motor nameplate current will be increased by 15 percent.

$$132 \times 1.15 = 151.8 \text{ amperes}$$

PROBLEM 3: A 150-horsepower, 3-phase squirrel-cage induction motor is connected to 440 volts. The motor nameplate lists the following information:

Motor amps 175   SF 1.25   NEMA code B

The conductors are to be copper with type THHN insulation. The short-circuit protective device is to be an inverse-time circuit breaker. The termination temperature rating is not known. Refer to Figure 10-3.

Solution: The conductor size is determined from the current listed in Table 10-4 (*NEC Table 430.250*) and then increased by 25 percent.

$$180 \times 1.25 = 225 \text{ amperes}$$

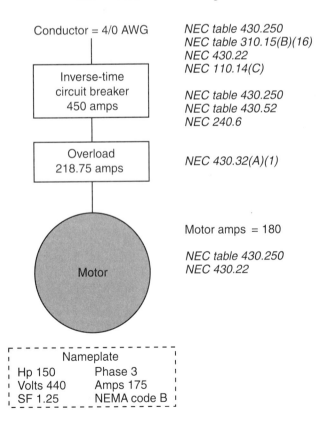

Conductor = 4/0 AWG

NEC table 430.250
NEC table 310.15(B)(16)
NEC 430.22
NEC 110.14(C)

Inverse-time circuit breaker 450 amps

NEC table 430.250
NEC table 430.52
NEC 240.6

Overload 218.75 amps

NEC 430.32(A)(1)

Motor

Motor amps = 180

NEC table 430.250
NEC 430.22

Nameplate
Hp 150   Phase 3
Volts 440   Amps 175
SF 1.25   NEMA code B

FIGURE 10-3   Values for Problem 3.
(Delmar/Cengage Learning)

Table 6-2 (*NEC Table 310.15(B)(16)*) is used to determine conductor size. Type THHN insulation is located in the 90°C (194°F) column. Because the motor nameplate indicates a NEMA code B, the conductor size will be selected from the 75°C (167°F) column. The conductor size will be 4/0 AWG.

The overload size is determined from the nameplate current and *430.32(A)(1)*. Because the motor has a marked service factor of 1.25, the motor nameplate current will be increased by 25 percent.

$$175 \times 1.25 = 218.75 \text{ amperes}$$

The percentage of full-load current listed in Table 10-7 is used to determine the circuit-breaker rating. The table indicates a factor of 250 percent for squirrel-cage motors with a NEMA design code B. The value of current from Table 10-4 is used in this calculation.

$$180 \times 2.50 = 450 \text{ amperes}$$

One of the standard circuit-breaker ratings listed in *240.6(A)* in Table 10-8 is 450 amperes. A 450-ampere inverse-time circuit breaker will be used as the short-circuit protective device.

MULTIPLE MOTOR CALCULATIONS

When several motors or loads are to be connected to a single branch circuit, *NEC 403.53(C)* requires that the branch circuit be protected by fuses or inverse time circuit breakers.

The main feeder short-circuit protective device and conductor size for a multiple motor connection are set forth in *430.62(A)* and *430.24*. In this example, three motors are connected to a common feeder. The feeder is 440 volts, 3-phase, and the conductors are to be copper with type THHN insulation. Each motor is to be protected with a dual-element time-delay fuse and a separate overload device. The main feeder is also protected by a dual-element time-delay fuse. The termination temperature rating of the connected devices is not known. The motor nameplate ratings are as follows:

Motor #1

Phase 3	HP 20
SF 1.25	NEMA code C
Volts 440	Amperes 23
Type Induction	

Motor #2

Phase 3	HP 60
Temp. 40°C	Code J
Volts 440	Amperes 72
Type Induction	

Motor #3

Phase 3	HP 100
Code A	Volts 440
Amperes 96	PF 90%
Type Synchronous	

Motor #1 Calculations

The first step is to calculate the values for motor amperage, conductor size, overload size, and short-circuit protective device size for each motor. These values for motor #1 are shown in Figure 10-4. The motor amperage rating from Table 10-4 is used to determine the conductor and fuse size. The amperage rating must be increased by 25 percent for the conductor size.

$$27 \times 1.25 = 33.75 \text{ amperes}$$

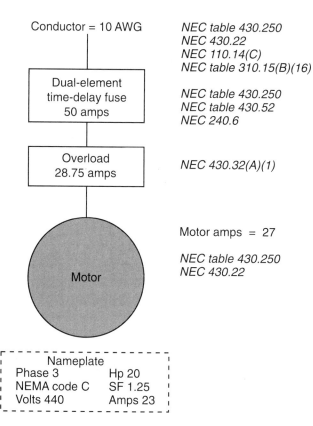

FIGURE 10-4   Values for motor #1.
(Delmar/Cengage Learning)

The conductor size is now chosen from Table 6-2. Although type THHN insulation is located in the 90°C (194°F) column, the conductor size will be chosen from the 75°C (167°F) column, per *110.14(C)(1)(d)*.

$$33.75 \text{ amperes} = 10 \text{ AWG}$$

The fuse rating is determined by using the motor current rating from Table 10-4 and the demand factor from Table 10-7. The percent of full-load current for a dual-element time-delay fuse protecting a squirrel-cage motor listed as Design C is 175 percent. The current listed in Table 10-4 will be increased by 175 percent.

$$27 \times 175\% = 47.25 \text{ amperes}$$

The next higher standard fuse rating listed in Table 10-8 is 50 amperes. A 50-ampere fuse will be used.

The overload size is calculated from the nameplate current. The demand factors in *430.32(A)(1)* are used for the overload calculation. Because this motor has a marked service factor of 1.25, the motor nameplate current will be increased by 25 percent.

$$23 \times 1.25 = 28.75 \text{ amperes}$$

Motor #2 Calculations

Figure 10-5 shows an example for the calculation of motor #2. Table 10-4 lists a full-load current of 77 amperes for this motor. This value of current is increased by 25 percent for the calculation of the conductor current.

$$77 \times 1.25 = 96.25 \text{ amperes}$$

Table 6-2 indicates that a 1 AWG conductor should be used for this motor connection. The conductor size is chosen from the 60°C (140°F) column because the circuit current is less than 100 amperes in accord with *110.14(C)*, and the motor nameplate does not indicate a NEMA design code.

The fuse size will be determined from Table 10-7. The value is to be increased by 175 percent for squirrel-cage motors other than design E.

$$77 \times 175\% = 134.75 \text{ amperes}$$

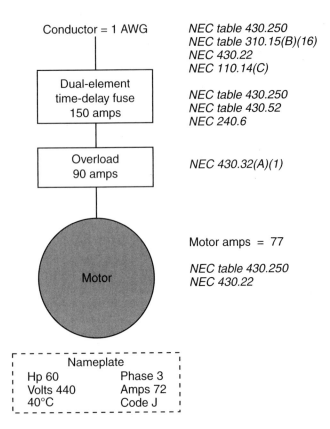

Conductor = 1 AWG

NEC table 430.250
NEC table 310.15(B)(16)
NEC 430.22
NEC 110.14(C)

Dual-element
time-delay fuse
150 amps

NEC table 430.250
NEC table 430.52
NEC 240.6

Overload
90 amps

NEC 430.32(A)(1)

Motor

Motor amps = 77

NEC table 430.250
NEC 430.22

Nameplate
Hp 60   Phase 3
Volts 440   Amps 72
40°C   Code J

FIGURE 10-5   Values for motor #2.
(*Delmar/Cengage Learning*)

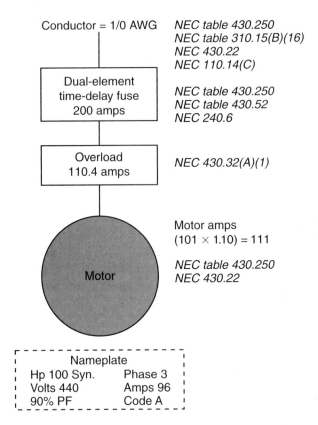

Conductor = 1/0 AWG

NEC table 430.250
NEC table 310.15(B)(16)
NEC 430.22
NEC 110.14(C)

Dual-element
time-delay fuse
200 amps

NEC table 430.250
NEC table 430.52
NEC 240.6

Overload
110.4 amps

NEC 430.32(A)(1)

Motor

Motor amps
(101 × 1.10) = 111

NEC table 430.250
NEC 430.22

Nameplate
Hp 100 Syn.   Phase 3
Volts 440   Amps 96
90% PF   Code A

FIGURE 10-6   Values for motor #3.
(*Delmar/Cengage Learning*)

The next higher standard fuse size listed in Table 10-8 is 150 amperes. Thus, 150-ampere fuses will be used to protect this circuit.

The overload size is determined from *430.32(A) (1)*. The motor nameplate lists a temperature rise of 40°C for this motor. The nameplate current will be increased by 25 percent.

$$72 \times 1.25 = 90 \text{ amperes}$$

Motor #3 Calculations

Motor #3 is a synchronous motor intended to operate with a 90 percent power factor. Figure 10-6 shows an example of this calculation. The note at the bottom of Table 10-4 indicates that the listed current is to be increased by 10 percent for synchronous motors with a listed power factor of 90 percent.

$$101 \times 1.10 = 111 \text{ amperes}$$

The conductor size is calculated by using this current rating and increasing it by 25 percent.

$$111 \times 1.25 = 138.75 \text{ amperes}$$

Table 6-2 indicates that a 1/0 AWG conductor will be acceptable for this circuit. Because the circuit current is over 100 amperes, the conductor is chosen from the 75°C (167°F) column.

The fuse rating is determined from Table 10-7. The percentage of full-load current for a synchronous motor is 175 percent.

$$111 \times 1.75 = 194.25 \text{ amperes}$$

The next highest standard rating fuse listed in Table 10-8 is 200 amperes. Thus, 200-ampere fuses will be used to protect this circuit.

This motor does not have a marked service factor or a marked temperature rise. The overload size will be calculated by increasing the nameplate current by 15 percent, as indicated in *430.32(A)(1)*.

$$96 \times 1.15 = 110.4 \text{ amperes}$$

Main Feeder Calculation

An example of the main feeder calculation is shown in Figure 10-7. The conductor size is

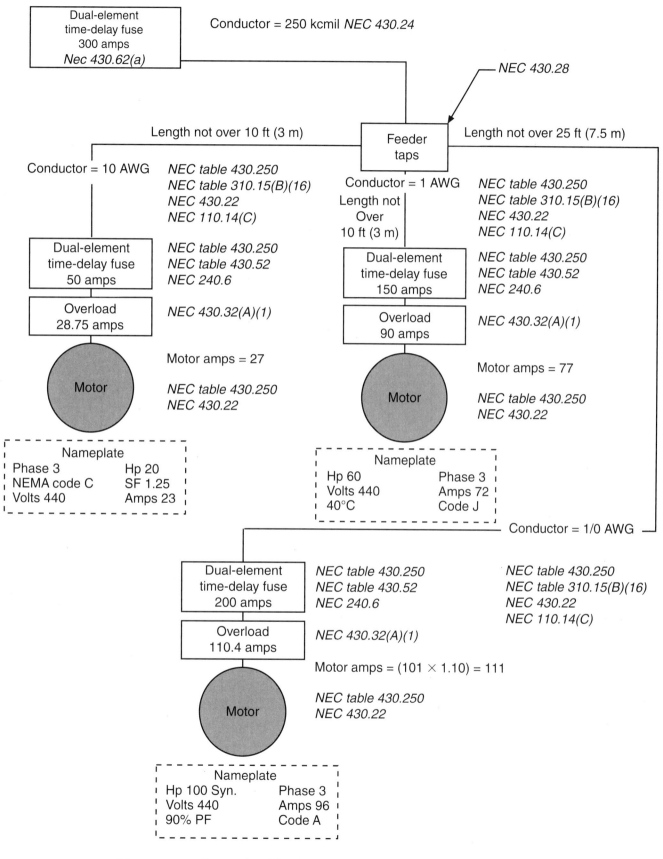

FIGURE 10-7   Example of feeder calculations. (*Delmar/Cengage Learning*)

calculated by increasing the highest amperage rating of the motors connected to the feeder by 25 percent and then adding the amperage ratings of the other motors to this amount. In this example, the 100-horsepower synchronous motor has the highest running current. This current will be increased by 25 percent, and then the running currents of the other motors as determined from Table 10-4 will be added.

$$111 \times 1.25 = 138.75 \text{ amperes}$$

$$138.75 + 77 + 27 = 242.75 \text{ amperes}$$

Table 6-2 indicates that 250 kcmil conductors are to be used as the main feeder conductors. The conductors were chosen from the 75°C (167°F) column.

The size of the short-circuit protective device is determined by *430.62(A)*. The *Code* states that the rating or setting of the short-circuit protective device shall not be greater than the highest rating or setting of the branch-circuit short-circuit and ground-fault protective device plus the sum of the full-load currents of the other motors. The highest fuse size was that of the 100-horsepower synchronous motor. The fuse calculation for this motor was 200 amperes. The running currents of the other two motors will be added to this value to determine the rating of the fuse for the main feeder.

$$200 + 77 + 27 = 304 \text{ amperes}$$

The highest-rated standard fuse listed in Table 10-8, without going over 304 amperes, is 300 amperes. Thus, 300-ampere fuses will be used as the short-circuit protective devices for this circuit.

REVIEW QUESTIONS

All answers should be written in complete sentences, calculations should be shown in detail, and *Code* references should be cited when appropriate.

Unless otherwise stated, all conductors are copper type THHN, and the supply is ac.

1. What is the full-load current of a 500-volt, dc, 20-horsepower motor? _____

_____

_____

2. Which *NEC* table is used to determine the full-load current of a torque motor? _____

_____

3. What is the rated full-load current of a ¾-horsepower, 208-volt single-phase motor?

_____

_____

4. What are the minimum allowable ampacities of the phase conductors and the neutral conductor that are to supply a 2-phase, 3-wire, 230-volt, 30-horsepower motor?

_____

_____

_____

5. What is the full-load current of a 230-volt, 3-phase, 125-horsepower synchronous motor when it is operated at an 80 percent power factor? _____

_____

_____

6. What is the full-load current of a 600-volt, 3-phase, 50-horsepower induction motor?

_____

_____

7. What is the minimum conductor ampacity, overload protection, and dual-element fuse rating for a circuit supplying a 560-volt, 3-phase, code J, 40°C, 125-horsepower motor with a nameplate full-load current of 115 amperes? _____

_____

_____

_____

8. What is the minimum conductor size, overload protection, and inverse-time circuit-breaker rating for a circuit supplying a 230-volt, 3-phase, code A, 40°C, 75-horse-power synchronous motor operating at an 80 percent power factor with a nameplate full-load current of 185 amperes? _____

_____

_____

_____

9. What size conductor would be required to supply three motors connected to a 440-volt, 3-phase branch circuit? Motor #1 is a 50-horsepower, code B induction motor; motor #2 is a 40-horsepower, code H induction motor; and motor #3 is a 50-horsepower, code J induction motor. _____

_____

_____

_____

10. What would be the rating of an inverse-time circuit breaker required for the branch circuit serving the three motors in the previous question? What would be the rating of a dual-element fuse?

Inverse-time circuit breaker: _____

_____

Dual-element fuse: _____

_____

CHAPTER

11

Power Factor

OBJECTIVES

After studying this chapter, the student should be able to

- define and use the concept of power factor.
- correct low power factor situations with a synchronous condenser.
- correct low power factor situations with capacitors.

184

LOADING ON ALTERNATING-CURRENT CIRCUITS

Three kinds of electrical loads may be placed on the lines of ac circuits:

- a resistive load consisting solely of resistance;

- an inductive load consisting of some resistance combined with a larger amount of inductance; and

- a capacitive load consisting almost entirely of capacitance.

All three of these loads are present in varying degrees in nearly all ac lines. The resistive part of the total load is due to the fact that metal conductors do not have a 100 percent efficiency in conducting electricity; thus, some losses will occur in the conductors. All metals used as conductors have some resistance that opposes the free flow of electrons. The resistance depends upon the kind of metal used as the conductor, the length of the conductor, the size or circular mil area of the cross section, and the temperature of the conductor.

When the load includes electromagnetic devices such as motors with windings formed from turns of wire wound on steel cores, then inductance is a factor in analyzing the total load. Inductance creates an additional opposition to the flow of current in an ac system.

The amount of opposition due to inductance, called inductive reactance, depends upon the amount of inductance present and the frequency (in hertz) of the ac system. The inductive reactance is determined by the following equation:

$$\text{Inductive reactance } (X_L) = 2\pi fL$$

where   f = the frequency of the system, in hertz
         L = the inductance, in henrys

Because two forces (resistance and reactance) are now opposing the current, it becomes necessary to combine the effect of the two forces to find the total opposition, which is called impedance. Each of these factors (resistance, reactance, and impedance) is measured in ohms.

Reactance can be either leading (capacitance) or lagging (inductance). When both are present, the larger value is reduced by the amount of the smaller value. In motor circuits the reactance is primarily inductive.

The inductive reactance adds to the resistance in a geometric manner, rather than in an arithmetic manner. That is, the inductive reactance is at right angles to the resistance. The impedance, as given by the following formula, is the resultant (hypotenuse of a right triangle) of these two forces. This formula is based on the Pythagorean theorem found in basic trigonometry.

$$\text{Impedance} = \sqrt{(\text{resistance})^2 + (\text{reactance})^2}$$

or

$$Z = \sqrt{R^2 + X_L^2}$$

For example, in the circuit shown in Figure 11-1, there is a 0.5-ohm resistance and a 0.6-ohm inductive reactance. Thus, the impedance is

$$Z = \sqrt{0.5^2 + 0.6^2} = 0.781 \text{ ohm}$$

When the circuit current lags behind the voltage, an angle is formed. The tangent of this angle is the inductive reactance divided by the resistance, or

$$\tan \theta = \frac{X_L}{R} = \frac{0.6}{0.5} = 1.2$$

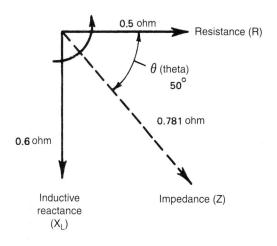

FIGURE 11-1 Vector diagram showing angular displacement of resistance, reactance, and impedance. (*Delmar/Cengage Learning*)

A table of trigonometric functions shows that the angle whose tangent is 1.2 is 50°. The vector diagram in Figure 11-1 shows the relationship among the resistance, the reactance, and the impedance.

Power Factor

The power factor is determined by the cosine of the angle between the voltage and the current. In Figure 11-1, the power factor is 0.643; this value is the cosine of 50°. Note that the angle between the resistance and the impedance in the figure is 50°. If this angle is zero, then the power factor would be 1.0. This indicates perfect conditions in that all of the current is active and useful. The power factor can also be expressed as the true power divided by the apparent power (watts/volt-amperes).

⎓⌁ POWER FACTOR MEASUREMENT

When an industrial plant has a lagging power factor, the value of the power factor should be maintained between 0.9 and 1.0, if possible. This condition is desirable because of a number of factors, including the need to reduce the reactive current to achieve more capacity for useful current on the mains, feeders, and subfeeders; the need for better voltage regulation and stability; and the desirability of obtaining lower power rates from the power company.

As shown in the following list, there are several ways of determining the power factor of an entire plant, a single feeder, or even a branch circuit. The power factor can be determined by the use of

1. A power factor meter

2. A kilovarmeter and kilowattmeter

3. Wattmeters or kilowattmeters in combination with voltmeters and ammeters

The last of the three methods listed is the most convenient one in terms of the connections to be made. However, the use of a permanently connected kilovarmeter is also convenient. When the power factor of the plant is 1.0 (unity), the kilovarmeter reads zero.

The instrument has a zero center scale. The needle indicates the number of reactive kilovolt amperes (kiloVARs) by which the system is lagging or leading. The electrician can tell at a glance whether it is necessary to supply either a larger or a smaller leading component to the system. This determination is a simple matter when a synchronous condenser is available. (See the following section on synchronous condensers.)

When a synchronous condenser is connected across the line, it supplies the leading component of the current needed to offset or counteract the existing lagging component of the current. The plans show the locations of the two synchronous condensers used to correct the power factor of feeder duct No. 2.

Power Factor Correction

Power or true power is expressed in kilowatts. The term kilovolt amperes, which means volts multiplied by amperes and divided by 1000, is called the apparent power. Reactive power is not power at all, but rather is a component that is 90° out of phase with the true power. Reactive power is measured in volt-amperes reactive (VAR).

The total motor load on feeder duct No. 2 is approximately 927 horsepower. This load requires nearly 1293 amperes per leg to supply the 3-phase, 480-volt, 60-hertz motors. The apparent power required for this load is found as shown by the following calculation:

$$\text{Apparent power} = \frac{1293 \times 480 \times 1.73}{1000}$$
$$= 1074 \text{ kVA}$$

The actual value of the apparent power is obtainable only when all of the motors are running and loaded. If the motors are running at less than the full rated horsepower, the power factor will be less than the value of 0.85 used in previous examples.

For the present example, assume that the power factor is 0.74. Recall that the power factor is the cosine of the angle of lag (u). When the value of 0.74 for the cosine is located on a table of functions, the angle u is found to be about 42°, Figure 11-2. The kilovar (kVAR) reactive (useless) component in

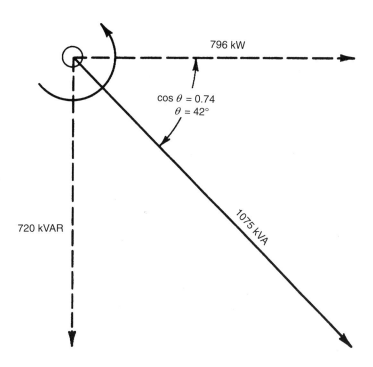

FIGURE 11-2   Angular displacement without synchronous condensers, power factor = 0.742. (*Delmar/Cengage Learning*)

the power group is equal to the value of the apparent power (in kVA) multiplied by the sine of θ:

$$kVAR = kVA \text{ sine } 42°$$
$$= 1074 \times 0.67$$
$$= 720 \text{ kilovolt-amperes reactive}$$

Because 180° separate the leading component supplied by the synchronous condenser and the lagging component caused by the inductive characteristics of the motor load (90° lead and 90° lag, respectively), the leading kVAR value needed to cancel the lagging kVAR value is 720. Two 350-kVAR synchronous condensers are to be installed to provide a leading power factor and reduce the lagging power factor produced by the induction motors and other inductive loads. It is not necessary to reduce the reactance to zero as shown in Figure 11-3. This illustration shows that the synchronous condenser is supplying 520 kVAR of leading reactance. This value reduces the original lagging reactance to 200 kVAR. However, the power factor is increased to 0.97 and the apparent power to 821 kVA. The kW value is unchanged, as it is established by the loading of the motors and the friction losses in the motor. Both of these quantities are actual loads.

THE SYNCHRONOUS CONDENSERS

The synchronous condenser is a rotating electrical machine, as shown in Figure 11-4. It is similar to a synchronous motor or an ac generator. However, in operation, there are no electrical or mechanical loads connected to it. The only power required for the operation of a synchronous condenser is the power needed to supply its own small losses.

The synchronous condenser has a stationary 3-phase armature winding rated at 480 volts and 60 hertz. The rotating field of the condenser is excited from a source of dc, usually a small dc generator mounted on the shaft of the synchronous condenser. The schematic diagrams of the controller and the control scheme for a synchronous condenser are shown in Figure 11-5.

The operation of the synchronous condenser is such that when the field winding is overexcited, the machine has a leading current. As the field excitation becomes stronger, there is no lead or lag by the condenser, and the current in the synchronous condenser branch is in phase with the voltage.

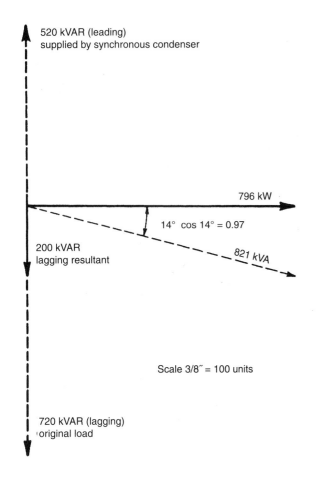

520 kVAR (leading)
supplied by synchronous condenser

796 kW

14° cos 14° = 0.97

200 kVAR
lagging resultant

821 kVA

Scale 3/8″ = 100 units

720 kVAR (lagging)
original load

FIGURE 11-3   Vector diagram with synchronous condenser, power factor = 0.97. (*Delmar/Cengage Learning*)

When the industrial plant is not operating at its full capacity, the excitation must be regulated to maintain a low kVAR value in the feeder. Periodic inspections and adjustments as necessary will ensure that a high power factor is maintained. The power factor is corrected or increased only from the point of attachment of the synchronous condenser back to the source of supply, Figure 11-6. Power factor corrections have no effect on the current in the plug-in ducts or the motor branch circuits.

Therefore, to cancel the lagging currents in the feeder duct, the synchronous condenser must be overexcited. To accomplish this, the strength of the field excitation is increased by adjusting the field rheostat on the control panelboard.

When the synchronous condenser is started from the control panelboard, it performs in the same manner as any synchronous motor with the exception that a load is not connected. Before the condenser starts, a normally closed contactor shorts out the field windings to prevent a high-voltage buildup. As soon as the machine is close to its synchronous speed, the contactor removes the short and connects the field to the exciter. As soon as the normal field excitation is achieved, the current decreases to a relatively low value. The ac ammeter on the control panelboard will rise rapidly

FIGURE 11-4   Synchronous condenser with directly connected exciter. (*Courtesy ARCO Electric Products*)

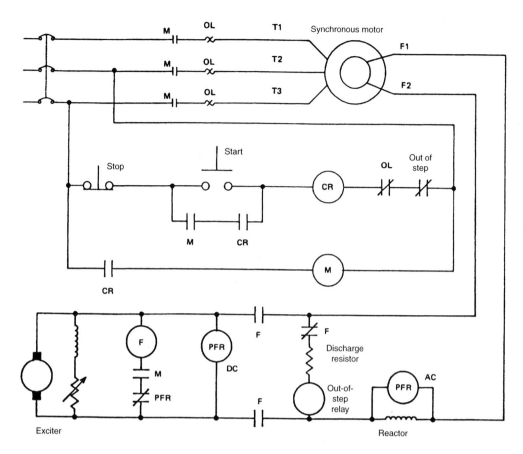

FIGURE 11-5   Basic synchronous motor controller circuitry. (*Delmar/Cengage Learning*)

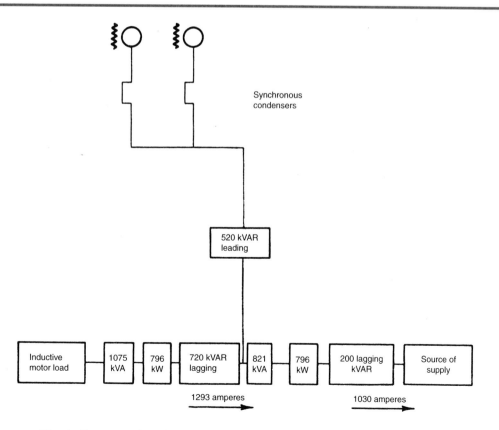

FIGURE 11-6   Block diagram showing how the power factor is corrected. (*Delmar/Cengage Learning*)

when the field rheostat is adjusted for either under-excitation or overexcitation.

When the machine is underexcited by a lower-than-normal field current, the current lags the voltage. But, when the field excitation is stronger than normal, the current leads the voltage and provides a leading kVAR value to counteract the lagging kVAR value in the feeder duct. This lagging value is caused by the inductive effect of the ac motor load. By using the field rheostat to regulate the field excitation, any value of leading kVAR is made available up to the rated output of the machine.

If the plant is running at full capacity, both synchronous condensers must be used. The condensers should be adjusted so that their kVAR outputs are equal. The total kVAR value should be such that the power factor is improved to a value close to unity.

THE TIE-IN

Each of the 350-kVAR synchronous condensers is connected to the No. 2 feeder busway. The wiring for these connections is sized according to *430.22*. The conductors must have an ampacity of at least 125 percent of the full-load current rating of the synchronous condenser that is supplied by the manufacturer. Using 370 amperes as an example, when 370 is multiplied by 1.25, the result is 463 amperes, the required ampacity of the conductors. To permit the use of smaller conductors and conduit, it was decided to parallel the feeders. Several rules must be followed when using parallel feeders (see *300.20* and *310.10(H)*):

- all phase conductors, and the neutral if used, as well as all equipment grounding conductors must be grouped in each raceway;

- the conductors must be 1/0 AWG or larger;

- the conductor in one grouping must be the same length, the same conductor material, the same size, have the same type of insulation, and be terminated in the same manner as the conductors in the other grouping(s); and

- the raceways containing the groups of conductors must have the same physical characteristics.

Because two parallel feeders are installed to each of the synchronous condensers, the conductors can be sized for half the value of the required ampacity, or 232 amperes. The connections are illustrated in Figure 11-7.

CORRECTING POWER FACTOR WITH CAPACITORS

Although this plant uses synchronous condensers for power factor correction, it is common practice in many industrial installations to use banks of capacitors, as shown in Figure 11-8, to perform this task. A capacitor is a device that has a leading current and, therefore, a leading power factor. When capacitors are connected in a circuit with inductors, the leading VARs of the capacitors act to cancel the lagging VARs of the inductors. In this way, power factor can be corrected.

To correct the power factor of a circuit or motor, the existing power factor must first be determined. In the example shown in Figure 11-9, a 3-phase wattmeter, an ammeter, and a voltmeter have been connected to a 3-phase circuit. It is assumed the meters indicate the values shown:

> Wattmeter: 13.9 kilowatts
>
> Ammeter: 25 amperes
>
> Voltmeter: 480 volts

To calculate the power factor, it will be necessary to first calculate the apparent power. In a 3-phase circuit, the apparent power can be calculated using the formula:

$$VA = \sqrt{3} \times volts \times amps$$
$$= 1.732 \times 480 \times 25$$
$$= 20.8 \, kVA$$

Now that both the apparent power and the true power of the circuit are known, the power factor can be determined by comparing the true power and apparent power.

$$PF = \frac{W}{VA} = \frac{13.9}{20.8} = 0.668 = 66.8\%$$

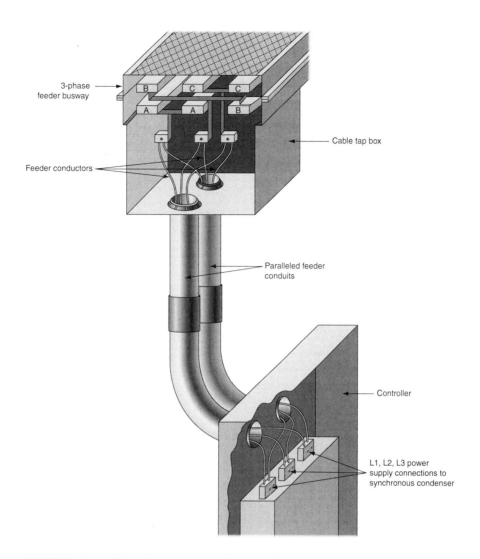

FIGURE 11-7   Synchronous condenser tie-in. (*Delmar/Cengage Learning*)

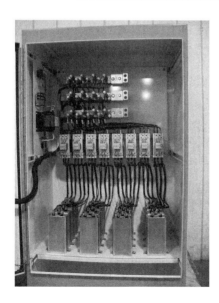

FIGURE 11-8   Capacitors used for power factor correction. (*Courtesy ARCO Electric Products*)

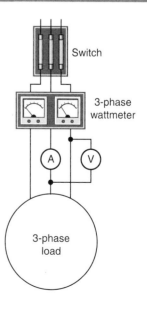

FIGURE 11-9   Measuring power factor. (*Delmar/Cengage Learning*)

The amount of reactive power in the circuit can now be calculated using the formula:

$$kVAR = \sqrt{kVA^2 - kW^2}$$
$$= \sqrt{20.8^2 - 13.9^2}$$
$$= \sqrt{432.64 - 193.21}$$
$$= \sqrt{239.43}$$
$$= 15.47 \, kVAR$$

To determine the capacitive kVARs needed to correct the power factor to 97 percent, first determine what the apparent power would be with a 97 percent power factor.

$$kVA = \frac{kW}{PF} = \frac{13.9}{0.97} = 14.3$$

Now determine the amount of inductive VARs necessary to produce an apparent power to 14.3 kVA.

$$kVAR = \sqrt{kVA^2 - kW^2}$$
$$= \sqrt{14.3^2 - 13.9^2}$$
$$= 3.36$$

Because the circuit presently contains 15.47 inductive kVARs, 12.11 capacitive kVARs (15.47–3.36) would be added to the circuit.

Another method for determining the capacitance needed to correct the power factor is to use Table 11-1. To find the amount of capacitance needed, calculate the power factor in the same manner described previously. The circuit in the previous example has a true power of 13.9 kilowatts and a power factor of 67 percent. To find the amount of capacitive VAR needed to correct the power factor to 97 percent, find 67 percent in the left-hand column. Follow this row across to the 97 percent column. The multiplication factor is 0.857. Multiply the true power value by the multiplication factor.

$$13.9 \, kW \times 0.857 = 11.9 \, kVARs$$

CORRECTING MOTOR POWER FACTOR

It is often desirable to correct the power factor of a single motor. The amount of capacitance needed can be determined in the same manner as shown previously, by connecting a wattmeter, ammeter, and voltmeter in the circuit. Charts similar to the ones shown in Table 11-2 can also be used. These two charts list the horsepower and synchronous speed of both U-frame and T-frame motors. The charts assume a correction factor of 93 to 97 percent. The values shown are the kVAR of capacitance needed to correct the motor power factor. For motors designed to operate on 208 volts, the kVAR value shown should be increased by 1.33. For motors designed to operate on 50 hertz, increase the chart values by a factor of 1.2.

INSTALLING CAPACITORS

NEC Article 460 covers the installation and protection of capacitor circuits. As stated in *460.8(A)*, conductors in capacitor circuits must be rated no less than 135 percent of the current rating of the capacitor. This section further states that if the capacitor is used in a motor circuit, the conductors connecting the capacitor cannot be less than one-third the rating of the motor current and in no case less than 135 percent of the rated current of the capacitor.

If capacitors are to be used to correct the power factor of a single motor, the manner in which the capacitors are installed can greatly influence the *Code* requirements. For example, *460.8(B)* states that an overcurrent device must be provided for each ungrounded conductor in a capacitor bank. The exception, however, states that a separate overcurrent device does not have to be provided if the capacitor bank is connected to the load side of the motor overload protective device, Figure 11-10. *NEC 460.8(C)* states that a capacitor must have a separate disconnecting means rated no less than 135 percent of the rated capacitor current. The exception, however, states that a separate disconnect means is not required if the capacitor is connected to the load side of the motor overload protective device, as shown in Figure 11-10. If the capacitor is connected ahead of the overload protective device, as shown in Figure 11-11, both a separate disconnecting means and an overcurrent protective device are required. Table 11-3 provides a list of sizes of wire, fuses, and switches for different kVAR capacitor ratings on different voltages of 3-phase systems.

TABLE 11-1

Kilowatt multipliers for determining capacitor kilovars. (*Courtesy ARCO Electric Products*)

Original PF	80	81	82	83	84	85	86	87	88	89	90	91	92	93	94	95	96	97
50	0.982	1.008	1.034	1.060	1.086	1.112	1.139	1.165	1.192	1.220	1.248	1.276	1.306	1.337	1.369	1.403	1.440	1.481
51	0.937	0.962	0.989	1.015	1.041	1.067	1.094	1.120	1.147	1.175	1.203	1.231	1.261	1.292	1.324	1.358	1.395	1.436
52	0.893	0.919	0.945	0.971	0.997	1.023	1.050	1.076	1.103	1.131	1.159	1.187	1.217	1.248	1.280	1.314	1.351	1.392
53	0.850	0.876	0.902	0.928	0.954	0.980	1.007	1.033	1.060	1.088	1.116	1.144	1.174	1.205	1.237	1.271	1.308	1.349
54	0.809	0.835	0.861	0.887	0.913	0.939	0.966	0.992	1.019	1.047	1.075	1.103	1.133	1.164	1.196	1.230	1.267	1.308
55	0.769	0.795	0.821	0.847	0.873	0.899	0.926	0.952	0.979	1.007	1.035	1.063	1.093	1.124	1.156	1.190	1.227	1.268
56	0.730	0.756	0.782	0.808	0.834	0.860	0.887	0.913	0.940	0.968	0.996	1.024	1.054	1.085	1.117	1.151	1.188	1.229
57	0.692	0.718	0.744	0.770	0.796	0.822	0.849	0.875	0.902	0.930	0.958	0.986	1.016	1.047	1.079	1.113	1.150	1.191
58	0.655	0.681	0.707	0.733	0.759	0.785	0.812	0.838	0.865	0.893	0.921	0.949	0.979	1.010	1.042	1.076	1.113	1.154
59	0.619	0.645	0.671	0.697	0.723	0.749	0.776	0.802	0.829	0.857	0.885	0.913	0.943	0.974	1.006	1.040	1.077	1.118
60	0.583	0.609	0.635	0.661	0.687	0.713	0.740	0.766	0.793	0.821	0.849	0.877	0.907	0.938	0.970	1.004	1.041	1.082
61	0.549	0.575	0.601	0.627	0.653	0.679	0.706	0.732	0.759	0.787	0.815	0.843	0.873	0.904	0.936	0.970	1.007	1.048
62	0.516	0.542	0.568	0.594	0.620	0.646	0.673	0.699	0.725	0.754	0.782	0.810	0.840	0.871	0.903	0.937	0.974	1.015
63	0.483	0.509	0.535	0.561	0.587	0.613	0.640	0.666	0.693	0.721	0.749	0.777	0.807	0.838	0.870	0.904	0.941	0.982
64	0.451	0.474	0.503	0.529	0.555	0.581	0.608	0.634	0.661	0.689	0.717	0.745	0.775	0.806	0.838	0.872	0.909	0.950
65	0.419	0.445	0.471	0.497	0.523	0.549	0.576	0.602	0.629	0.657	0.685	0.713	0.743	0.774	0.806	0.840	0.877	0.918
66	0.388	0.414	0.440	0.466	0.492	0.518	0.545	0.571	0.598	0.626	0.654	0.682	0.712	0.743	0.775	0.809	0.846	0.887
67	0.358	0.384	0.410	0.436	0.462	0.488	0.515	0.541	0.568	0.596	0.624	0.652	0.682	0.713	0.745	0.779	0.816	0.857
68	0.328	0.354	0.380	0.406	0.432	0.458	0.485	0.511	0.538	0.566	0.594	0.622	0.652	0.683	0.715	0.749	0.786	0.827
69	0.299	0.325	0.351	0.377	0.403	0.429	0.456	0.482	0.509	0.537	0.565	0.593	0.623	0.654	0.686	0.720	0.757	0.798
70	0.270	0.296	0.322	0.348	0.374	0.400	0.427	0.453	0.480	0.508	0.536	0.564	0.594	0.625	0.657	0.691	0.728	0.769
71	0.242	0.268	0.294	0.320	0.346	0.372	0.399	0.425	0.452	0.480	0.508	0.536	0.566	0.597	0.629	0.663	0.700	0.741
72	0.214	0.240	0.266	0.292	0.318	0.344	0.371	0.397	0.424	0.452	0.480	0.508	0.538	0.569	0.601	0.635	0.672	0.713
73	0.186	0.212	0.238	0.264	0.290	0.316	0.343	0.369	0.396	0.424	0.452	0.480	0.510	0.541	0.573	0.607	0.644	0.685
74	0.159	0.185	0.211	0.237	0.263	0.289	0.316	0.342	0.369	0.397	0.425	0.453	0.483	0.514	0.546	0.580	0.617	0.658
75	0.132	0.158	0.184	0.210	0.236	0.262	0.289	0.315	0.342	0.370	0.398	0.426	0.456	0.487	0.519	0.553	0.590	0.631
76	0.105	0.131	0.157	0.183	0.209	0.235	0.262	0.288	0.315	0.343	0.371	0.399	0.429	0.460	0.492	0.526	0.563	0.604
77	0.079	0.105	0.131	0.157	0.183	0.209	0.236	0.262	0.289	0.317	0.345	0.373	0.403	0.434	0.466	0.500	0.537	0.578
78	0.052	0.078	0.104	0.130	0.156	0.182	0.209	0.235	0.262	0.290	0.318	0.346	0.376	0.407	0.439	0.473	0.510	0.551
79	0.026	0.052	0.078	0.104	0.130	0.156	0.183	0.209	0.236	0.264	0.292	0.320	0.350	0.381	0.413	0.447	0.484	0.525
80	0.000	0.026	0.052	0.078	0.104	0.130	0.157	0.183	0.210	0.238	0.266	0.294	0.324	0.355	0.387	0.421	0.458	0.499
81		0.000	0.026	0.052	0.078	0.104	0.131	0.157	0.184	0.212	0.240	0.268	0.298	0.329	0.361	0.395	0.432	0.473
82			0.000	0.026	0.052	0.078	0.105	0.131	0.158	0.186	0.214	0.242	0.272	0.303	0.335	0.369	0.406	0.447
83				0.000	0.026	0.052	0.079	0.105	0.132	0.160	0.188	0.216	0.246	0.277	0.309	0.343	0.380	0.421
84					0.000	0.026	0.053	0.079	0.106	0.134	0.162	0.190	0.220	0.251	0.283	0.317	0.354	0.395
85						0.000	0.027	0.053	0.080	0.108	0.136	0.164	0.194	0.225	0.257	0.291	0.328	0.369
86							0.000	0.026	0.053	0.081	0.110	0.137	0.167	0.198	0.230	0.264	0.301	0.342
87								0.000	0.027	0.055	0.083	0.111	0.141	0.172	0.204	0.238	0.275	0.316
88									0.000	0.028	0.056	0.084	0.114	0.145	0.177	0.211	0.248	0.289
89										0.000	0.028	0.056	0.086	0.117	0.149	0.183	0.220	0.261
90											0.000	0.028	0.058	0.089	0.121	0.155	0.192	0.233

Original Power-Factor Percentages

TABLE 11-2

Correcting the power factor of motors. (Abbreviated table: *Courtesy ARCO Electric Products*)

Kilovolt-Amperes Reactive for U-Frame Motors
Suggested Capacitor Ratings for Approximately 93% to 97% Power Factor

HP	3600 RPM	1800 RPM	1200 RPM	900 RPM	720 RPM	600 RPM
5	2.5	2.5	2.5	3	4	4
7½	2.5	2.5	3	4	5	6
10	3	3	3	5	6	7.5
15	4	4	5	6	8	8
20	5	5	6	7.5	8	10
25	6	6	7.5	8	10	13
30	8	8	8	10	13	15
40	10	10	10	13	15	20
50	12	10	13	15	18	23
60	13	13	15	18	21	26
75	17	15	18	21	26	35
100	21	21	25	26	35	40
125	26	26	30	30	40	50
150	30	30	35	37	50	50
200	40	37	40	50	60	60
250	50	45	50	60	70	75
300	60	50	60	60	80	90
350	60	60	75	75	90	95

Kilovolt-Amperes Reactive for T-Frame Motors
Suggested Capacitor Ratings for Approximately 94% to 97% Power Factor

HP	3600 RPM	1800 RPM	1200 RPM	900 RPM	720 RPM	600 RPM
5	2.5	2.5	3	4	4	5
7½	2.5	3	4	5	5	6
10	4	4	5	6	7.5	8
15	5	5	6	7.5	8	10
20	6	6	7.5	9	10	12
25	7.5	7.5	8	10	12	18
30	8	8	10	14	15	20.5
40	12	14	16	18	20	25
50	15	18	20	21	23	30
60	18	20	23	25	30	35
75	20	25	25	30	35	40
100	23	30	30	35	40	45
125	25	35	35	40	45	50
150	30	40	40	50	50	60
200	35	50	50	70	70	90
250	40	60	60	80	90	100
300	45	70	75	100	100	
350	50	75	90			

For maximum benefit, locate at motor. For a 208-volt system, use 33% larger.

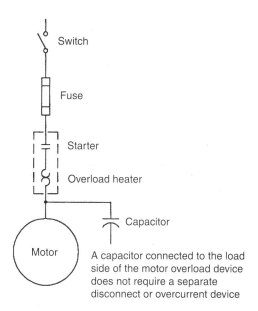

FIGURE 11-10 Capacitor connected to load side of motor. (*Delmar/Cengage Learning*)

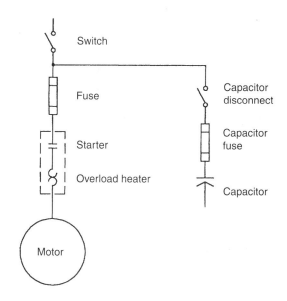

FIGURE 11-11 Capacitor connected ahead of overload protective device. (*Delmar/Cengage Learning*)

TESTING CAPACITORS

WARNING: Capacitors are among the most dangerous components in the electrical field. A charged capacitor has the ability to deliver an almost infinite amount of current. Use caution when testing or working with capacitors. NEVER charge a capacitor and hand it to someone as a joke. Capacitors can cause the heart to go into fibrillation under the right conditions.

To understand how to test a capacitor, it is necessary to first understand what a capacitor is. Most capacitors used in industry, especially for power factor correction, are called ac or nonpolarized capacitors. This simply means that the capacitor is not sensitive to which polarity of voltage is connected to which plate. These capacitors are generally constructed of two metal plates separated by an insulating material called the dielectric, as illustrated in Figure 11-12. To accurately test a capacitor, two measurements must be made. One is to measure the capacitance value of the capacitor to determine whether it is the same or approximately the same as the rate value. The other is to test the strength of the dielectric.

The first test should be made with an ohmmeter. With the power disconnected, connect the terminals of an ohmmeter directly across the capacitor terminals as shown in Figure 11-13. (It is a good practice to discharge the capacitor, by touching the leads together, before connecting to the ohmmeter.)

This test determines whether the dielectric is shorted. When the ohmmeter is connected, the needle should swing up-scale and return to infinity. The amount of needle swing is determined by the capacitance of the capacitor. Then reverse the ohmmeter connection; the needle should move twice as far up-scale and return to the infinity setting.

If the ohmmeter test is successful, the dielectric must be tested at its rated voltage. This is called a dielectric strength test. To make this test, a dielectric test set must be used. This device is often referred to as a HIPOT because of its ability to produce a

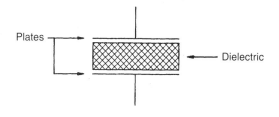

FIGURE 11-12 Basic capacitor. (*Delmar/Cengage Learning*)

TABLE 11-3

Three-phase wiring and fusing for capacitor installations. (*Courtesy ARCO Electric Products*)

	240 VOLTS				480 VOLTS				600 VOLTS			
KVAR	NOM. AMPS.	SIZE THW WIRE	FUSE	Switch	NOM. AMPS.	SIZE THW WIRE	FUSE	Switch	NOM. AMPS.	SIZE THW WIRE	FUSE	Switch
1	2.4	14	5	30	1.2	14	5	30				
2.5	6.0	14	10	30	3.0	14	5	30	2.4	14	5	30
3	7.2	14	15	30	3.6	14	10	30	3.0	14	5	30
4	9.6	14	20	30	4.8	14	10	30	3.8	14	10	30
5	12.0	12	20	30	6.0	14	10	30	4.8	14	10	30
6	14.4	12	25	30	7.2	14	15	30	5.7	14	10	30
7.5	18.0	10	30	30	9.0	14	20	30	7.0	14	15	30
8	19.2	10	35	60	9.6	14	20	30	7.6	14	15	30
10	24.0	8	40	60	12.0	12	25	30	9.5	14	20	30
13	31.2	6	50	60	15.6	10	30	30	12.2	12	20	30
15	36.0	6	60	60	18.0	10	30	30	14.2	12	25	30
18	43.4	4	80	100	21.7	10	35	60	17.3	10	30	30
20	48.0	4	80	100	24.0	8	40	60	19.0	10	35	60
21	50.5	4	80	100	25.2	8	40	60	20.1	10	40	60
23	55.2	3	90	100	27.6	8	50	60	22.8	10	40	60
25	60.0	2	90	100	30.0	6	60	60	23.8	8	40	60
26	62.5	2	90	100	31.2	6	60	60	24.8	8	40	60
30	72.0	2	125	200	36.0	6	60	60	28.8	8	50	60
33	79.2	1	150	200	39.6	6	80	100	31.3	6	60	60
35	84.0	1	150	200	42.0	4	80	100	33.6	6	60	60
37	88.8	1/0	150	200	44.4	4	80	100	35.1	6	60	60
40	96.0	1/0	175	200	48.0	4	80	100	38.0	6	80	100
45	108.0	2/0	200	200	54.0	3	90	100	42.7	4	80	100
50	120.0	2/0	200	200	60.0	2	90	100	47.6	4	80	100
55	132.0	3/0	225	400	66.0	2	100	100	52.4	3	90	100
60	144.0	3/0	250	400	72.0	2	125	200	57.6	3	90	100
65	156.0	3/0	250	400	78.0	1	150	200	62.4	2	90	100
70	168.0	4/0	300	400	84.0	1	150	200	66.2	2	100	100
75	180.0	250	300	400	90.0	1/0	150	200	71.0	2	125	200
80					96.0	1/0	175	200	77.0	1	150	200
85					102.0	1/0	175	200	81.0	1	150	200
90					108.0	2/0	200	200	85.5	1/0	150	200
95					114.0	2/0	200	200	90.0	1/0	150	200
100					120.0	2/0	200	200	95.0	1/0	175	200
125					150.0	3/0	250	400	119.0	2/0	200	200

Wire sized at 135% of rated current. See *NEC 460.8(A)*.

high voltage or high potential. The dielectric test set contains a variable voltage control, a voltmeter, and a microammeter. To use the HIPOT, connect its terminal leads to the capacitor terminals. Increase the output voltage until rated voltage is applied to the capacitor. The microammeter indicates any

Ohmmeter

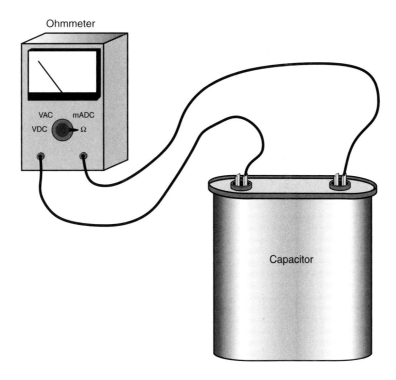

FIGURE 11-13   Testing a capacitor with an ohmmeter. (*Delmar/Cengage Learning*)

Ammeter

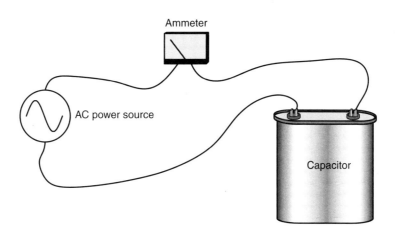

FIGURE 11-14   Determining the capacitance of a capacitor. (*Delmar/Cengage Learning*)

current flow between the plates and the dielectric. If the capacitor is good, the microammeter should indicate zero current.

The capacitance value must be measured to determine whether there are any open plates in the capacitor. To measure the capacitance value of the capacitor, connect, as shown in Figure 11-14, some value of ac voltage across the plates of the capacitor. This voltage must not be greater than the rated capacitor voltage. Then measure the amount of current in the circuit. Now that the voltage and current are known, the capacitive reactance can be calculated using the formula:

$$X_C = \frac{E}{I}$$

After the capacitive reactance has been determined, the capacitance can be calculated using the formula:

$$C = \frac{1}{2\pi f\, X_C}$$

REVIEW QUESTIONS

All answers should be written in complete sentences, calculations should be shown in detail, and *Code* references should be cited when appropriate.

1. List the three kinds of electrical loads that are connected to an ac circuit. _____

   _____

   _____

   _____

2. An electrician uses a clamp-on ammeter and a voltmeter to measure the current and voltage of a motor.

   a. If the two values are multiplied, is the product the true power or the apparent power?

      _____

      _____

   b. What would you need to know to be able to calculate the other power value?

      _____

      _____

3. The ballast in some of the older fluorescent luminaires was said to have a low power factor. What device was added to the circuit to improve the power factor? _____

   _____

   _____

4. What effect does the power factor of a motor have on the branch circuit supplying power to the motor? _____

   _____

   _____

   _____

5. Inductive reactance, resistance, and impedance are all measured in ohms. What is the common characteristic that makes this possible?_____

   _____

   _____

   _____

6. Inductive reactance, resistance, and impedance are always present in circuit-serving motors. Which of the three actually determines the required conductor size? _____

   _____

   _____

   _____

7. We are told that 1 horsepower is equivalent to 746 watts, but a 1-horsepower, 200-volt single-phase motor, according to Table 10-2 (*NEC Table 430.248*), has a current of 9.2 amperes. Explain this apparent discrepancy. _____

   _____

   _____

   _____

8. A 240-volt, 3-phase circuit has a current of 116 amperes. A wattmeter connected to the circuit indicates a load of 34.7 kilowatts.

   a. What is the power factor? _____

      _____

      _____

   b. How many capacitive VAR would be needed to correct the power factor to 95 percent?

      _____

      _____

9. When a 60-volt, 60-hertz supply is connected to a capacitor, the current reads 0.6 ampere. What is the capacitance of the capacitor?_____

   _____

   _____

   _____

Ventilating, Air Conditioning, and Other Facilities

OBJECTIVES

After studying this chapter, the student should be able to

- explain the ventilator circuits.
- determine the requirements for heating controls.
- determine the electrical needs for air conditioning.

The plans and specifications for the industrial building indicate that there are still several other circuits to be installed. Although these circuits are smaller in scale than those covered in previous chapters, they are important nonetheless.

THE VENTILATOR AND EXHAUST SYSTEMS

According to the plans and specifications, there are six ventilating units to be installed and connected for operation. Four of these units are located on the roof of the manufacturing area. The two remaining units are located on the roof of the office structure (see Sheet E-3 of the plans). Each ventilating unit consists of a steel housing designed for mounting on a flat surface and a blower unit enclosed in the steel housing.

The exhaust blower unit consists of a 3-horsepower, 3-phase, 208-volt motor driving a propeller-type fan through a V-belt drive, Figure 12-1. This arrangement results in a quieter mode of operation than is obtainable using a direct drive unit. The motor and fan assembly are cushion-mounted to absorb vibration. As a result, the unit is almost noiseless during operation except for the sound of the rush of air. The fan rotates at a speed of 905 rpm and is rated at 17,300 cubic feet per minute (cfm) at a static air pressure of 0 inches.

The two ventilating blowers located on the roof of the office structure are also rated at 3 horsepower, Figure 12-2. These blowers exhaust the air from the toilets and washrooms located on the first and second floors as well as the basement washroom and locker room. The fan has a speed of 913 rpm and should be used with a static air pressure of less than 1 inch.

As shown on Sheet E-3, the two ventilators and the four exhaust blowers are supplied from a special power panel, P-11. This power panelboard is shown in Figure 12-3. The panelboard contains a fusible disconnect means for each of the motor circuits. Six-inch wireways are installed to the right and left from the panelboard. Manual motor controllers are supplied from these wireways.

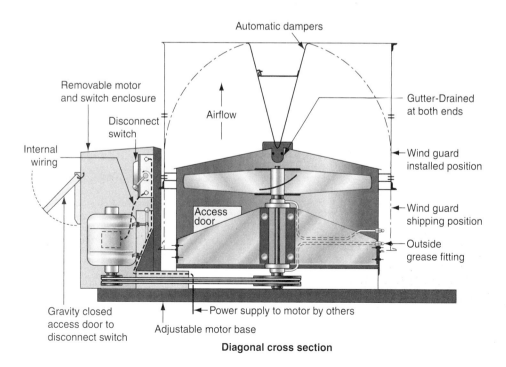

FIGURE 12-1 Exhaust blower unit. (*Delmar/Cengage Learning*)

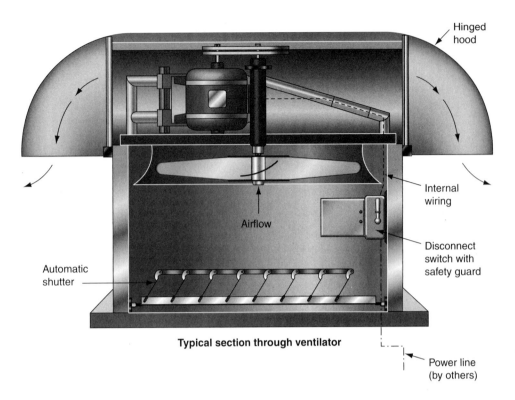

FIGURE 12-2 Ventilating blower. (*Delmar/Cengage Learning*)

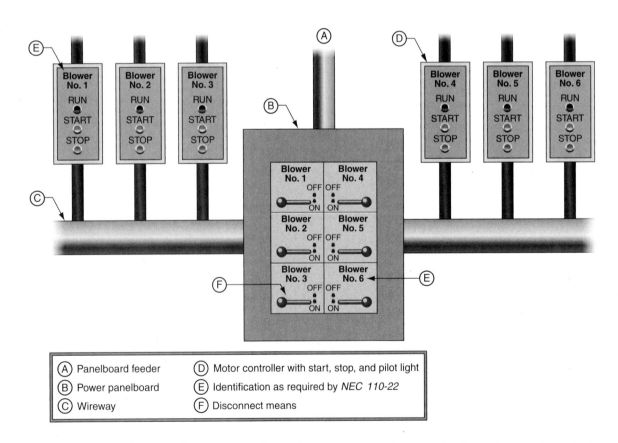

(A) Panelboard feeder	(D) Motor controller with start, stop, and pilot light
(B) Power panelboard	(E) Identification as required by *NEC 110-22*
(C) Wireway	(F) Disconnect means

FIGURE 12-3 Layout of power panelboard and motor controllers. (*Delmar/Cengage Learning*)

SPECIAL TERMINOLOGY

Most of the following phrases are used repeatedly in *NEC Article 440*.

Maximum Continuous Current

The *maximum continuous current* is determined by the manufacturer of the hermetic refrigerant motor-compressor under specific test conditions. The maximum continuous current is needed to properly design the unit. The electrician need not know this information, and it is not placed on the nameplate.

Rated-Load Current

The *rated-load current* is determined by the manufacturer of the hermetic refrigerant motor-compressor by testing at rated refrigerant pressure, temperature conditions, and voltage. In most instances, the rated-load current is at least equal to 64.1 percent of the hermetic refrigerant motor-compressor's maximum continuous current.

Branch-Circuit Selection Current

Some hermetic refrigerant motor-compressors are designed to operate continuously at currents greater than 156 percent of the rated-load current. In such cases, the unit's nameplate is marked with branch-circuit selection current. The *branch-circuit selection current* will be at least 64.1 percent of the maximum continuous current rating of the hermetic refrigerant motor-compressor.

Minimum Circuit Ampacity

The manufacturer of an air-conditioning unit is required to mark the nameplate with the *minimum circuit ampacity*. This is important information for the electrician. The manufacturer determines the minimum circuit ampacity by multiplying the rated-load current, or the branch-circuit selection current of the hermetic refrigerant motor-compressor, by 125 percent. The current ratings of all other concurrent loads, such as fan motors, transformers, relay coils, and so on, are then added to this value.

Maximum Overcurrent Protective Device

The manufacturer is required to mark the *maximum overcurrent protective device* on the nameplate. This value is determined by multiplying the rated-load current, or the branch-circuit selection current of the hermetic refrigerant motor-compressor, by 225 percent and then adding all concurrent loads such as electric heaters, motors, and so on.

THE COOLING EQUIPMENT

The cooling equipment for the industrial building consists of three liquid chillers and ten fan coil units as shown on Sheet E-3 of the plans. The water-circulating pumps are located in the boiler room and are not shown in the plans. Although the actual installation of the cooling equipment is the responsibility of another contractor, the electrician should be familiar with and understand the basic operation of the cooling system.

Each of the three liquid chillers is capable of cooling nearly 100 gallons of water per minute to a temperature of 46°F (8°C) from a return water temperature of 52°F (11°C). This water is circulated through a piping system to the fan coil units, Figure 12-4, located at various points in the building. Cool air is then blown from these units into the immediate area. Each serves to maintain a comfortable air temperature.

Each fan coil unit is equipped with a 2-horsepower induction motor driving a squirrel-cage-type fan. The fan moves nearly 3000 cubic feet of air per minute through a fin tube coil, Figure 12-5. The movement of air through the coil removes the heat from the air, which is then forced through ductwork to the proper area to be cooled. Whenever the specific area is cooled to the desired temperature, a thermostat opens the control circuit to the motor controller and the air movement stops until the area again requires cooling. The controller for each fan coil unit is located adjacent to the cooling unit. Three circuits from panelboard P-14 are used to supply the ten fan coil cooling units.

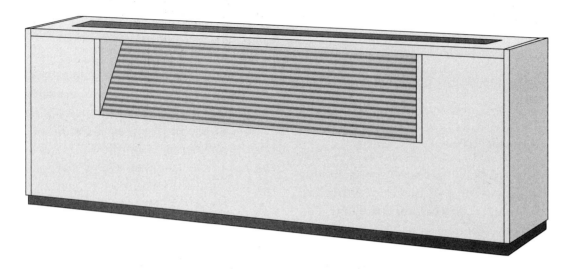

FIGURE 12-4 Fan coil unit. (*Delmar/Cengage Learning*)

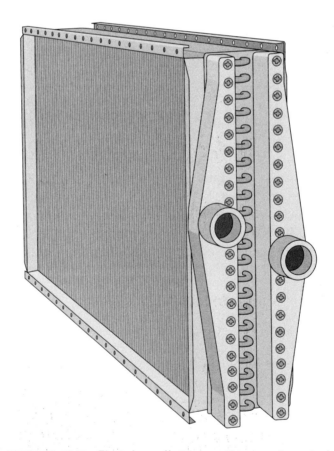

FIGURE 12-5 Fin tube coil. (*Delmar/Cengage Learning*)

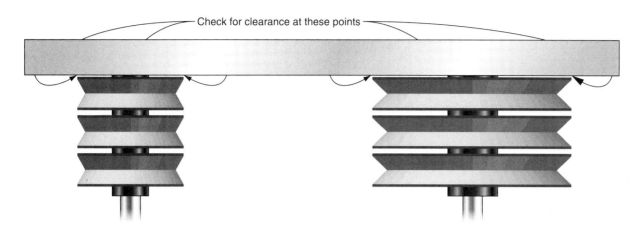

FIGURE 12-6 Aligning sheaves with straightedge. (*Delmar/Cengage Learning*)

The motors of the fan coil units can be expected to run for continuous periods of time exceeding 3 hours. Thus, these motors are considered to be continuous-duty motors. Overload protection must be provided for the motors and is sized according to *430.32(A)(1)*. In summary, this section requires

- a separate overload device for each motor;
- a device that is selected as follows:

  —for motors with a service factor of not less than 1.15, the device rating is 125 percent Full Load Amps (FLA).

  —for motors with a temperature rise of not over 40°C, the device rating is 125 percent FLA.

  —for all other motors, the device rating is 115 percent FLA.

For example, the protection for a 50°C motor having a service factor of 1 is sized to trip at a value not to exceed 115 percent of the full-load rating of the motor. For the fan coil cooling units, this trip value is

$$7.5 \text{ amperes} \times 1.15 = 8.625 \text{ amperes}$$

When the raceway is installed from the controller to the motor, several sections of the *NEC* must be applied. In particular, *430.242*, *430.245(A)*, *250.112(C)*, *250.112(D)*, and *250.134* are relevant to the installation. In addition, because the motor must be movable to adjust the tension in the drive belt, *NEC Article 350* will apply.

A typical installation includes a section of flexible metal conduit slightly less than 3 ft (900 mm)

long, with an equipment grounding conductor installed in the raceway to ground the motor.

After the raceway is installed, the electrician usually is responsible for aligning the motor and adjusting the drive belts. The drive for the cooling unit requires three V-belts. It is recommended that the three belts be purchased as a set so that they all have the same length. The belts should be installed only after the motor is loosened from its base and moved closer to the fan sheave. The belts should not be pulled on over the sheaves because the belt fabric can be damaged. The motor should be placed so that the sheaves are in perfect alignment, Figure 12-6. The motor is then moved to tighten the belts.

The correct tension for a belt can be determined from the motor manufacturer's literature. A measurement can be made to find the force that is required to deflect the belt a distance equal to $\frac{1}{64}$ of the belt span, Figure 12-7. Depending upon the class of the belt cross section, representative values of the deflection are shown in Table 12-1.

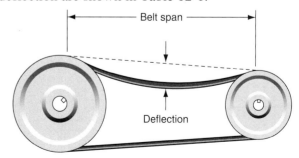

FIGURE 12-7 Measuring the correct belt tension. (*Delmar/Cengage Learning*)

TABLE 12-1

Values of belt deflections.

Belt Cross Section	Deflection Force in Pounds	
	Minimum	Maximum
A	2	3½
B	2½	6
C	6	12
D	13	25
E	25	36

LIQUID CHILLERS

The cooling installation includes three liquid chillers, Figure 12-8, which provide cooled water to the fan coil units. Each chiller is connected to a separate circuit from panelboard P-14. Any two of the three chillers are adequate to meet the cooling needs of the industrial building. As a result, one unit acts as

a reserve to ensure that adequate cooling is always available.

The chillers each have the equivalent of two 12½-horsepower motors in the form of hermetic motor-compressors (see *NEC Article 440*). A hermetic motor is an integral part of the refrigeration system. The refrigerant medium provides a cooling effect as it passes through the motor, Figure 12-9. The motor is labeled with the rated-load current and the locked-rotor current. The rated-load current is used to size the heater elements for the overload relays. According to *440.52(A)(1)*, these relays are selected to trip at not more than 140 percent of the rated-load current.

$$37.4 \text{ amperes} \times 1.40 = 52.36 \text{ amperes}$$

When the disconnect means is selected, both the locked-rotor current and the rated-load current are considered. For example, if a hermetic compressor has a rated-load current of 37.4 amperes and a locked-rotor current of 250 amperes, the requirements of *440.12* are applied as follows to determine

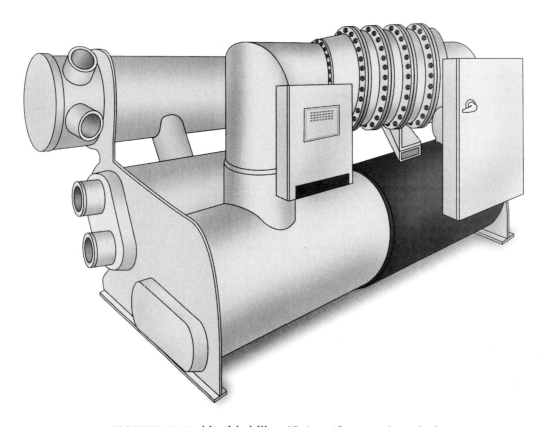

FIGURE 12-8 Liquid chiller. (*Delmar/Cengage Learning*)

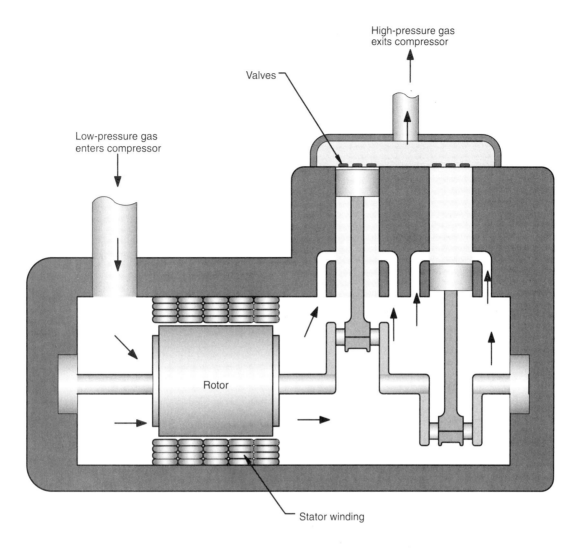

Low-pressure gas
enters compressor

Valves

High-pressure gas
exits compressor

Rotor

Stator winding

FIGURE 12-9   Low-pressure refrigerant flows through the stator winding to provide cooling for the motor of a hermetically sealed compressor. (*Delmar/Cengage Learning*)

the equivalent horsepower required for selecting the correct disconnect means.

1. From Table 10-4 (*NEC Table 430.250*):
    A 208-volt, 3-phase motor with a rated-load current of 37.4 amperes is equivalent to a 230-volt, 3-phase motor with a rated-load current of 42 amperes. The table shows that this value exceeds that given for a 10-horsepower motor. As a result, a disconnect means with a 15-horsepower rating is acceptable according to this table.

2. From Table 10-4 [*NEC Table 430.251(B)*]:
    A locked-rotor current of 250 amperes is found to be less than the value given for

a 15-horsepower motor; thus, a disconnect means rated at 15 horsepower is acceptable according to this table.

3. Because a 15-horsepower disconnect means is the minimum acceptable by *NEC Table 430.251(B)*, this is the minimum required for installation.

The two hermetic motors on each of the chiller units are connected so that they cannot start at the same time. A timer is installed to delay the start of the second motor until 15 seconds after the start of the first motor. This delay reduces the current surge on the supply conductors.

THE PRECIPITATION UNIT

Oil mist is present in the air of most large machine shops. The mist consists of tiny, almost microscopic, droplets of oil or coolant. High-speed machine tools such as boring mills, grinders, and turret lathes tend to pollute the surrounding air with an oil mist. A single high-speed grinder can give off nearly 38 liters (10 gallons) of coolant oil in the form of mist in an 8-hour period.

The oil mist in the air lowers visibility within the manufacturing area and leaves a coating or residue on any surrounding machinery and equipment. In addition, the mist may be the cause of skin and eye irritations, as well as throat and lung ailments among the workers.

One method of removing this pollutant is by the use of precipitation units such as the one shown in Figure 12-10. These units are manufactured in sizes large enough to be used with groups of machines or in a small package unit that can be installed at the individual machine causing the oil mist. Precipitation units can be installed directly behind, above, or at the side of a machine; however, a site at the rear of the machine is the most common location.

The specifications for the industrial building call for individual precipitation units to be installed at the rear of each vertical boring mill, turret lathe, and cylindrical grinder. As a result, 23 precipitation units must be installed and connected.

Each precipitation unit occupies a floor area of 18 in. by 30 in. (45.7 cm by 76.2 cm). The unit is 36 in. (91.4 cm) in height and is mounted on a specially constructed stand. The unit has an air-handling capacity of 600 cu. ft per minute.

The basic construction of the precipitation unit is as follows: Mounted at or near a high-speed tool, it draws the contaminated air from around the cutting or grinding operation; removes the oil mist, smoke, and odor; and returns the cleaned air to the shop space. The salvaged coolant oil is returned to the machine coolant supply reservoir and reused.

This air-cleaning unit consists of a rugged steel cabinet, which is easily installed. The entire unit is accessible from the side. Hinged doors permit ready access to the unit. The cabinet contains collector cells, an ionizer, a power pack, and the fan assembly. The air inlet is located near the bottom of the cabinet. This inlet is connected to the source of the oil mist or the hood of the machine. Flexible or fixed piping similar to an aluminum stovepipe is used to make the connection. A connecting flange provides a means of attachment between the machine and the air inlet. The air outlet is located at the back of the precipitation unit and is provided with a grill.

FIGURE 12-10   The precipitator unit.
(*Courtesy Trion Indoor Air Quality*)

Basic Construction of a Precipitator

The main parts of a precipitator are the ionizer, collector cells, and power supply. The ionizer consists of tungsten wires that are electrically connected together. The entire ionizer unit is insulated from the metal case. The ionizer is connected to one

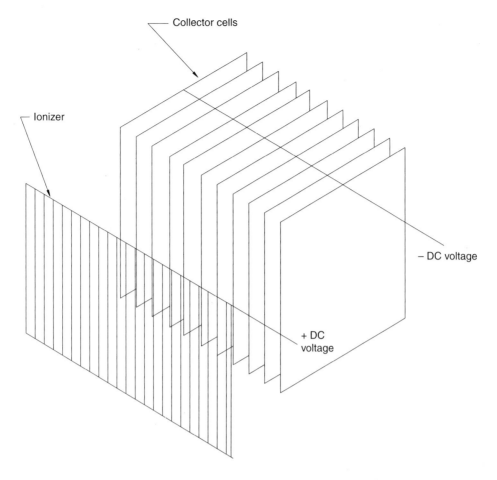

FIGURE 12-11   Basic construction of a precipitator. (*Delmar/Cengage Learning*)

conductor of a high-voltage dc source. The ionizer generally is connected to the positive voltage.

Collector plates are flat metal plates that are electrically connected together, and the entire assembly is also insulated from the metal case. The collector plates are connected to the negative terminal of the high-voltage dc source, Figure 12-11. The power supply consists of a high-voltage transformer, a bridge rectifier to convert the ac into dc, and a filter capacitor, Figure 12-12.

A blower fan draws air across the ionizer. Microscopic particles of oil, dust, pollen, and so on, receive a positive charge as they flow across the tungsten wires. The positively charged particles are then attracted to the negative collector plates. To clean the precipitator, the power is turned off and the plates are removed for cleaning. Large amounts of oil, however, will accumulate and flow down the plates, where they are returned to the machine cooling tank.

Portable Precipitators

The plant also uses portable precipitators that can be moved to any desired location, Figure 12-13. The portable unit contains a flexible hose that can be positioned in the desired location. The unit is connected to a 120-volt power source by a cord. Although precipitators require a very high voltage for their operation, they actually consume very little energy. The ionizer and collector plates are insulated from each other, and there is no direct path for current flow. The current draw of a precipitator is very low.

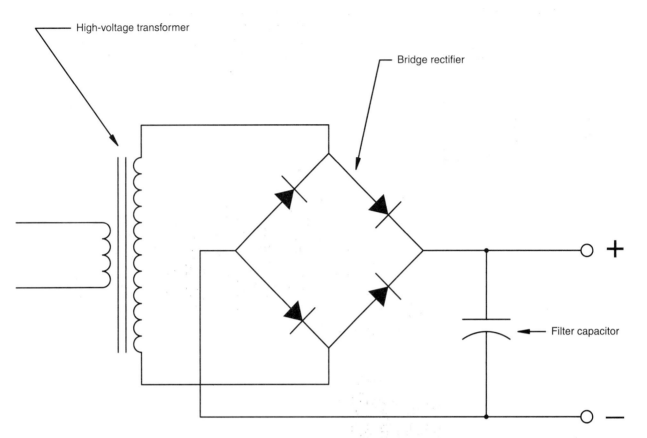

High-voltage transformer

Bridge rectifier

+

Filter capacitor

−

FIGURE 12-12 A rectifier converts high-voltage ac into high-voltage dc. (*Delmar/Cengage Learning*)

FIGURE 12-13 Portable precipitator unit. (*Courtesy Trion Indoor Air Quality*)

All answers should be written in complete sentences, calculations should be shown in detail, and *Code* references should be cited when appropriate.

1. How many conductors should be installed from the disconnect means in the power panelboard to the motor controller? What size would they be?

_____

_____

_____

For the following questions, the nameplate information for a hermetic refrigerant motor-compressor unit is given. In each case, the branch-circuit protection is located in an equipment room remote from the cooling unit. The conductor size and type must be determined, along with the disconnect switch rating. Voltage drop is a critical consideration and should be kept within recommended limits.

2. The system voltage is a 208Y/120-volt single-phase. The branch-circuit conductors will be 75 feet long.

   a. Conductor size: c. Switch rating:

   b. Conductor type: d. Voltage drop:

Voltage 208 to 230	Phase 1
Use copper conductors only.	
Minimum Circuit Ampacity 23.3 Amps	
Compressor RLA 17.6	LRA 87
Fuse Max. Amps 40	Hz 60
Max. HACR Circuit Breaker	40-ampere
Fan Motor FLA 1.3	Hp ⅙

3. The system voltage is a 208Y/120-volt 3-phase. The branch-circuit conductors will be 100 feet long.

   a. Conductor size: c. Switch rating:

   b. Conductor type: d. Voltage drop:

Voltage 208 to 230	Phase 1 Hz 60
Minimum Circuit Amps	38
Compressor RLA 29.1	LRA 141
Branch-Circuit Selection Current 29.1	
Maximum Fuse 60/50 Amps	
Maximum HACR Circuit Breaker 60/50 Amps	
Fan Motor FLA 1.9	HP ¼

4. Give the purpose and describe the function of the precipitation unit.

_____

_____

_____

_____

System Protection

OBJECTIVES

After studying this chapter, the student should be able to

- identify the devices used to provide system protection.

- explain the operation of circuit breakers, fuses, and ground-fault protective devices.

- make the proper adjustments of those devices with adjustable elements.

- determine when selective coordination is achieved.

⊣〰⊢ SYSTEM PROTECTION

The previous chapters of this text described numerous devices and methods of providing system protection. This chapter evaluates the complete power protective system to determine whether it complies fully with the recommendations of the *NEC*.

Ground-Fault Protection in a Coordinated System

The *NEC* specifies in *230.95* that ground-fault protection shall be provided on certain electrical equipment. In addition, the *NEC* lists several other applications as follows for this type of protective device:

Required

- for 1000 ampere or larger, solidly grounded wye service of more than 150 volts to ground but less than 600 volts phase to phase

- for service devices, ground-fault protection set to operate at a maximum of 1000 amperes

Recommended

- for feeder and branch-circuit protection

- for service of less than 1000 amperes where it is solidly grounded and for more than 150 volts to ground not exceeding 600 volts phase to phase

Exception

- provisions of this section of the *Code* shall not apply in a continuous industrial process where a nonorderly shutdown will introduce or increase hazards.

Of equal importance to the requirement for ground-fault protection is the need to coordinate the proper selective overcurrent protection. To achieve coordination, the electrician must be knowledgeable with regard to the operating characteristics of the various types of protective devices. Thus, after the proper selection of equipment is made, the electrician must be able to check and make adjustments where necessary to achieve coordination, Figure 13-1.

Selective coordination means that when an overload or fault condition occurs, only the part of the electrical system that is in jeopardy is disconnected. For example, a fault on a branch circuit causes the branch-circuit protective device to open. At the same time, all of the other protective devices remain closed. Similarly, an overload on a feeder causes only the feeder overcurrent protective device to open.

The three basic types of devices involved in selective coordination are circuit breakers, fuses, and ground-fault protectors. Circuit breakers and fuses are installed in the ungrounded conductors

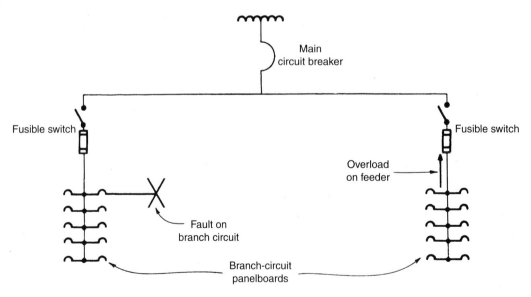

FIGURE 13-1   Selective overcurrent protection coordination. (*Delmar/Cengage Learning*)

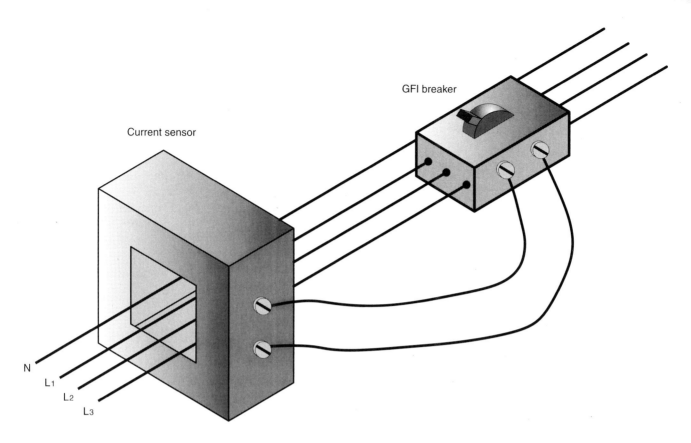

Current sensor

GFI breaker

N

L₁

L₂

L₃

FIGURE 13-2   Ground-fault protection for a 3-phase system. (*Delmar/Cengage Learning*)

of an electrical system to protect the system by monitoring the current in those conductors. These protective devices disconnect the conductors from the power source if a specified abnormal condition occurs, Figure 13-2.

A ground-fault protective device consists of a ground-fault sensor (current transformer) and a relay. All of the phase conductors and the neutral of the system are installed through the center of the sensor. As long as the current in these conductors is balanced (the normal condition), the relay is static. However, if one of the conductors makes contact with ground, the resulting current through the sensor is unbalanced.

If this fault has sufficient magnitude and lasts for a long enough period, the relay sends a signal to the circuit protective device, which then opens the circuit. The following detailed description of the operating characteristics of these devices is presented to help the student gain an understanding of their operation.

CIRCUIT BREAKERS

Circuit breakers are categorized by the method employed to interrupt the circuit current (extinguish the arc) when the contacts open. The three major types of circuit breakers are air, oil, and vacuum. Regardless of the method employed to extinguish an arc, circuit breakers sense circuit current in one of two ways. One method of sensing circuit current is through the production of heat. These breakers are often referred to as thermal circuit breakers. Thermal circuit breakers generally use some type of heating element inserted in series with the load, Figure 13-3. The heater is located close to a bimetallic strip. The bimetallic strip is mechanically connected to the movable contacts of the circuit breaker.

When there is a current through the heater, it causes the bimetallic strip to bend or warp. If the current is higher than a predetermined limit, the bimetallic strip will warp far enough to cause the contacts to snap open.

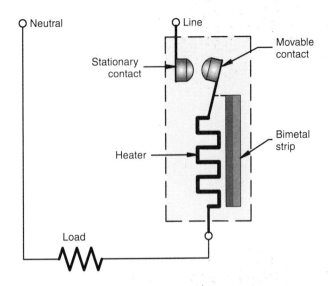

FIGURE 13-3 The thermal circuit breaker senses circuit current by inserting a heating element in series with the load. (*Delmar/Cengage Learning*)

FIGURE 13-5 Schematic symbol used to represent a single-pole thermal circuit breaker. (*Delmar/Cengage Learning*)

symbol generally used to represent a thermal-type circuit breaker is shown in Figure 13-5.

Magnetic Circuit Breakers

The second method of sensing circuit current is accomplished by connecting a coil in series with the load, Figure 13-6. As current flows through the circuit, a magnetic field is established around the coil. The magnetic field attracts the metal arm of a solenoid. If the magnetic field becomes intense enough, the metal arm mechanically opens the contacts of the circuit breaker. Circuit breakers that operate on this principle are referred to as magnetic circuit breakers. A 3-pole magnetic circuit breaker is shown in Figure 13-7.

Because the action of the circuit breaker depends on heating a bimetallic strip, there is some amount of time delay before the circuit opens. The amount of time delay depends on the amount of overcurrent. If the amount of overcurrent is small, it may take several minutes before the circuit breaker opens its contacts. A large overcurrent will cause the contacts to open much faster. A single-pole thermal circuit breaker is shown in Figure 13-4. The schematic

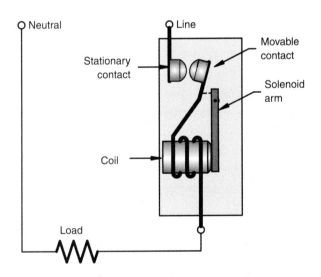

FIGURE 13-6 The magnetic circuit breaker senses circuit current by inserting a coil in series with the load. (*Delmar/Cengage Learning*)

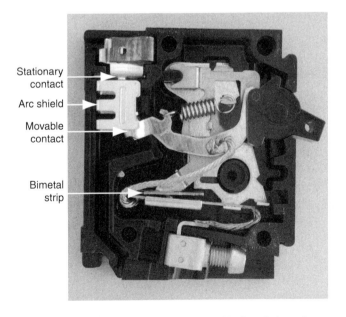

FIGURE 13-4 Single-pole thermal circuit breaker. (*Courtesy of General Electric*, www.geindustrial.com)

FIGURE 13-7   Three-pole magnetic circuit breaker.
(*Courtesy of General Electric*, www.geindustrial.com)

The internal construction of the magnetic circuit breaker is shown in Figure 13-8. In Figure 13-9, one of the solenoids has been removed. This permits the series coil to be seen. The schematic symbol generally used to represent a magnetic circuit breaker is shown in Figure 13-10.

Because magnetic-type circuit breakers do not depend on heating a bimetallic strip, there is very little time delay in the opening of the contacts when an overload occurs. For this reason, they are often referred to as instantaneous circuit breakers. *NEC Table 430.52* lists the maximum rating or setting of motor branch-circuit, short-circuit, and ground-fault protective devices. One of the protection devices listed is the instantaneous trip breaker.

There are some types of circuit breakers that employ both thermal and magnetic current sensors. These circuit breakers are known as thermomagnetic circuit breakers. The schematic symbol generally used to denote the use of a thermomagnetic breaker is shown in Figure 13-11.

Circuit-Breaker Current Ratings

Circuit breakers actually have two different current ratings. One is the *trip* rating, and the other is the *interrupt* rating. The trip current rating is the amount of current that should cause the circuit breaker to open its contacts when it is exceeded. Standard trip current ratings for inverse-time circuit breakers (thermal circuit breakers) and fuses are listed in *240.6*. Trip current ratings range from 15 to 6000 amperes.

The interrupt rating indicates the maximum amount of current a circuit breaker is intended to interrupt when its contacts open. The amount of current that will flow during a short-circuit condition is determined by two factors:

1. The circuit voltage
2. The circuit impedance

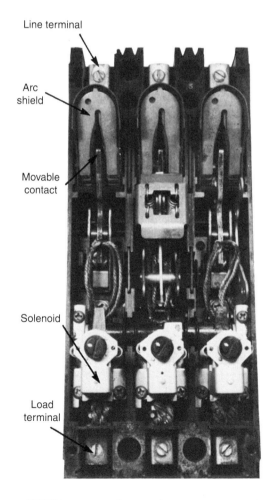

FIGURE 13-8   Internal construction of a
3-pole magnetic circuit breaker.
(*Courtesy of General Electric*, www.geindustrial.com)

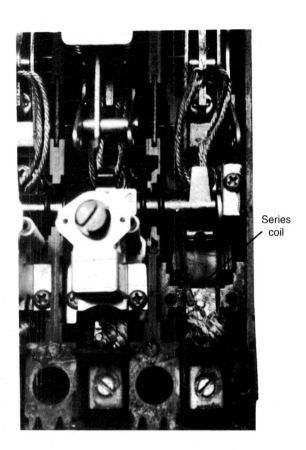

FIGURE 13-9 Series coil of a magnetic circuit breaker. (*Courtesy of General Electric, www.geindustrial.com*)

FIGURE 13-11 Schematic symbol generally used to represent a thermomagnetic circuit breaker. (*Delmar/Cengage Learning*)

FIGURE 13-10 Schematic symbol generally used to represent a magnetic circuit breaker. (*Delmar/Cengage Learning*)

The circuit impedance is determined by factors such as the kVA capacity of the transformers supplying power to the branch circuit, the size of wire used in the circuit, contact resistance of connections, and so on.

When a short or grounded circuit occurs, the circuit breaker must be capable of interrupting the current. Assume, for example, that a circuit

breaker has a trip current rating of 100 amperes. Now assume that a short circuit occurs and there is a current of 15,000 amperes in the circuit. Because the circuit breaker has a trip current rating of 100 amperes, it will open its contacts almost immediately.

For the circuit breaker to stop the current, it must be capable of interrupting a current of 15,000 amperes. Circuit breakers commonly have an interrupt rating of 5000 amperes. Because the interrupt rating can be very important in the event of a short circuit, *240.83(C)* states that circuit breakers having an interrupt rating other than 5000 amperes must have the rating marked on the circuit breaker, Figure 13-12.

When it is necessary to replace a circuit breaker, always make sure of the interrupt rating. If a circuit breaker with an interrupt rating of 5000 amperes is used to replace a breaker with an interrupt rating of 10,000 amperes, a short circuit could cause a great deal of damage to both equipment and individuals.

Shunt Trips and Auxiliary Switches

Some circuit breakers contain a small solenoid coil known as a *shunt trip*. Shunt trips are used to open the circuit breaker contacts by energizing the solenoid from an external source.

Assume, for example, that it is desirable to disconnect the power to a circuit if the temperature

FIGURE 13-12 Circuit-breaker interrupt rating of 10,000 amperes. (*Courtesy of General Electric, www.geindustrial.com*)

FIGURE 13-14 Circuit breaker with shunt trip and auxiliary switch. (*Courtesy of General Electric, www.geindustrial.com*)

rises above a certain level. If the circuit breaker protecting the circuit contains a shunt trip, a thermostat can be connected in series with the solenoid. If the temperature rises above the desired level, the thermostat contact will close and energize the coil, Figure 13-13. When the coil energizes, the circuit-breaker contacts will open and disconnect power to the circuit. A 3-pole circuit breaker con-

taining a shunt trip and auxiliary switch is shown in Figure 13-14. The shunt trip connection for the circuit breaker is shown in Figure 13-15.

Some circuit breakers contain an auxiliary switch as shown in Figure 13-14. The auxiliary switch is a

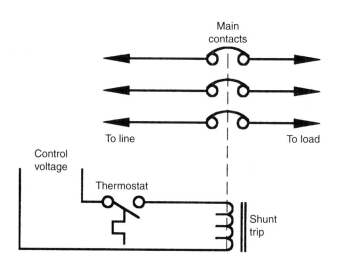

FIGURE 13-13 A thermostat is used to disconnect power to the circuit if the temperature rises to a certain point. (*Delmar/Cengage Learning*)

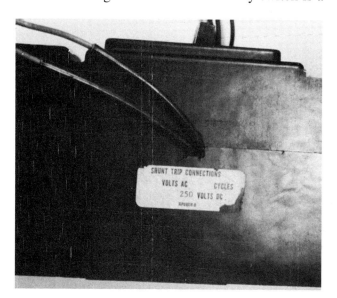

FIGURE 13-15 Shunt trip connection. (*Courtesy of General Electric, www.geindustrial.com*)

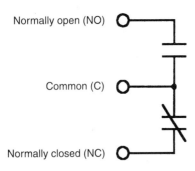

FIGURE 13-16 Auxiliary switch contacts. (*Delmar/Cengage Learning*)

FIGURE 13-17 Auxiliary switch connection. (*Courtesy of General Electric*, www.geindustrial.com)

small micro-limit switch. Its contacts are controlled by the action of the circuit breaker. The auxiliary switch generally contains a set of normally open and normally closed contacts connected to a common terminal, as shown in Figure 13-16. The contacts are shown in the position they will be in when the circuit breaker is turned off, or open. When the circuit breaker is turned on, or closed, the auxiliary switch contacts will change position. The normally closed contact will open and the normally open contact will close.

Auxiliary switch contacts can be used for a variety of purposes. In some instances, if the circuit breaker should open, it may be desirable to disconnect power to some other control device on a different circuit. In another application, if the breaker should open, the auxiliary contacts may be used to illuminate an indicator light on the operator's panelboard. The auxiliary switch connection for a 3-pole circuit breaker is shown in Figure 13-17.

Air Circuit Breakers

Air circuit breakers are so named because they use air as the insulating medium to break the arc when contacts open. They can be divided into three basic types:

1. Molded-case circuit breakers

2. Low-voltage power circuit breakers

3. Medium-voltage circuit breakers

Regardless of the type of air circuit breaker employed, all have one similar characteristic: They use air as a medium to extinguish an arc. When

contacts separate to interrupt the current, an arc is produced that contains a great deal of heat. The farther apart the contacts become, the longer the arc becomes and the greater the cooling effect. Convection airflow will cause the arc to rise, Figure 13-18.

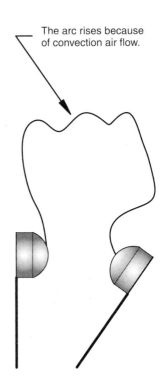

The arc rises because of convection air flow.

FIGURE 13-18 An arc is produced when contacts open. (*Delmar/Cengage Learning*)

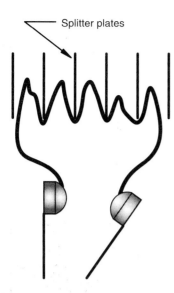

FIGURE 13-19 Splitter plates lengthen the arc, which helps extinguish it. (*Delmar/Cengage Learning*)

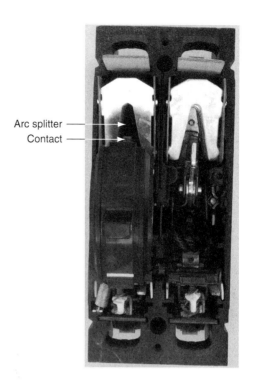

FIGURE 13-20 Two-pole molded-case circuit breaker with arc splitter. (*Courtesy of General Electric*, www.geindustrial.com)

Molded-Case Circuit Breakers

Molded-case circuit breakers are used in low-voltage (600 volts or less), low-current circuits. They are characterized by the use of a molded case, which results in minimum space requirements. They are used to protect small motor, lighting, and appliance circuits.

Circuit breakers intended for lower voltage and current ratings often depend on the distance between the contacts being sufficient to stretch the arc far enough to extinguish it. Circuit breakers intended for use on higher-voltage circuits often employ other devices to help extinguish an arc. One of these devices is the *splitter*. A splitter consists of metal or insulated plates located at the top of the contacts, Figure 13-19. Their function is to permit hot gases to escape but lengthen the path of the arc so the cooling effect is increased. The longer arc path weakens the arc to the point that it is eventually extinguished. A 2-pole molded-case circuit breaker with an arc splitter is shown in Figure 13-20. The arc splitter is shown outside the circuit breaker in Figure 13-21.

Low-Voltage Power Circuit Breakers

Low-voltage power circuit breakers are generally constructed with a metal case. They can

be obtained in case sizes that range from 100 to 6000 amperes and can have trip current ratings that range from 15 to 6000 amperes. Because they are intended to interrupt higher currents, the contact

FIGURE 13-21 The arc splitter is constructed of individual plates. (*Courtesy of General Electric*, www.geindustrial.com)

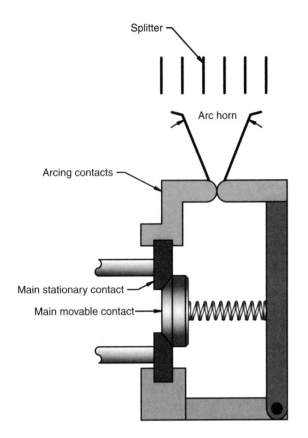

FIGURE 13-22 Low-voltage power circuit breakers generally contain two sets of contacts and an arc horn. (*Delmar/Cengage Learning*)

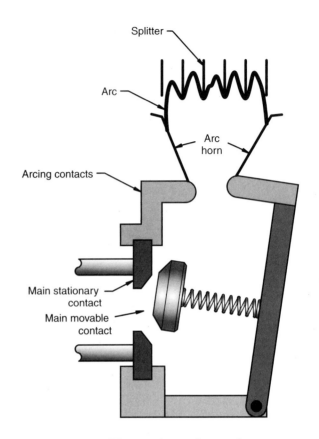

FIGURE 13-23 The arc horn draws the arc away from the arcing contacts. (*Delmar/Cengage Learning*)

arrangement is generally different from that of molded-case circuit breakers.

Low-voltage power circuit breakers commonly have two separate sets of contacts. One set is the main contacts and is used to connect the line and load together. The second set, the arcing contacts, is used to direct the arc away from the main contacts, Figure 13-22. The arcing contacts are further assisted by an *arc horn*, which aids in drawing the arc away from the arcing contacts and also helps stretch the arc. A splitter, generally located above the arc horn, breaks the arc into pieces to extinguish it, Figure 13-23.

Medium-Voltage Air Circuit Breakers

Medium-voltage air circuit breakers are intended to operate on system voltages that range from 600 volts to 15 kilovolts. They are constructed with a metal case and generally contain *blow-out coils* and a *puffer*, as well as a splitter and an arc horn to help extinguish the arc.

Blow-out coils are connected in series with the arcing contacts so that current flows through them when the main contacts open. The current through the coils produces a magnetic field, which attracts the arc and helps to move it into the splitter. The puffer is constructed by attaching a small piston to the operating lever of the breaker. The piston is located inside a cylinder. When the circuit breaker opens, the piston is forced to move through the cylinder, sending a puff of air in the direction of the arc. This puff of air helps move the arc into the splitter.

Oil Circuit Breakers

Oil circuit breakers are often used in substations to interrupt voltages as high as 230 kilovolts. They use oil as a dielectric or insulator. The contacts are located under the oil. When the contacts open, the heat of the arc causes the surrounding oil to decompose and form a gas. The gas extinguishes the arc.

There are two basic types of oil circuit breakers, the *full tank* or *dead tank* type and the *low*

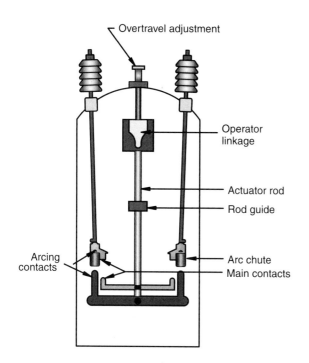

FIGURE 13-24 Typical dead tank oil circuit breaker. (*Delmar/Cengage Learning*)

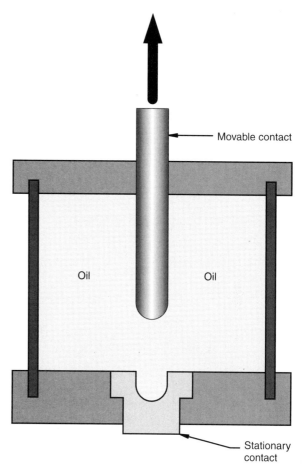

FIGURE 13-25 Plain-break low oil–type circuit breaker. (*Delmar/Cengage Learning*)

oil or *oil poor* type. The dead tank type is the oldest and is generally used for voltages above 13.8 kilovolts. The construction of a typical dead tank circuit breaker is shown in Figure 13-24. The dead tank circuit breaker receives its name from the fact that the tank is at ground potential and insulated from the live parts by the dielectric oil. The circuit breaker shown is a double-break type containing a set of main contacts and arcing contacts. The movable parts of both the main and arcing contacts are controlled by an actuator rod, which is manually operated.

Oil poor circuit breakers are manufactured in several different styles. The *plain-break* type relies on the surrounding oil and the pressure generated by the production of gas to control the arc when the contacts open, Figure 13-25. The pressure is eventually vented between the case and the movable contact. Another type of low oil circuit breaker is often referred to as a *vented* type and is designed with vents that permit the pressure produced by the formation of gas to exit the arc chamber, Figure 13-26. Another type of low oil circuit breaker intended for use on higher voltages, called the *double-break*

type, employs a double-break contact arrangement, as shown in Figure 13-27.

Vacuum Circuit Breakers

An understanding of the operation of vacuum circuit breakers begins with an understanding of the mechanics of an electric arc occurring in air. When an electric arc occurs in air, gas molecules in the air become ionized. These ionized molecules form a conducting path for the flow of electrons. It is the ionization of gas molecules that makes the job of extinguishing an electric arc in air so difficult.

In the vacuum circuit breaker, the contacts are contained inside a sealed enclosure, Figure 13-28. If the air could be completely removed from the container, no arc could occur because there would be no molecules to ionize. Although it is not possible to obtain a perfect vacuum, very few air molecules are left in the chamber, and any arc produced by the

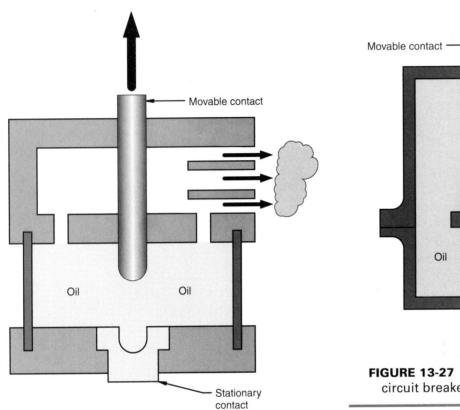

FIGURE 13-26 Typical vented–type low oil circuit breaker. (*Delmar/Cengage Learning*)

FIGURE 13-27 Typical double-break low oil circuit breaker. (*Delmar/Cengage Learning*)

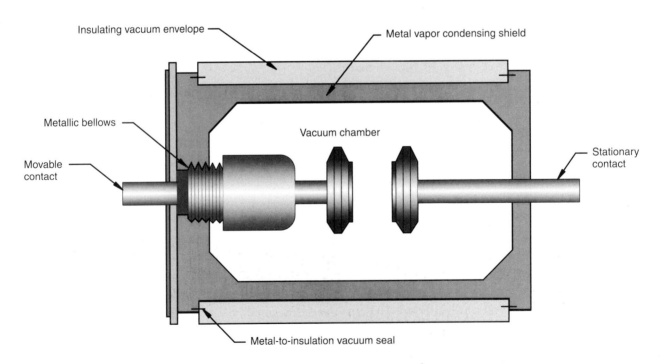

FIGURE 13-28 Typical vacuum circuit breaker. (*Delmar/Cengage Learning*)

opening of the contacts will be extremely small. This small arc is extinguished by the distance between the contacts. Most vacuum circuit breakers require only about one-half to three-quarters of an inch clearance between the contacts to control voltages over 13.8 kilovolts. A metallic bellows is connected to the movable contact. The bellows permits movement of the contact while maintaining the vacuum. Vacuum circuit breakers are being used to replace the older oil circuit breakers because they are smaller in size and require very little maintenance.

CIRCUIT-BREAKER TIME-CURRENT CHARACTERISTIC CHARTS

Time-current characteristic charts are published for most protective devices, Figure 13-29. A log–log grid is used for the chart with time on the vertical axis and current on the horizontal axis. In general, time is given in seconds and current is in amperes. Terms used with circuit-breaker curves are *trip coil rating, frame size, long-time delay, long-time pickup,*

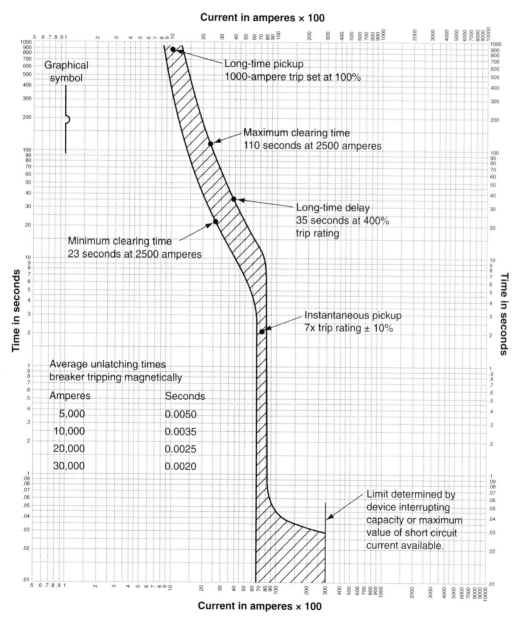

FIGURE 13-29   A circuit-breaker time-current curve. (*Delmar/Cengage Learning*)

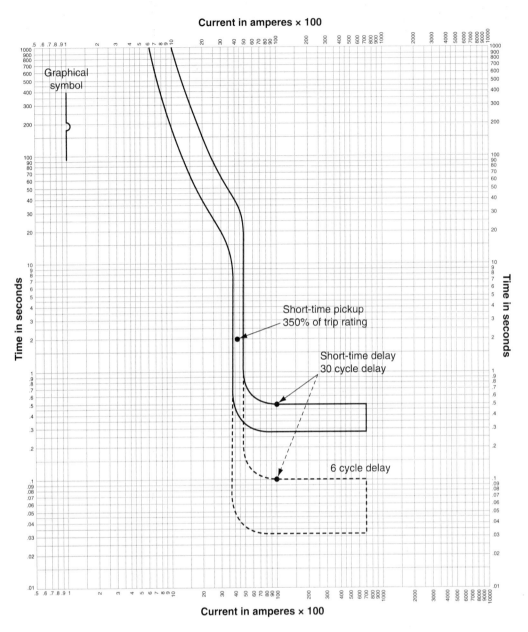

FIGURE 13-30 Characteristic curve for a circuit breaker with adjustable short-time delay.
(*Delmar/Cengage Learning*)

instantaneous pickup current, short-time delay, short-time pickup, unlatching time, and *interrupting rating.*

The *trip coil rating* is also known as the breaker rating. That is, a 150-ampere breaker is a breaker with a 150-ampere trip coil. This rating is not adjustable in molded-case circuit breakers. However, in some breaker models, the physical construction is such that breakers of various ratings are interchangeable. Because of this, a breaker of a different size (rating) can be installed to meet specific protec-

tive needs. Most air-type circuit breakers are adjustable. The rating can be changed from 80 percent to 160 percent of the trip coil rating. An adjustment above 100 percent of the trip coil rating should be made only in those installations where motors or other surge-generating loads are factors.

The value of current at which the trip coil operates is called the *long-time pickup*. A 200-ampere trip coil adjusted to 120 percent has a long-time pickup of 240 amperes.

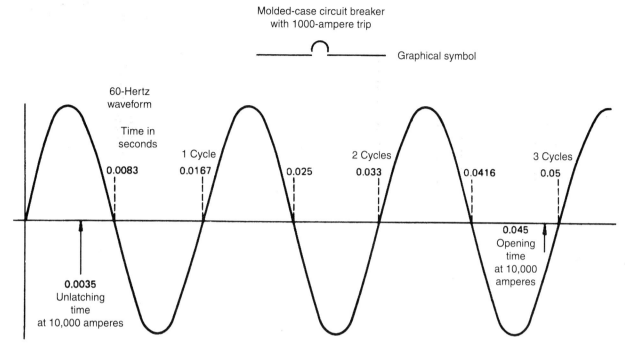

FIGURE 13-31 A comparison of latching and opening times for a circuit breaker tripping magnetically.
(*Delmar/Cengage Learning*)

The *frame size* indicates the maximum size of trip unit that a specific breaker can accommodate. Thus, a breaker with a 100-ampere frame size will accept a trip unit of any standard rating ranging from 15 to 100 amperes.

The *long-time delay* portion of the curve indicates the operating characteristic of a breaker under overload conditions. For molded-case circuit breakers, this delay is usually controlled by a device sensitive to thermal changes. If a breaker has long-time delay adjustment, the time value may be set to a low value for lighting and resistive loads. However, a high time-value setting is required for motor starting and other surge-generating loads.

The *instantaneous pickup current* is the point at which a breaker responds to a short-circuit current through a magnetically actuated trip arrangement. This value is adjustable in many breakers.

The *instantaneous opening* is the time required for the breaker to open when no intentional delay is added. However, when the trip must be delayed intentionally, a short-time delay is added, Figure 13-30. This feature is available only on more sophisticated

breakers. *Short-time pickup* is that value of current at which short-time delay is initiated.

Unlatching time is the point beyond which the opening action of the breaker is irreversible, Figure 13-31.

The characteristic curve of a circuit breaker is a band that represents the range of time or current through which the breaker can be expected to operate. The upper limit of the band indicates the maximum value; the lower band limit is the minimum value. In Figure 13-29, the curve indicates that a 250 percent load (2500 amperes) can be cleared in no less than 23 seconds or in no more than 110 seconds.

Each protective device that is designed to open a circuit under fault conditions must be able to interrupt the maximum current that can flow in that circuit (*110.9*). The *interrupt rating* of a protective device indicates the maximum current that the device can interrupt. For currents above this value, an arc may be sustained across the contact gaps after they open. This arc continues to supply current to the fault and damages the protective device. Devices are available with interrupting capacities ranging from 5000 to 200,000 amperes.

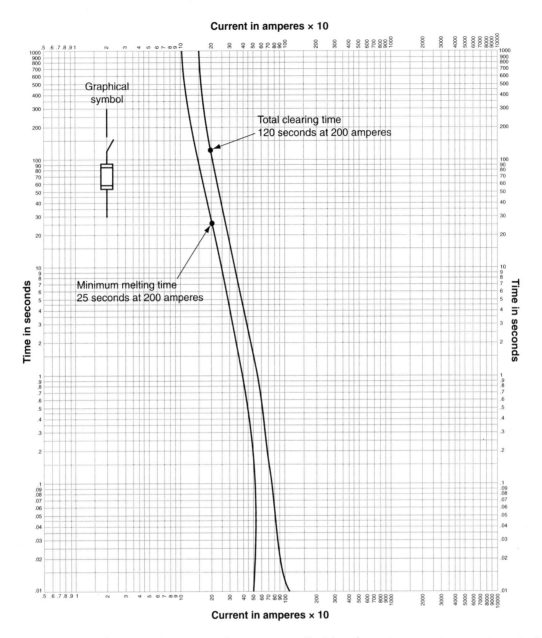

FIGURE 13-32 Characteristic curve for a current-limiting fuse. (*Delmar/Cengage Learning*)

FUSE TIME-CURRENT CHARACTERISTIC CHARTS

A fuse has a highly predictable performance, which is usually represented on a chart by a single curve similar to the right-hand line in Figure 13-32. A curve of this type is called the total clearing time-current characteristic curve. Another value is also significant in determining the selectivity of fuses. The minimum melting time is that value of time-current at which the opening of the fuse becomes irreversible.

GROUND-FAULT PROTECTOR TIME-CURRENT CHARACTERISTIC CHARTS

Ground-fault protector curves indicate the reaction of the device at a specific time-delay setting, Figure 13-33. A ground-fault sensor is always used in conjunction with another protective device that can respond to a signal. Thus, for a given ground-fault current, the time required for the clearance of its circuit is the sum of the time delay of the

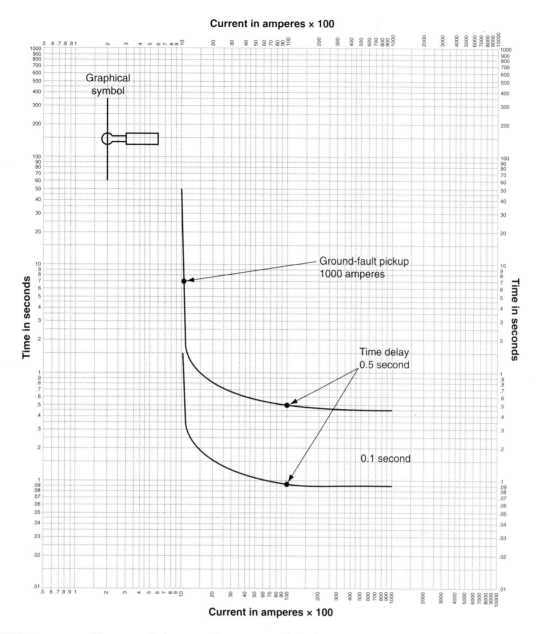

FIGURE 13-33   Characteristic curve for a ground-fault protector with adjustable time delay.
(*Delmar/Cengage Learning*)

ground-fault sensor and the time required for the associated protective device to open.

 ## COORDINATION

Each of the three types of protective devices described previously has distinct operating characteristics. The addition of these devices to a coordinated system requires that the proper selection of the rating be made. Figure 13-34 shows the character-

istic curves of two circuit breakers, a 1600-ampere main and an 800-ampere feeder. The cross-hatching in the figure indicates areas in which the breakers do not coordinate. For this situation, if a 5000-ampere overload continues for 20 seconds, there is a high probability that both breakers will open. Thus, instead of protecting a feeder circuit, the open breakers will cause an entire building to be without power. The same end result will occur for a short circuit of more than 18,000 amperes. This problem can be minimized by the proper selection, adjustment, and

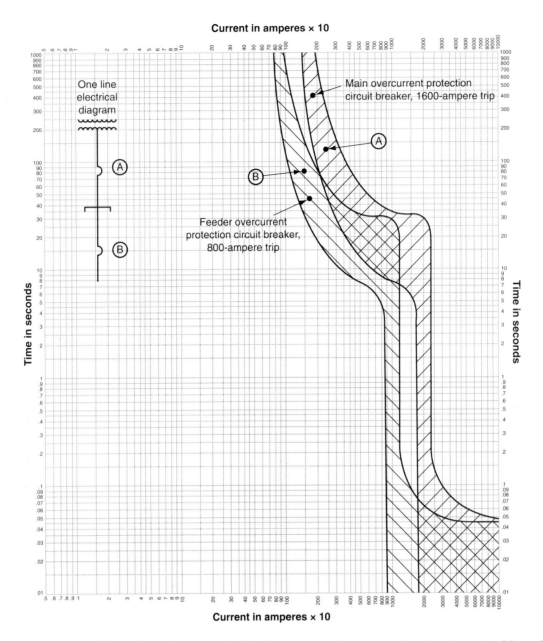

FIGURE 13-34 Circuit-breaker curve indicating areas where coordination is not achieved.
(*Delmar/Cengage Learning*)

maintenance of the protective devices. However, it must be noted that circuit breakers in general are difficult to coordinate. In particular, molded-case circuit breakers are almost impossible to coordinate, except at low overload current values.

All circuit breakers in a system, except the branch-circuit breakers, must have a short-time delay feature if coordination is to be achieved. The insertion of a delay in the magnetic tripping of the feeder and main devices makes it possible to

achieve coordination, as indicated in Figure 13-35. There is a problem with this method of obtaining coordination: A condition is established where a fault may not be opened for several cycles. This increases the possibility of damage that may occur as a result of the faulting condition.

Coordination can be accomplished easily, as shown in Figure 13-36, by using fuses alone or in combination with circuit breakers. The 800-ampere fuse coordinates with the 1600-ampere circuit

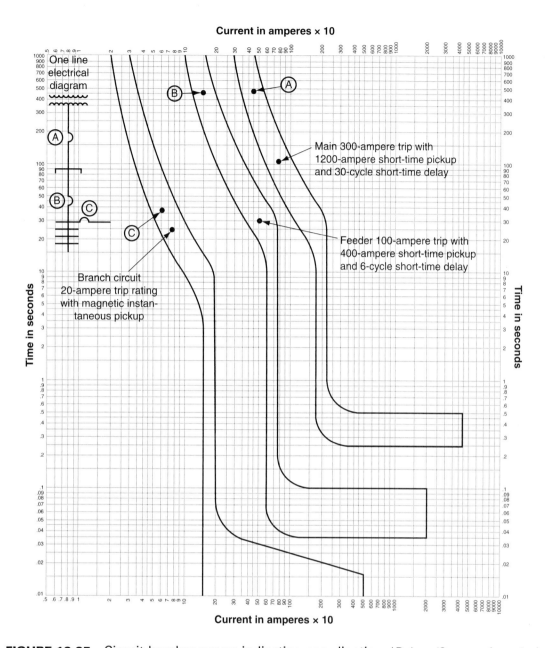

Current in amperes × 10

One line electrical diagram

Main 300-ampere trip with 1200-ampere short-time pickup and 30-cycle short-time delay

Feeder 100-ampere trip with 400-ampere short-time pickup and 6-cycle short-time delay

Branch circuit 20-ampere trip rating with magnetic instantaneous pickup

Time in seconds

Current in amperes × 10

FIGURE 13-35 Circuit-breaker curves indicating coordination. (*Delmar/Cengage Learning*)

breaker, and coordination is achieved—with one possible exception. The unlatching time of the breaker may, under certain conditions, exceed the speed of the fuse, in which case coordination will not be obtained.

A system containing only fuses is the easiest situation to coordinate. When the fuses are selected according to the manufacturer's recommendation, complete coordination can be achieved. That is, fuses are used with a certain ratio to the upstream protective device. For example, a 1600-ampere current-limiting fuse used for a main will coordinate on a 2:1 basis with another current-limiting fuse or on a 4:1 basis with a time-delay fuse (such as the type used with motors). If ground-fault protection is required after the overcurrent protective devices are coordinated, it should be added to the system without disrupting the coordination.

If more than one ground-fault protector is installed, a double problem is posed: Both protectors should coordinate with the overcurrent system and with one another, Figure 13-37.

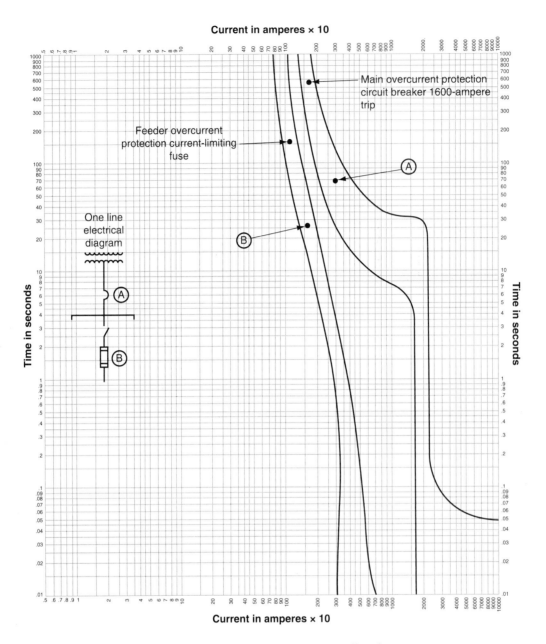

FIGURE 13-36 A circuit breaker and a fuse-achieving coordination. (*Delmar/Cengage Learning*)

Coordination with overcurrent protective devices can be achieved by selecting and adjusting the ground-fault protector so that its characteristic curve is above the total clearing curve of the next downstream overcurrent protective device. Coordination with other ground-fault protectors is achieved by using a lower trip setting and progressively shorter time settings on each of the downstream devices, or by making interlock connections between the devices so that the device that first senses a fault locks the upstream device in until the time setting of the downstream device is exceeded.

In summary, to coordinate a system, proper selection of the protective devices must be accompanied by the correct sizing of the various components followed by careful adjustment of these devices. In addition, the proper maintenance of the devices after they are placed in operation will help to ensure that coordination, once achieved, will be maintained.

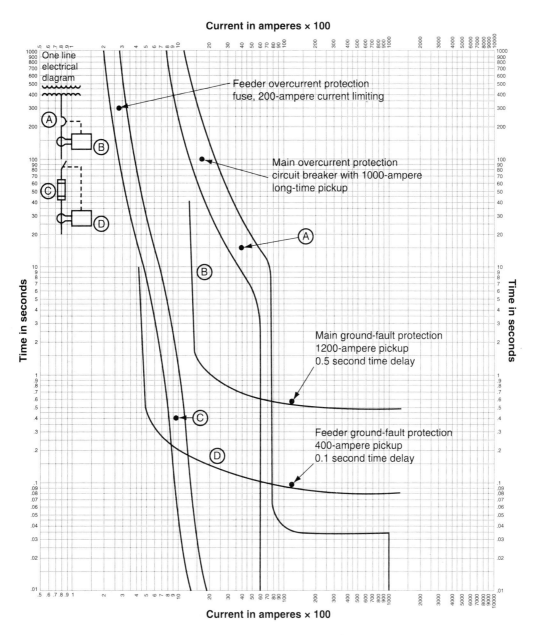

FIGURE 13-37   Selective coordination of a system with ground-fault protection. (*Delmar/Cengage Learning*)

REVIEW QUESTIONS

All answers should be written in complete sentences, calculations should be shown in detail, and *Code* references should be cited when appropriate.

1. List the service types that are required by the *NEC* to have ground-fault protection.

_____

_____

_____

2. List the service types that are recommended to have ground-fault protection. _____

_____

_____

_____

3. How many conductors of a 3-phase, 4-wire system must be installed through the sensor of a ground-fault protective device? _____

_____

_____

4. What is the range of the trip current for an elapsed time of 1 minute (refer to Figure 13-29)?

_____

_____

5. What is the range of trip time for a current of 3000 amperes (refer to Figure 13-29)?

_____

_____

6. In referring to circuit breakers, what is meant by the *instantaneous opening time*? _____

_____

_____

_____

7. In referring to circuit breakers, what is meant by the *long-time delay*? _____

_____

_____

_____

8. What is meant by the phrase *minimum melting time*? _____

_____

_____

_____

9. For a current of 300 amperes, the opening time of the fuse is how many seconds (refer to Figure 13-32)? _____

_____

_____

_____

10. For a current of 300 amperes, the opening action of the fuse is irreversible after how many seconds (refer to Figure 13-32)? _____

_____

_____

_____

11. What is the difference between a short-circuit fault and a ground fault? _____

_____

_____

_____

12. Explain what is meant by *selective coordination* and list three types of devices that are involved. _____

_____

_____

_____

13. If a 1000-ampere fault were to occur on the system represented in Figure 13-34, what would probably happen? _____

_____

_____

_____

14. What are the maximum current and time-delay settings of ground-fault protection as set forth by the *NEC*? _____

_____

_____

_____

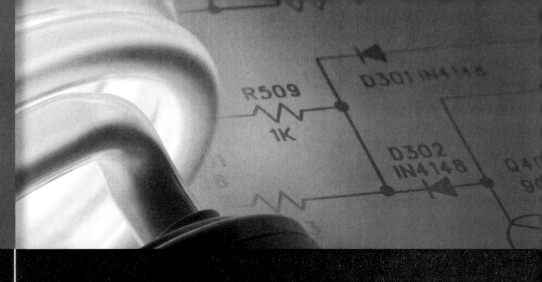

CHAPTER
14

Lightning Protection

OBJECTIVES

After studying this chapter, the student should be able to

- describe the lightning process.
- identify the requirements for protecting a building.
- list lightning safety rules.

Lightning is simultaneously a fascinating, awesome, and mysterious phenomenon. It is beautiful to witness but destructive and fatal to experience. Yet, it is a natural, necessary occurrence. Lightning is electricity on exhibit; and, as with the electrical power in our homes, businesses, and factories, specific precautions are to be taken or we must expect to suffer the consequences.

To understand the intricacies of lightning protection, it is necessary to also understand, or at least accept, the rudiments of atomic structure and what occurs within that structure. For that reason, this study of lightning protection begins with a brief presentation of the theory of atomic structure.

ATOMIC STRUCTURE

All matter is made up of *atoms*. For example, a single drop of water contains about 100 billion atoms. Each atom has at its center a *nucleus* that is composed of *protons* and *neutrons*. The nucleus is considered to have a positive charge equal to the number of protons. Under normal conditions, the nucleus is surrounded by a number of *electrons*, each having a negative charge, equal to the number of protons in the nucleus.

The exact numbers of protons or electrons, under normal conditions, are different for each element and are stated as the atomic number of that element. Hydrogen has an atomic number of 1, copper 29, lead 82, and so on.

If a force is exerted on an atom to the extent that an electron is dislodged or added, the atom is said to become an *ion*. The atom that loses an electron has a net positive charge and thus is called a *positive ion*; the atom that gains an electron is a *negative ion*. This phenomenon of being able to remove electrons from atoms, or *ionizing* them, makes it possible to store electrical power and to transfer that power from place to place.

Electrons at Work

For the purpose of further explanation, assume a hypothetical situation in which a large number of positive ions are collected at Point A. At another point (call it Point B) located in space are the collected electrons that have been dislodged from the atoms at Point A. If 6,250,000,000,000,000,000 (6.25×10^{18}) electrons are transferred from Point A to Point B, the quantity of charge is referred to as a *coulomb*.

For a moment, digress from the hypothetical situation and turn your attention to the basic nature of charges. It is a fundamental law that charged bodies of unlike charges attract each other, and charged bodies of like charges repel each other. The force of attraction or repulsion between charges is directly proportional to the square of the distance between these charges. This brings us to another definition of the coulomb. A coulomb (C) is that quantity of charge that, when placed 1 meter (m) from a like charge, repels it with a force of 9,000,000,000 (9×10^9) newtons (N). It follows that it takes work (force $\times$ distance) to collect charges at a point, for, as 1 coulomb of charge is formed, a force of 9,000,000,000 (9×10^9) newtons must be exerted over a distance of 1 meter. The work done (energy released) by a force of 1 newton, acting over a distance of 1 meter, is expressed as 1 joule (J).

Going back to Points A and B, it should now be apparent that, as electrons are forced from Point A to Point B, work is done and an energy differential is established. In electrical studies, this is called a *potential difference*, which is measured in volts (V). A *volt* is that potential energy a charge gains when 1 joule of work is done on 1 coulomb of charge.

The Ionosphere

We now expand on the hypothetical example by giving locations to Points A and B. Because the subject is lightning, it is reasonable to locate Point B on Earth and Point A in a region called the ionosphere. The *ionosphere* is located at altitudes of 40 miles or so above the earth, where the atmosphere contains more ions than neutral atoms, Figure 14-1.

The earth has a surplus of electrons and is about 300,000 (3×10^5) volts negative with respect to the ionosphere. This means that a person of average height, while standing, is covered from foot

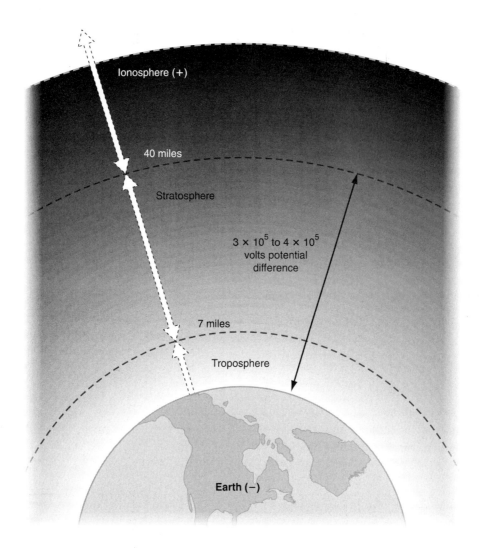

FIGURE 14-1 Location of the ionosphere in relation to the earth. (*Delmar/Cengage Learning*)

to head by a potential gradient of about 260 volts, Figure 14-2. The typical reaction to this statement is "Why doesn't the person feel a shock?" Electrical shock is measured in terms of current, and the current in this case is infinitesimal. (The air–Earth current is calculated to be between 1400 amperes and 1800 amperes or about 0.000,009 [10^{-6}] ampere per square mile.) But the total current from Earth is sufficient to upset nature's balance, and lightning is thought to be a natural way of restoring the balance. It is estimated that, on the average, the earth is struck by lightning 100 times per second, and in about 90 percent of these events, electrons flow to the earth. However, there is not full agreement on how this comes about.

HOW LIGHTNING IS GENERATED

Lightning is generated by the ominous-looking cumulonimbus cloud, or thundercloud. Here the action is so violent that charges may be formed on the lower portion of the cloud. Most of these charges are negative with respect to the upper portion of the cloud and to the earth below the cloud, Figure 14-3. Now, we essentially have a Point B at a high negative charge and several Points A at high positive charges. What happens depends upon how high the potential differences become and the impedance between the points. There may be lightning strokes from cloud to cloud, from points of high potential

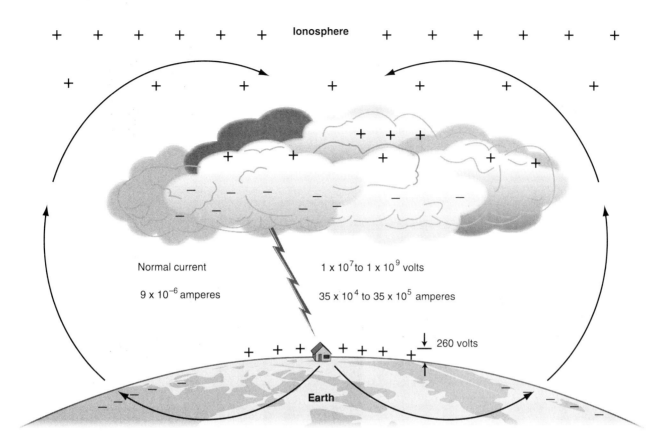

FIGURE 14-2 Person standing on Earth is surrounded by a potential gradient of about 260 volts. (*Delmar/Cengage Learning*)

Ionosphere

Normal current

9×10^{-6} amperes

1×10^{7} to 1×10^{9} volts

35×10^{4} to 35×10^{5} amperes

260 volts

Earth

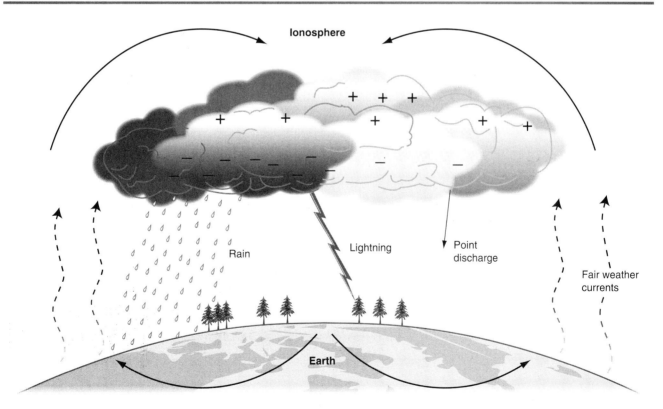

FIGURE 14-3 Lightning strokes are generated between the thundercloud and the Earth. (*Delmar/Cengage Learning*)

Ionosphere

Rain

Lightning

Point discharge

Fair weather currents

Earth

difference within a cloud, or between the thunder-cloud and the earth. We, of course, are primarily interested in the cloud-to-Earth strokes.

These strokes are most likely to occur at a high point—that is, where a tree, a tall building, or some other extension of the earth rises upward, thereby reducing the distance and, thus, the impedance between the charge on Earth and the cloud charge.

Lightning Strokes

According to the National Center for Health Statistics, each year lightning in the United States alone kills about 150 people, injures another 250, and causes damage in excess of $250 million. These disasters are all caused by the enormous energy transfer that takes place during a cloud-to-Earth strike. The current may rise to as high as 200,000 (2×10^5) amperes, and the potential difference may be as much as 100 million volts. Yet, the stroke lasts only a fraction of a second. Our goal in lightning protection is to provide a low-impedance path from the high point, where the stroke makes contact with the earth, so that the current may be dissipated over a large area.

MASTER LABEL

The installation of lightning protection systems for buildings, trees, and other structures in the open should be guided by the requirements of Master Label Service, which is endorsed by UL, and/or by the Lightning Protection Institute Installation Code. The systems advocated by these organizations are based upon the basic principle of providing a low-impedance path for the stroke to follow to Earth, while minimizing the possibility of damage, fire, and personal injury or death as the stroke follows that path.

BUILDING PROTECTION

Lightning protection systems have three basic components: air terminals, lightning conductors, and grounding connections.

Air Terminal (Lightning Rod)

The air terminal, or lightning rod, is the highest element of a lightning protection system. It is a solid or tubular rod made of copper, bronze, or aluminum. Usually sharp pointed, it is available with a safety tip, as shown in Figure 14-4. These are installed where they would not create a hazard to personnel, such as on a flat roof.

The terminal attracts lightning, but does not prevent lightning, as originally supposed by Benjamin Franklin. The lightning stroke is attracted to the terminal because it is a part of the path that offers the least impedance to ground. In general, air terminals should

- extend above the object to be protected, not less than 10 in. (250 mm) or more than 36 in. (900 mm);
- be placed on ridges of gable, gambrel, and hip roofs;
- be placed on the perimeter of flat roofs at intervals not exceeding 20 ft (6 m) and within 2 ft (600 mm) of the edge;
- be placed in the center of a roof area at intervals not exceeding 50 ft (15 m);
- be placed on dormers and chimneys, except when these projections are protected by other terminals.

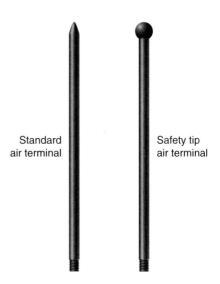

Standard air terminal

Safety tip air terminal

FIGURE 14-4 Air terminals.
(Delmar/Cengage Learning)

Lightning Conductors

Lightning conductors are installed to connect and interconnect to the earth, the air terminals, and other metal parts of the object to be protected. They are made of copper or aluminum. In general, lightning conductors should

- not be bent to a radius of less than 8 inches;
- not be bent to an angle of more than 90°;
- have a cross-sectional area of not less than 59,000 circular mils for copper or 98,500 circular mils for aluminum;
- maintain a horizontal or downward course;
- provide at least two paths for current flow from an air terminal to Earth;
- provide a ground path for every 100 ft (30 m) of perimeter;
- connect all metal bodies, such as exhaust fans and roof vents, to the protection system;
- be securely fastened to the air terminals, the grounds, and the structure; and
- not be concealed in metal conduit unless the conductor is securely bonded to the conduit at both ends.

Grounding Connections

The grounds are usually copper-clad steel rods at least 10 feet long and ½ inch in diameter. However, where the topsoil is very shallow, groundplates may be used. In general, grounding connections should be

- made with rods driven into the earth, at least 2 ft (600 mm) from the protected object so that the rod top is 1 ft (300 mm) under grade;
- made to underground metal water pipes or well casings; and
- interconnected with driven ground electrodes for the electric or telephone system.

──⌇⊢ SAFETY RULES

Following the specific requirements of the organizations referred to previously will result in a structure that is free from the hazards of lightning.

But because humans are also vulnerable to lightning, we should be aware of various safety rules that will help us avoid the shocking experience of a lightning stroke. The United States Department of Commerce, National Oceanic and Atmospheric Administration, has compiled the following safety rules for human protection when lightning threatens:

- Stay indoors and do not venture outside unless absolutely necessary.
- Stay away from open doors and windows, fireplaces, radiators, stoves, metal pipes, sinks, and plug-in electrical appliances.
- Do not use plug-in electrical equipment such as hair dryers, electric toothbrushes, or electric razors during the storm.
- Do not use the telephone during the storm—lightning may strike telephone lines outside.
- Do not take laundry off the clothesline.
- Do not work on fences, telephone or power lines, pipelines, or structural steel fabrication.
- Do not use metal objects such as fishing rods and golf clubs. Golfers wearing cleated shoes are particularly good lightning rods.
- Do not handle flammable materials in open containers.
- Stop tractor work and dismount, especially when the tractor is pulling metal equipment. Tractors and other implements in metallic contact with the ground are often struck by lightning.
- Get out of the water and off small boats.
- Stay in your automobile if you are driving. Automobiles offer lightning protection.
- Seek shelter in a building. If a building is unavailable, seek protection in a cave, a ditch, a canyon, or under head-high clumps of trees in open forest glades.
- When there is no shelter, avoid the highest object in the area. If only isolated trees are nearby, the best protection is to crouch in the open, keeping twice as far away from isolated trees as the trees are high.

- Avoid hilltops, open spaces, wire fences, metal clotheslines, exposed sheds, and any electrically conductive elevated objects.

- Should you feel the electrical charge—if your hair stands on end or your skin tingles—lightning may be about to strike you. Drop to the ground immediately.

Persons struck by lightning suffer a severe electrical shock and may be burned, but they carry no electrical charge and can be handled safely. A person thought to be killed by lightning can often be revived by prompt CPR, cardiac massage, and prolonged artificial respiration. In a group struck by lightning, the apparently dead should be treated first. Those who show vital signs will probably recover spontaneously, although burns and other injuries may require treatment. Recovery from lightning strokes is usually complete, except for possible impairment or loss of sight or hearing.

If you are ever present when a person is struck by lightning, immediately begin cardiopulmonary resuscitation (CPR). If you do not know how to apply this basic life-support technique, contact your American Heart Association for instruction. Encourage your friends to learn too, for if lightning ever strikes you, it may save your life to have friends who are able to come to your aid.

REVIEW QUESTIONS

All answers should be written in complete sentences, calculations should be shown in detail, and *Code* references should be cited when appropriate.

1. In your own words, write two definitions of a coulomb. _____

2. In your own words, explain what causes lightning. _____

3. List the general rules for placement of air terminals. _____

4. List the general rules for grounding connections. _____

5. If you were asked to evaluate a lightning protection installation, what would you look for? _____

6. If you were attending a Little League baseball game when nearby lightning was observed, what actions would you advise others to take? _____

_____

_____

_____

REFERENCES

Lightning Protection Institute. Arlington Heights, IL: [Home page: http://www.lightning.org/ e-mail: strike@lightning.org]

National Fire Protection Association. (1992). *Lightning Protection Code,* 1992 Ed. NFPA 780. Quincy, MA: Author.

Site Lighting

OBJECTIVES

After studying this chapter, the student should be able to

- list the important considerations in lamp selection for site lighting.
- select illuminance values for site lighting.
- calculate the power limit and power demand for site lighting.
- locate luminaires for site lighting.
- list control options for site lighting.

LAMP SELECTION

In choosing an appropriate lamp for site lighting, at least three factors should be considered: (1) The amount of power the lamp requires to provide the needed light, (2) the color the lamp creates, and (3) the maintenance requirements of the lamp.

At first thought, the study of site lighting could be perceived as a limited subject. However, a few experiences will reveal there is an array of options that can be applied in the lighting of a large area site. To examine these options, it is useful to divide the system into three parts: power control, light sources, and light distribution.

Power Control

A toggle switch may be adequate if the load is small and the control requirements are minimal. As the load increases, it is common to install a contactor that will, on signal, turn on or off a large number of light sources. A contactor is similar to a motor starter in that completing the control circuit will energize a coil that closes a set of contacts. The circuit controlled may connect directly to the light sources, or it may energize a panelboard where many circuits are connected to an array of light sources. The operation of the system may be manual, such as the toggle switch, or it may be fully automated. The advantage to using a toggle switch is initial cost; the disadvantage is that a person must perform the operation, which can be costly and likely unreliable. The automated system has a high initial cost but only a minimal maintenance cost, plus it is reliable. For site lighting, it would need to be sensitive to the day and night cycle. The simplest option is an astronomical time clock. These clocks would be set for the time and the date, then they would automatically adjust the on period for the seasonal changes in nighttime hours. A control device with greater reliability would be a photocell that could be adjusted to a precise intensity of light for on and off control. On a cloudy day, it would probably energize the lighting earlier than the time clock.

Light Sources

The light sources (lamps) can be placed in three categories: incandescent, fluorescent, and high-intensity discharge (HID). It is recommended that the person selecting or installing the light source have at hand a lamp specification and application guide (this is available from most electrical distribution centers or can be requested from any of the major lamp manufacturers), for there are hundreds of lamp types available, all with different characteristics. Following are abbreviated descriptions of several types of lamps.

Incandescent   There are two styles of incandescent lamps. The first to be discussed is the filament, or Edison, type, and the second is the tungsten halogen lamp. The *Edison*-type lamp uses a wire filament enclosed in a glass bulb to produce the light. The bulb may be vacuumed or filled with an inert gas. The filament is a coil of wire that emits light when heated. The light output varies from 100 to 10,000 lumens and the lamp wattage varies from 3 to 1500 watts. Both the light output and the lamp life are sensitive to the voltage applied. A small increase in voltage above the rated value will result in a higher intensity of light and a much shorter lamp life. In practice, where lamp life is an important factor, lamps rated for 130 volts are installed on 120-volt systems, which will almost double the lamp life. The *halogen*-type lamp uses a tightly wound tungsten filament coil in a small quartz tube. The light is produced at a higher temperature; thus, it has superior color-rendering properties when compared with the Edison lamps. The lamp is costly but life is longer than, sometimes double, that of the Edison lamp. The luminaires are smaller and could provide superior control of the light distribution.

Fluorescent   The fluorescent is a tubular type of lamp with a filament at both ends of the tube. The tube size and length vary greatly from a few inches (millimeters) to 8 ft (2.5 m) in length and from $\frac{5}{8}$ to $\frac{12}{8}$ in. (15.87 to 38 mm) in diameter. Light output varies from slightly over 400 lumens to near 6000 lumens. The light is produced by electrons bombarding phosphors; thus, the color rendition properties of the light can be varied by the selection of phosphors. They have a long life and a fair tolerance for voltage swings. Their application in site lighting has been limited because the lamp light output drops with the ambient temperature, and because

TABLE 15-1

Lamp performance data.

Lamp Type	Watts Rating	Lumen Rating	Total Watts	Lamp Life (hours)	Lamp Length	(Lamps) Load*
Incandescent	1000	17,700	1000	1000	13 in.	(17) 929 kW
Halogen	1000	21,000	1000	3000	10 in.	(14) 24 kW
Fluorescent (800 ma)	60	4050	100	12,000	48 in.	(74) 15.5 kW
Mercury	250	12,100	285	24,000	8.5 in.	(25) 11.7 kW
Metal halide	250	20,500	285	17,000	8.5 in.	(15) 6.8 kW
High-pressure sodium	250	28,500	310	28,500	10 in.	(11) 5.6 kW
Low-pressure sodium	135	22,500	180	18,000	20 in.	(13) 3.9 kW

*The approximate number of lamps and the resultant load required to produce 300,000 lumens.
All values are generic approximations; manufacturer's data should be consulted for specific information.

of the length, the luminaires must be strongly supported if there is the possibility of high-velocity winds.

HID There are four styles of HID lamps: mercury, metal halide, high-pressure sodium, and low-pressure sodium. The vast majority of these lamps have a single screw-type base. The *mercury lamp* is the oldest style but is now in diminished use. It is usually available with power ratings of 100 to 1000 watts with a light output of 2850 to 63,000 lumens. Light depreciation over the lamp's life is high and the color rendition poor. The *metal halide lamp* is available in power ratings from 39 to 1000 watts with the light output varying from 2300 to 125,000 lumens and lamp life from 10,000 to 24,000 hours. The lamps vary in length from 4 to 15 in. (100 to 375 mm). They are usually preferred over the mercury lamp.

The *high-pressure sodium lamp* is available with power ratings from 35 to 1000 watts, with the light output varying from 1250 to 140,000 lumens. Lamp life varies from 10,000 to 24,000 hours. Color rendition is poor, but color discrimination is possible.

The *low-pressure sodium lamp* has a power rating that varies from 18 to 180 watts and light output from 1800 to 33,000 lumens. Lamp life ranges from 14,000 to 18,000 hours. The light output and the life are excellent, but color discrimination is void, everything appearing the same color: yellow. People parking in the lot would not be able to identify their automobiles by color.

Table 15-1 provides a comparison of lamps that might be selected for illuminating the site at the industrial building. The operational data for the listed lamps is provided in columns two through six. In the last column, a comparison is made by choosing a quantity of light to be produced, then calculating the number of lamps and the electrical power required. As can be seen in the table, the low-pressure sodium will be the least expensive to operate. This does not indicate it will be the lamp of choice. When all other factors such as color discrimination, lamp life, and number of poles required are considered, the metal halide will be a strong contender.

Light Distribution

The distribution of the light after it has been created in the lamp is affected by three factors: the lamp, reflectors, and lens. Many halogen and all the fluorescent lamps are tubular in shape, thus emitting light in 360° the entire length of the tube. In most cases, a reflector is installed to redirect 50 percent or more of the light. When light is reflected by a surface, there is absorption dependent on the reflectivity of the surface. When selecting a reflector, keep in mind that white has the highest reflectivity. This surface should be cleaned before installation and every time the lamps are replaced. Although the distribution from the other types of lamps is more directed, most likely there will be a reflector to redirect the light in a specific pattern. Lenses are made of clear glass or plastic and are designed to redirect the light. They should be cleaned regularly and replaced if they begin to yellow as is the tendency of some plastics. With most luminaires, charts are available to detail the light distribution. These should be studied to determine whether the luminaire has the proper distribution pattern for the application. For example, a luminaire designed for illuminating book shelves would be a poor choice for an assembly room.

Lamp Efficacy

To be fair in comparing the power requirement of lamps, it is assumed that the light output of the lamps is the same. Because this is rarely the case, a technique has been developed that compares the light output of a lamp (lumen) with the power requirement (watt). By dividing the lumen rating of a lamp by the watts rating, a value called the *efficacy* of the lamp is determined. A high efficacy means that the lamp produces a great amount of light for each watt of power. When evaluating lamps that use ballasts, the watts rating of the ballast should also be included.

Typical lamp types and their efficacies (lumens per watt) are shown in Table 15-1. A review of Table 15-1 would indicate that the low-pressure sodium lamp is preferred, when *power only* is considered; the high-pressure sodium lamp is second; and the metal-halide lamp and the high-output fluorescent lamp are tied for third. The incandescent lamp and the mercury lamp are so poor for this purpose that their use would be difficult to justify.

Lamp Color Characteristics

The colors that a person perceives when viewing a building and its surrounding area can be very important. The image of the structure, the automobiles in the parking lot, the materials in the storage yard, and the people entering the building will all have a different appearance when illuminated by light with different color content.

The perceived color of any of these objects is affected by the color of the object itself and the

TABLE 15-2

Lamp color characteristics.

Lamp Type	Color Characteristics
Incandescent	Is accepted by many as the color standard; is considered a warm light; blues will appear somewhat gray
Mercury deluxe white	Has a cool, greenish light; reds and oranges will appear somewhat gray
Fluorescent	A wide variety of lamp color types makes any degree of color rendering possible
Metal halide	Similar to mercury
Low-pressure sodium	Has an orange-yellow light; permits good color discrimination, except with some reds and blues
High-pressure sodium	A yellow monochromatic light that permits no color discrimination, except with yellow objects

color of the light used to illuminate the object. A common example of this is often observed in parking lots where automobiles take on an entirely different look at nighttime when the parking lot lighting is turned on.

When it is important that good color discrimination be possible, then lamps must be selected to make this possible. If exact color discrimination is required, the objects to be seen should be viewed under the lamps before the lamps are selected for installation. However, for general usage, the listing in Table 15-2 can be used as a guide.

A comparison of the characteristics in Table 15-2, along with the performance information in Table 15-1, would indicate the low-pressure sodium lamp to be the best lamp if color discrimination is unimportant. The low-pressure sodium lamp would be acceptable for the majority of exterior uses, and the others would be used in cases where color is more important.

Lamp Maintenance

The maintenance of any lighting system is strongly sensitive to how often the lamp can be expected to fail and how difficult it is to replace. Looking back at Table 15-1, it can be noticed that the incandescent lamp has, by far, the shortest life expectancy, followed by the fluorescent and the metal halide; all the others have a life expectancy of more than 18,000 hours (over a 4-year period at 50 percent burning time). It should also be noted that the fluorescent lamp is 4 ft (1.2 m) in length, which makes it most difficult to handle on pole installations.

Again, the high-pressure sodium lamp and the low-pressure sodium lamp are given the highest rating. For the industrial building, the need to identify automobiles in the parking lot, and materials in the storage area, was justification to use the high-pressure sodium lamp as the light source for all the site areas.

ILLUMINANCE SELECTIONS

The amount of light, or illuminance, needed for exterior areas is dependent upon the type of activity that is to take place in a specific area. The Illuminating Engineering Society of North America (IES) has published recommendations for various activity areas. Unless otherwise specified, the illuminance value given is the footcandle, which can be measured on the horizontal surface of the area, on the pavement in a parking lot or roadway, or on the sidewalk where people walk.

For the areas on the industrial building site, the recommended illuminances are as shown in Table 15-3.

Note that the storage area is recommended to have a considerably higher lighting level than the other areas. This is because it is considered a work area, as compared with the other areas where only walking or driving takes place.

POWER LIMITATION

Since the oil embargo of the early 1970s, many states have legislated measures that restrict the power that can be dedicated to lighting exterior (and interior) areas. Several procedures are common, but the one adopted by the majority of the states is "ASHRAE/IES Standard 90, Energy Conservation in New Building Construction."

For the areas on the industrial building site, the power allowances are shown in Table 15-4. The procedure dictates that the power values determined for each of the areas be totaled and only the total value be considered the power limit.

For the industrial building site, the power limits are shown in Table 15-5. The actual connected

TABLE 15-3

Recommended illuminance.

Site Activity	Illuminance (footcandles)
Parking	0.5
Roadway	0.6
Storage	20
Pedestrian	0.9

TABLE 15-4

Power allowances.

Site Area	Power Allowance (watts)
Parking	30 watts per space
Roadway	2 W/ft (6.5 W/m)
Storage	0.4 W/ft^2 (4.3 W/m^2)
Pedestrian	30 W/ft (100 W/m)

TABLE 15-5

Industrial lighting power limits.

Site Activity		Power Allowance	Area Watts
Parking	90 spaces	30 watts per space	2700
Roadway	425 ft	2 watts per linear ft	850
	130 m	6.5 watts per m	
Storage	7900 sq. ft	0.4 watts per sq. ft	3160
	735 sq. m	43 watts per sq. m	
Pedestrian	60 ft	30 watts per linear ft	1800
	18 m	10 watts per m	
	Power limit (watts)		8510

TABLE 15-6

Site-lighting power requirements.

Site Activity	Number of Luminaires	Watts per Luminaire	Total Watts
Parking	12	310	3720
Roadway	6	125	750
Storage	6	480	2880
Pedestrian	18	65	1170
	Total watts		8520

load is shown in Table 15-6. Because the total connected load is not significantly greater than the power limit, this installation complies with the typical energy code.

LUMINAIRE PLACEMENT

The location and selection of luminaires, and the sizing of the lamp, are facilitated by drawings similar to that shown in Figure 15-1. This type of drawing, which is provided by the luminaire manufacturers, is called iso-illuminance curves or, in this specific instance, isofootcandle curves. Each curve represents a line, if drawn on the horizontal surface being illuminated, where the illuminance is at the designated value. In Figure 15-1, the egg-shaped

curve next to the center symbol shows that 5.6 footcandles are present along that line. The center symbol designates that this is for a twin luminaire. The labeling also indicates that it is for a 250-watt, high-pressure sodium lamp in each luminaire, and that the luminaires are mounted 30 ft (9 m) above the surface being illuminated.

Parking Lighting

In practice, this curve is combined with other similar and dissimilar curves to form a pattern such as that shown in Figure 15-2. This is the layout for the industrial building parking lot, showing three of the twin luminaires on the median and a single luminaire installed next to the street. The values assigned to the curves are additive; thus, this layout indicates that there will be a little more than 2.2 footcandles where the light patterns overlap. Because the recommendation was for 2 footcandles, this value is acceptable. This same technique was used to locate the remaining luminaires.

Pedestrian Lighting

The lighting of the sidewalks and the entryway was placed so that it is close to the walk surface. The luminaires are located using the same technique described earlier but are designed into the landscaping to be as unnoticeable as possible. The outer sidewalk, shown in Figure 15-3, is designed for use with a 50-watt high-pressure sodium lamp at a 36-in. (900-mm) mounting height. The center entry walk, shown in Figure 15-4, is designed

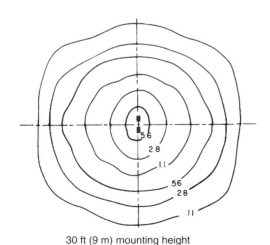

30 ft (9 m) mounting height

FIGURE 15-1   Isofootcandle curves.
(*Delmar/Cengage Learning*)

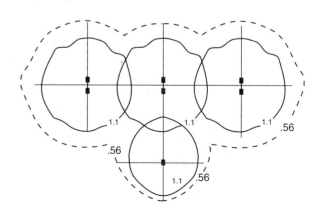

FIGURE 15-2   Determining footcandle values.
(*Delmar/Cengage Learning*)

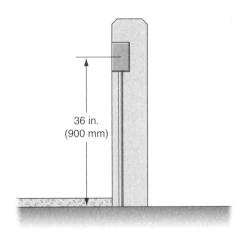

FIGURE 15-3 Sidewalk illumination.
(*Delmar/Cengage Learning*)

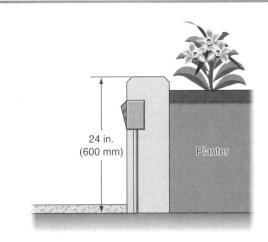

FIGURE 15-4 Center entry walk illumination.
(*Delmar/Cengage Learning*)

with 50-watt high-pressure sodium lamps at less than 24 in. (600 mm) mounting height built into the planter.

ELECTRICAL INSTALLATION

Direct-burial conductors, Type UF, are used to service the lighting installation. The *NEC* requirements discussed in Chapter 1 were followed and,

in addition, raceways were installed wherever the conductors had to run under sidewalk.

Two very important considerations dictated the selection of conductor size and the circuit arrangement. First, the conductors were sized to ensure a low voltage drop. This is particularly important because of the long distances that the conductors must run. The second important consideration was to be able to have selective control of the lighting in the various activity areas.

Lighting Control

Because of the high energy cost, the use of advanced electric lighting control for site lighting is cost-effective in many instances.

Photocell control is essentially a must for all site lighting. It can be installed on individual lighting units or at master control points. The photocell is used to ensure that the site lighting is turned off during the daylight hours.

Time clocks are often used in conjunction with photocells to deactivate the lighting when there is no longer a use for it. The industrial building uses a photocell to activate the pedestrian lighting at twilight, then a time clock is used to deactivate that lighting after the last worker has left.

Dimmers for site lighting are rapidly gaining in popularity. Dimming has the advantage, over other control systems, of being able to lower the lighting and the energy use while maintaining uniform lighting throughout the area. Energy use can be reduced by 20 percent, or more, of the original value. A dimming system is to be installed on the storage area and the parking lot. A photocell will signal when to turn the lights on; then, a time clock will initiate the dimmers when the high level of lighting is no longer needed. Override switches will be located in convenient places if there is a need to have full lighting at any time.

All answers should be written in complete sentences, calculations should be shown in detail, and *Code* references should be cited when appropriate.

1. List three common devices used to control illumination systems.

    1. _____ 2. _____ 3. _____

2. The three major light source types are

    1. _____ 2. _____ 3. _____

3. Arrange the lamps listed in Table 15-1 in order of their efficacy based on ratings, starting with the lowest efficacy, indicating the calculated efficacy.

    1. _____ (_____)   5. _____ (_____)

    2. _____ (_____)   6. _____ (_____)

    3. _____ (_____)   7. _____ (_____)

    4. _____ (_____)

4. Explain the difference between the terms *efficacy* and *efficiency*, and give examples of each.

    _____

    _____

    _____

5. When would color rendition be important in a parking lot and when would it not be important? _____

    _____

    _____

6. Explain why a storage area might require a higher level of lighting than a parking area.

    _____

    _____

    _____

7. List the factors that should be considered in selecting a lamp type for a parking lot.

    _____

    _____

    _____

8. Explain an isofootcandle curve. _____

    _____

    _____

    _____

9. List some devices that can be used with area lighting to reduce energy usage. _____

    _____

    _____

CHAPTER

16

Programmable Logic Controllers

OBJECTIVES

After studying this chapter, the student should be able to

- list the principal parts of a programmable logic controller (PLC).

- describe differences between PLCs and other types of computers.

- discuss differences between the input/output (I/O) track, central processing unit (CPU), and program loader.

- draw a diagram of how the input and output modules work.

PLCs were first used by the automotive industry in the late 1960s. Each time a change was made in the design of an automobile, it was necessary to change the control system operating the machinery. This consisted of physically rewiring the control system to make it perform the new operation. Rewiring the system was, of course, very time-consuming and expensive. What the industry needed was a control system that could be changed without the extensive rewiring required to change relay control systems.

DIFFERENCES BETWEEN PROGRAMMABLE LOGIC CONTROLLERS AND PERSONAL COMPUTERS

One of the first questions generally asked is "Is a PLC a computer?" The answer to that question is "yes." The PLC is a special type of computer designed to perform a special function. Although the PLC and the personal computer (PC) are both computers, there are some significant differences. Both generally employ the same basic type of computer and memory chips to perform the tasks for which they are intended, but the PLC must operate in an industrial environment. Any computer that is intended for industrial use must be able to withstand extremes of temperature, ignore voltage spikes and drops on the power line, withstand shock and vibration, and survive in an atmosphere that often contains corrosive vapors, oil, and dirt.

PLCs are designed to be programmed with schematic or ladder diagrams instead of common computer languages. An electrician who is familiar with ladder logic diagrams can generally learn to program a PLC in a few hours, as opposed to the time required to train a person how to write programs for a standard computer.

BASIC COMPONENTS

PLCs can be divided into four primary parts:

1. The power supply
2. The CPU
3. The programming terminal or program loader
4. The I/O (pronounced "eye-oh") rack

The Power Supply

The function of the power supply is to lower the incoming ac voltage to the desired level, rectify it to dc, and then filter and regulate it. The internal logic of a PLC generally operates on 5 to 24 volts dc, depending on the type of controller. This voltage must be free of voltage spikes and other electrical noise and be regulated to within 5 percent of the required voltage value. Some manufacturers of PLCs build a separate power supply, and others build the power supply into the CPU.

The CPU

The CPU is the "brains" of the PLC. It contains the microprocessor chip and related integrated circuits to perform all of the logic functions. The microprocessor chip used in most PLCs is the same as that found in most home and business PCs.

The CPU, Figure 16-1, often has a key located on the front panel. This switch must be turned on before the CPU can be programmed. This is done to prevent the circuit from being changed accidentally. Plug connections on the CPU provide connection for the programming terminal and I/O racks, Figure 16-2. CPUs are designed so that once a program has been developed and tested, it can be stored on some type of medium such as tape, disk, CD, or other storage device. As a result, if a CPU fails and has to be replaced, the program can be downloaded from the storage medium. This eliminates the time-consuming process of having to reprogram the unit by hand.

The Programming Terminal

The programming terminal or loading terminal is used to program the CPU. The type of terminal used depends on the manufacturer and often on the preference of the consumer. Some are small, handheld devices that use a liquid crystal display or LEDs to show the program, Figure 16-3. Some of these small units display one line of the program at a time, and others require the program to be entered in a language called Boolean.

Another type of programming terminal contains a display and keyboard, Figure 16-4. This

FIGURE 16-1 A CPU. (*Courtesy Rockwell Automation, Inc.*)

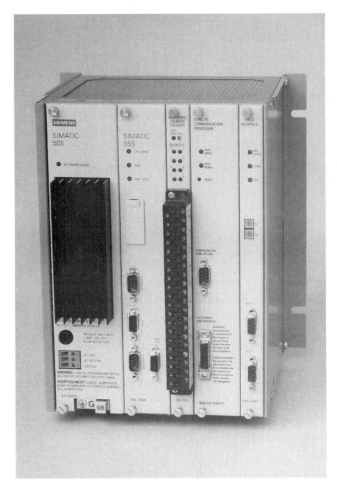

FIGURE 16-2 Plug connections located on the CPU. (*Courtesy Siemens*)

FIGURE 16-3 A handheld programming terminal and small PLC. (*Courtesy Eaton Corporation*)

FIGURE 16-4 A programming terminal. (*Courtesy Rockwell Automation, Inc.*)

type of terminal generally displays several lines of the program at a time and can be used to observe the operation of the circuit as it is operating.

Many industries prefer to use a notebook or laptop computer for programming, Figure 16-5. An interface that permits the computer to be connected to the input of the PLC and software program is generally available from the manufacturer of the PLC.

The terminal is used not only to program the PLC but also to troubleshoot the circuit. When the terminal is connected to the CPU, the circuit can be examined while it is in operation. Figure 16-6

illustrates a circuit typical of those seen on the display. Notice that this schematic diagram is different from the typical ladder diagram. All of the line components are shown as normally open or normally closed contacts. There are no NEMA symbols for push-button, float switch, limit switches, and so on. The PLC recognizes only open or closed contacts. It does not know whether a contact is connected to a push button, a limit switch, or a float switch. Each contact, however, does have a number. The number is used to distinguish one contact from another.

In this example, coil symbols look like a set of parentheses instead of a circle as shown on most ladder diagrams. Each line ends with a coil, and each coil has a number. When a contact symbol has the same number as a coil, it means that the contact is controlled by that coil. The schematic in Figure 16-6 shows a coil numbered 257 and two contacts numbered 257. When coil 257 is energized, the PLC interprets both contacts 257 to be closed.

A characteristic of interpreting a diagram when viewed on the screen of most loading terminals is that when a current path exists through a contact, or if a coil is energized, that coil or contact will be highlighted on the display. In the example shown in Figure 16-6 for coil 257, contact 16 and contact 18 are drawn with dark heavy lines, illustrating that they are highlighted or illuminated on the display. Highlighting a contact does not mean that it has changed from its original state. It means that there is a complete circuit through that contact. Contact 16 is highlighted, indicating that coil 16 has energized

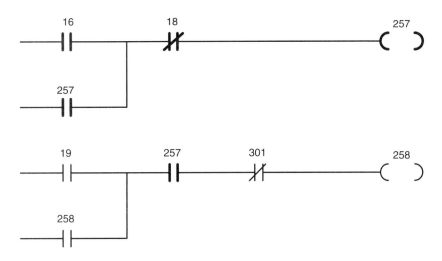

FIGURE 16-6 Analyzing circuit operation with a terminal. (*Delmar/Cengage Learning*)

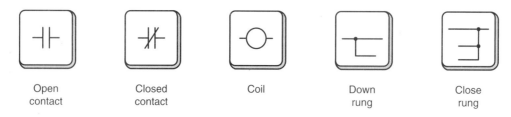

| Open contact | Closed contact | Coil | Down rung | Close rung |

FIGURE 16-7 Symbols are used to program the PLC. (*Delmar/Cengage Learning*)

and contact 16 is closed, providing a complete circuit. Contact 18, however, is shown as normally closed. Because it is highlighted, coil 18 has not been energized, as a current path still exists through contact 18. Coil 257 is shown highlighted, indicating that it is energized. Because coil 257 is energized, both 257 contacts are now closed, providing a current path through them.

When the loading terminal is used to load a program into the PLC, contact and coil symbols on the keyboard are used, Figure 16-7. Other keys permit specific types of relays, such as timers, counters, or retentive relays, to be programmed into the logic of the circuit. Some keys permit parallel paths, generally referred to as down rungs, to be started and ended. The method employed to program a PLC is specific to the make and model of the controller. It is generally necessary to consult the manufacturer's literature if you are not familiar with the specific PLC.

The I/O Rack

The I/O rack is used to connect the CPU to the outside world. It contains input modules that carry information from control sensor devices to the CPU and output modules that carry instructions from the CPU to output devices in the field. I/O racks are shown in Figure 16-8A and Figure 16-8B. Input and output modules contain more than one input or output. Any number from four to sixteen is common, depending on the manufacturer and model of PLC. The modules shown in Figure 16-8A each can handle sixteen connections. This means that each input module can handle sixteen different input devices such as push buttons, limit switches, proximity switches, and float switches. The output modules can each handle sixteen external devices such as pilot lights, solenoid coils, and relay coils. The operating voltage can be either ac or dc, depending on the make and model of controller, and is generally either 120 or 24 volts. The I/O rack shown in Figure 16-8A can handle ten modules. Because each module can handle sixteen input or output devices, the I/O rack is capable of handling 160 input and output devices. Many PLCs are capable of handling multiple I/O racks.

FIGURE 16-8A An I/O rack with input and output modules. (*Courtesy General Electric, www.geindustrial.com*)

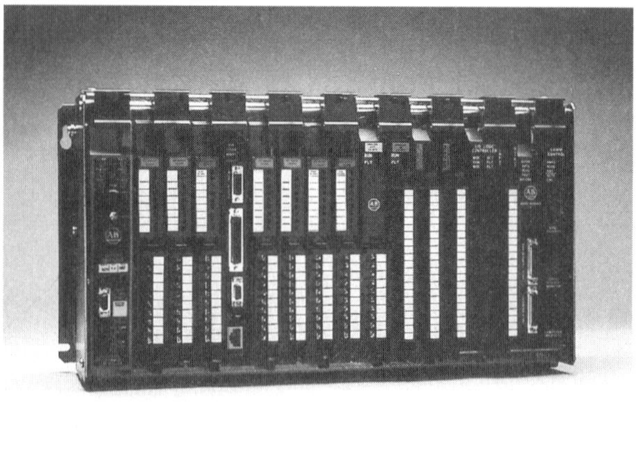

FIGURE 16-8B An I/O rack with input and output modules. (*Courtesy General Electric, www.geindustrial.com*)

FIGURE 16-9 A CPU with I/O racks. (*Courtesy Struthers-Dunn*)

I/O Capacity

One factor that determines the size and cost of a PLC is its I/O capacity. Many small units may be intended to handle as few as sixteen input and output devices. Large PLCs can generally handle several hundred. The number of input and output devices the controller must handle also affects the processor speed and amount of memory the CPU must have. A CPU with I/O racks is shown in Figure 16-9.

The Input Module

The CPU of a PLC is extremely sensitive to voltage spikes and electrical noise. For this reason, the I/O input module uses opto-isolation to electrically separate the incoming signal from the CPU. Figure 16-10 shows a typical circuit used for the input. A metal-oxide varistor (MOV) is connected across the ac input to help eliminate any voltage spikes that may occur on the line. The MOV is a voltage-sensitive resistor. As long as the voltage across its terminals remains below a certain level, it exhibits a very high resistance. If the voltage should become too high, the resistance almost instantly changes to a very low value. A bridge rectifier changes the ac voltage into dc. A resistor is used to limit current to an LED. When power is applied to the circuit, the LED turns on. The light is detected by a phototransistor, which signals the CPU that there is a voltage present at the input terminal.

When the module has more than one input, the bridge rectifiers are connected together on one side to form a common terminal. On the other side the rectifiers are labeled 1, 2, 3, and 4. Figure 16-11 shows four bridge rectifiers connected together to form a common terminal. Figure 16-12 shows a limit switch connected to input 1, a temperature switch connected to input 2, a float switch connected to input 3, and a normally open push button connected to input 4. Notice that the pilot devices complete a circuit to the bridge rectifiers. If any switch closes, 120 volts ac will be connected to a bridge rectifier, causing the corresponding LED to turn on and signal the CPU that the input has voltage applied to it. When voltage is applied to an input, the CPU considers that input to be at a high level.

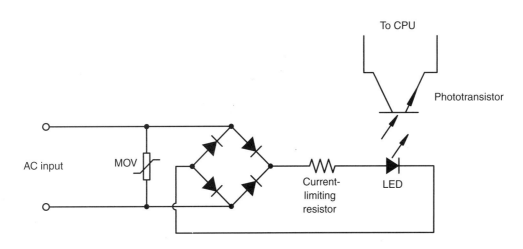

FIGURE 16-10 An input circuit. (*Delmar/Cengage Learning*)

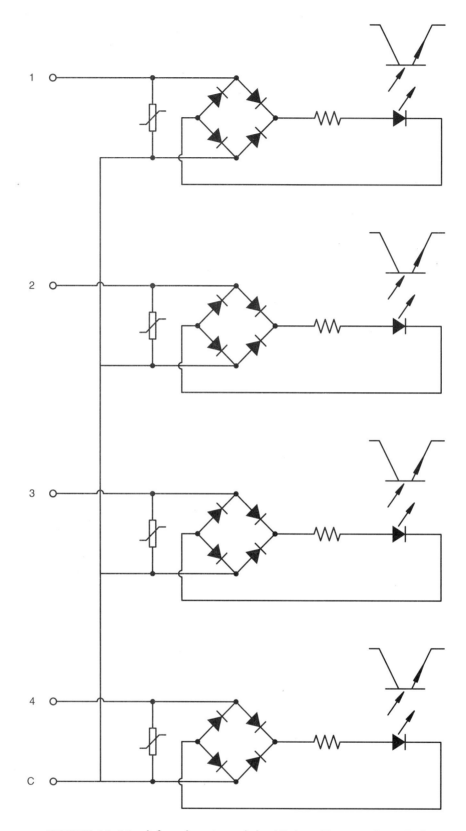

FIGURE 16-11 A four-input module. (*Delmar/Cengage Learning*)

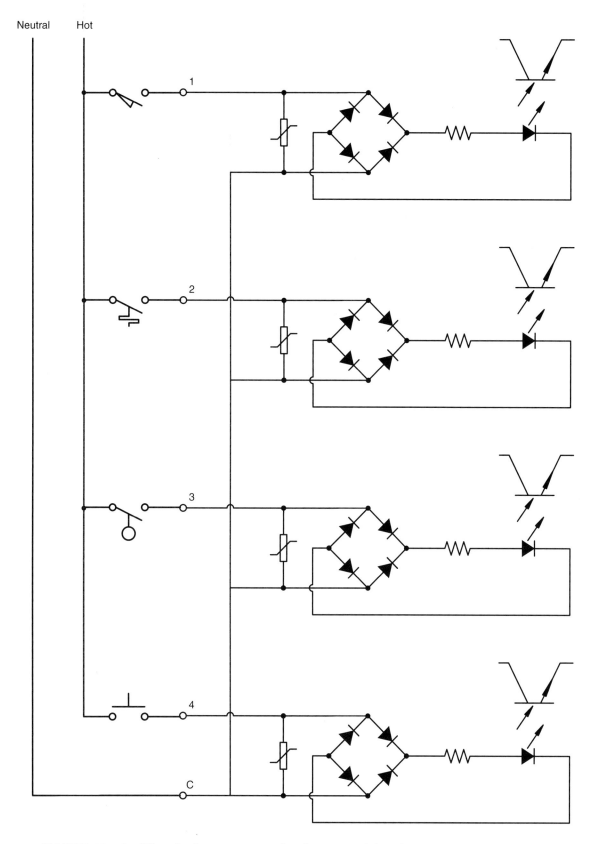

FIGURE 16-12 Pilot devices connected to input modules. (*Delmar/Cengage Learning*)

The Output Module

The output module is used to connect the CPU to the load. Output modules provide line isolation between the CPU and the external circuit. Isolation is generally provided in one of two ways. The most popular is with optical isolation, very similar to the input modules. In this case, the CPU controls an LED. The LED is used to signal a solid-state device to connect the load to the line. If the load is operated by dc, a power phototransistor is used to connect the load to the line, Figure 16-13. If the load is an ac device, a triac is used to connect the load to the line, Figure 16-14. Notice that the CPU is separated from the external circuit by a light beam. No voltage spikes or electrical noise can be transmitted to the CPU.

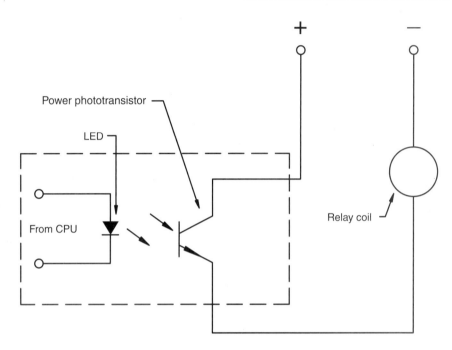

FIGURE 16-13 A power phototransistor connects a dc load to the line. (*Delmar/Cengage Learning*)

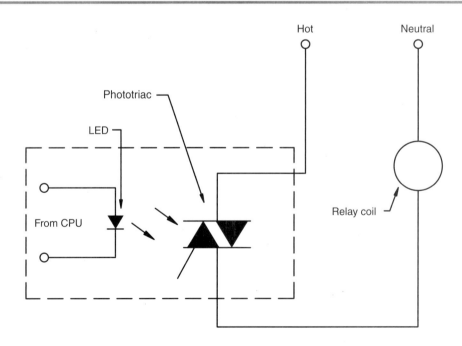

FIGURE 16-14 A triac connects an ac load to the line. (*Delmar/Cengage Learning*)

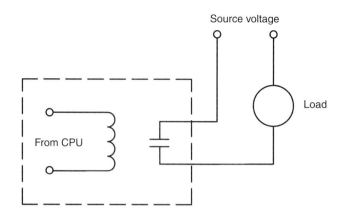

FIGURE 16-15 A relay connects the load to the line. (*Delmar/Cengage Learning*)

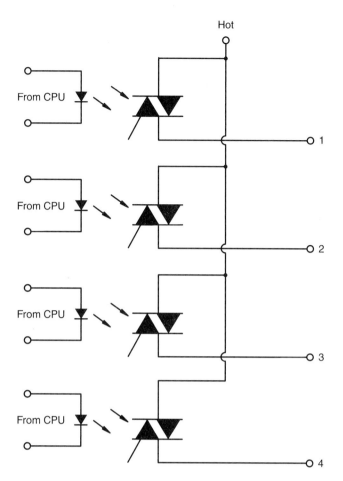

FIGURE 16-16 A multiple output module. (*Delmar/Cengage Learning*)

The second method of controlling the output is with small relays, Figure 16-15. The CPU controls the relay coil. The contacts connect the load to the line. The advantage of this type of output module is that it is not sensitive to whether the voltage is ac or dc and can control 120- or 24-volt circuits. The disadvantage is that it does contain moving parts that can wear. In this instance, the CPU is isolated from the external circuit by a magnetic field instead of a light beam.

If the module contains more than one output, one terminal of each output device is connected together to form a common terminal, similar to a module with multiple inputs, Figure 16-16. Notice that one side of each triac has been connected together to form a common point. The other side of each triac is labeled 1, 2, 3, or 4. If power transistors are used as output devices, the collectors or emitters of each transistor would be connected to form a common terminal. Figure 16-14 shows a relay coil connected to the output of a triac. Notice that the triac is used as a switch to connect the load to the line. The power to operate the load must be provided by an external source. Output modules *do not* provide power to operate external loads.

The amount of current an output can control is limited. The current rating of most outputs can range from 0.5 to about 3 amperes, depending on the make of the controller and the type of output being used. Outputs are intended to control loads that draw a small amount of current such as solenoid coils, pilot lights, and relay coils. Some outputs can control

motor starter coils directly, and others require an interposing relay. Interposing relays are employed when the current draw of the load is above the current rating of the output.

Internal Relays

The actual logic of the control circuit is performed by *internal relays*. An internal relay is an imaginary device that exists only in the logic of the computer. It can have any number of contacts from one to several hundred, and the contacts can be programmed normally open or normally closed. Internal relays are programmed into the logic of the PLC by assigning them a certain number. Manufacturers provide a chart that lists which numbers can be used to program inputs and outputs, internal relay coils, timers, counters, and so on. When a coil is entered at the end of a line of logic and is given a number that corresponds to an internal relay, it will act like a physical relay. Any contacts given the same number as that relay will be controlled by that relay.

Timers and Counters

Timers and counters are also internal relays. There is no physical timer or counter in the PLC. They are programmed into the logic in the same manner as any other internal relay, by assigning them a number that corresponds to a timer or counter. The difference is that the time delay or number of counts must be programmed when they are inserted into the program. The number of counts for a counter is entered using numbers on the keys on the load terminal. Timers are generally programmed in 0.1-second intervals. Some manufacturers provide a decimal key, whereas others do not. If a decimal key is not provided, the time delay is entered as 0.1-second intervals. If a delay of 10 seconds is desired, for example, the number 100 would be entered. One-hundred-tenths of a second equals 10 seconds.

Off-Delay Circuit

Some PLCs permit a timer to be programmed as on or off delay, but others permit only on-delay timers to be programmed. When a PLC permits only on-delay timers to be programmed, a simple circuit can be used to permit an on-delay timer to perform the function of an off-delay timer, Figure 16-17. To understand the action of the circuit, recall the operation of an off-delay timer. When the timer coil is energized, the timed contacts change position immediately. When the coil is de-energized, the contacts remain in their energized state for some period of time before returning to their normal state.

In the circuit shown in Figure 16-17, it is assumed that contact 400 controls the action of the timer. Coil 400 is an internal relay coil located somewhere in the circuit. Coil 12 is an output and controls some external device. Coil TO-1 is an on-delay timer set for 100-tenths of a second. When coil 400 is energized, both 400 contacts change position. The normally open 400 contact closes and provides a current path to coil 12. The normally closed 400 contact opens, which prevents a circuit from being completed to coil TO-1 when coil 12 energizes. Note that coil 12 turns on immediately when contact 400 is closed. When coil 400 is de-energized, both 400 contacts return to their normal position. A current path is maintained to coil 12 by the now closed 12 contact, in parallel with the normally open 400 contact. When the normally closed 400 contact returns to its normal position, a current path is established to coil TO-1 through the now closed 12 contact. This starts the time sequence of timer TO-1. After a delay of 10 seconds, the normally closed TO-1 contact opens and de-energizes coil 12, returning the two 12 contacts to their normal position. The circuit is now back in the state shown in Figure 16-17. Note the action of the circuit. When coil 400 was energized, output coil 12 turned on

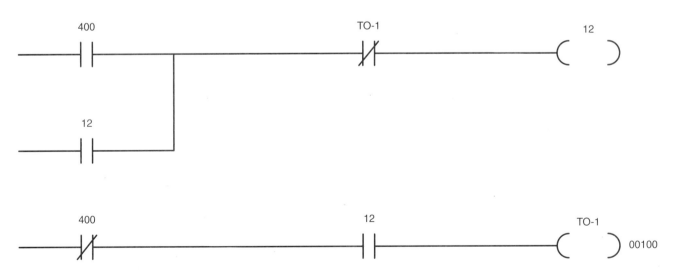

FIGURE 16-17 An off-delay timer circuit. (*Delmar/Cengage Learning*)

FIGURE 16-18 A direct-current drive unit controlled by a PLC. (*Courtesy Reliance Electric*)

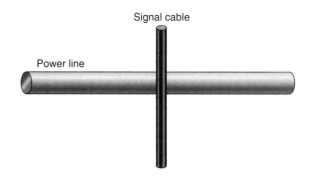

FIGURE 16-19 Signal cable crosses power line at right angle. (*Delmar/Cengage Learning*)

immediately. When coil 400 was de-energized, output 12 remained on for 10 seconds before turning off.

The number of internal relays and timers contained in a PLC is determined by the memory capacity of the computer. As a general rule, PLCs that have a large I/O capacity have a large amount of memory. The use of PLCs has steadily increased since their invention in the late 1960s. A PLC can replace hundreds of relays and occupy only a fraction of the space. The circuit logic can be changed easily and quickly without requiring extensive hand rewiring. It has no moving parts or contacts to wear out, and its downtime is less than that for an equivalent relay circuit. When replacement is necessary, it can be reprogrammed from a media storage device.

The programming methods presented in this text are general because it is impossible to include examples of equipment from each specific manufacturer. The concepts, however, are common to all programmable controllers. A PLC used to control a dc drive is shown in Figure 16-18.

INSTALLING PROGRAMMABLE LOGIC CONTROLLERS

In installing PLCs, several general rules should be followed. These rules are basically common sense and are designed to help reduce the amount of electrical noise that can be induced into the input cables. Electrical noise is such a problem in some installations that several manufacturers are using optical cables instead of wires for connection to the I/O rack.

Keep Wire Runs Short

Try to keep the wire runs as short as possible. A long wire run has more surface area of wire to pick up stray electrical noise.

Plan the Route of the Signal Cable

Before starting, plan how the signal cable should be installed. *Never run signal wire in the same conduit with power wiring.* Try to run signal wiring as far away from power wiring as possible. When it is necessary to cross power wiring, install the signal cable so that it crosses at a right angle, as shown in Figure 16-19.

Use Shielded Cable

Shielded cable is used for the installation of signal wiring. One of the most common types, shown in Figure 16-20, uses twisted wires with a Mylar foil shield. The ground wire must be grounded if the shielding is to operate properly. This type of shielded cable can provide a noise reduction ratio of about 30,000:1.

Another type of signal cable uses a twisted pair of signal wires surrounded by a braided shield. This type of cable provides a noise reduction ratio of about 300:1.

Use of common coaxial cable should be avoided. This cable consists of a single conductor surrounded by a braided shield. This type of cable offers very poor noise reduction.

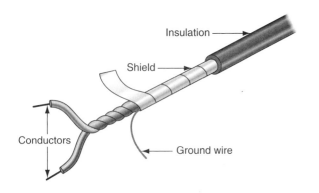

FIGURE 16-20 Shielded cable.
(*Delmar/Cengage Learning*)

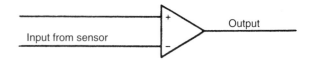

FIGURE 16-22 The differential amplifier detects a difference in signal level.
(*Delmar/Cengage Learning*)

Grounding

Ground is generally thought of as being electrically neutral or zero at all points. However, this may not always be the case in practical application. It is not uncommon to find that different pieces of equipment have ground levels that are several volts apart, Figure 16-21. To overcome this problem, large cable is sometimes used to tie the two pieces of equipment together. This forces them to exist at the same potential. This method is sometimes referred to as the brute-force method.

Where the brute-force method is not practical, the shield of the signal cable is grounded at only one end. The preferred method is generally to ground the shield at the sensor.

THE DIFFERENTIAL AMPLIFIER

An electronic device that is often used to help overcome the problem of induced noise is the differential amplifier. This device, as illustrated in Figure 16-22, detects the voltage difference between the pair of signal wires and amplifies this difference. Because the induced noise level should be the same in both conductors, the amplifier will ignore the noise. For example, assume a sensor is producing a 50-millivolt signal. This signal is applied to the input module, but induced noise is at a level of 5 volts. In this case the noise level is 100 times greater than the signal level. The induced noise level, however, is the same for both of the input conductors. Therefore, the differential amplifier ignores the 5-volt noise and amplifies only the voltage difference, which is 50 millivolts.

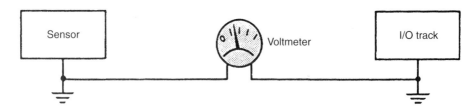

FIGURE 16-21 All grounds are not equal. (*Delmar/Cengage Learning*)

REVIEW QUESTIONS

All answers should be written in complete sentences, calculations should be shown in detail, and *Code* references should be cited when appropriate.

1. What industry first started using PLCs? _____

_____

_____

_____

2. Name the four basic sections of a PLC. _____

_____

_____

_____

3. In what section of the PLC is the actual circuit logic performed? _____

_____

_____

_____

4. What device separates the PLC from the outside circuits? _____

_____

_____

_____

5. If an I/O output module controls an ac voltage, what electronic device is used to actually control the load? _____

_____

_____

_____

6. What is opto-isolation? _____

_____

_____

_____

7. Why should signal wire runs be kept as short as possible? _____

_____

_____

_____

8. Why is shielded wire used for signal runs? _____

_____

_____

_____

9. What is the brute-force method of grounding? _____

_____

_____

_____

10. Explain the operation of a differential amplifier. _____

_____

_____

_____

Developing a Program for a PLC

OBJECTIVES

After studying this chapter, the student should be able to

- develop a program for a programmable logic controller using a schematic diagram.

- connect external devices to input and output terminals of the PLC.

Control circuits are generally drawn as standard schematic or ladder diagrams. These circuits are then converted into a logic diagram that can be loaded into the memory of a programmable logic controller. The circuit shown in Figure 17-1 will be converted for programming into a PLC. This circuit is used to control two well pumps. A housing development contains one pressure tank that supplies water to the development. There are two separate deep wells, however, that supply water to the tank. It is desired that the wells be used equally. The circuit in Figure 17-1 will cause the pumps to alternate running each time the pressure switch closes. A selector switch can be set to any of three operating modes. In the auto mode, the circuit will operate automatically and permit the pumps to alternate running each time the pressure switch closes. The selector switch can also be set to permit only one of the pumps to operate each time the pressure switch closes in the event that one pump fails. An on–off switch can be used to stop all operation of the circuit.

Before a program can be developed from a ladder diagram, it is first necessary to determine the number of input and output devices. In the circuit shown in Figure 17-1, there are actually three input devices: the on–off switch, the pressure switch, and the selector switch. The selector switch, however, requires three separate inputs. Therefore, there will be five inputs to the PLC. Only two outputs are required for motor starter coils 1M and 2M. Coils TR and CR are internal relays that exist only in the logic of the PLC.

The first step in developing a program is to assign the external input and output devices to specific inputs and outputs. In this example, it is assumed that the PLC to be used has sixteen inputs and eight outputs.

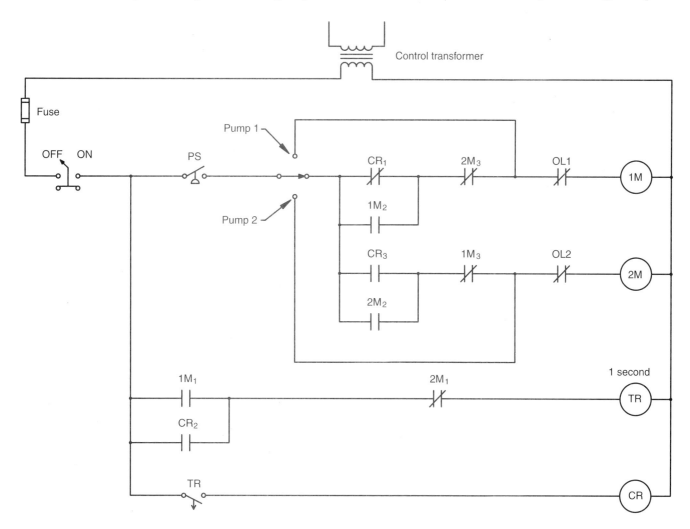

FIGURE 17-1 Circuit used to alternate the operation of two well pumps. (*Delmar/Cengage Learning*)

TABLE 17-1

Coil numbers associated with particular internal relays.

Inputs	1–16
Outputs	17–24
Internal Relays	100–175
Timers	200–225

Table 17-1 lists the numbers associated with inputs, outputs, internal relays, and timers. The table indicates that terminals 1 through 16 are inputs and terminals 17 through 24 are outputs. This PLC can have as many as seventy-five internal relays. Internal relay coil numbers will range from 100 through 175. There can be a total of twenty-five timers programmed into this controller. Coils numbered 200 through 225 are used for timers. It will also be assumed that the internal clock that controls the operation of timers operates in 0.1-second intervals. Therefore, it will be necessary to program a value of 10 to produce the 1-second time delay for timer TR, as indicated on the schematic.

ASSIGNING INPUTS AND OUTPUTS

In this example, the on–off switch is assigned to input 1, the pressure switch is assigned to input 2, the auto terminal of the selector switch is assigned to input 4, the pump 1 terminal is assigned to input 3, and the pump 2 terminal is assigned to input 5, Figure 17-2. Note that one side of each input device has been connected to the hot or ungrounded power terminal. The other side of each input device is connected to the appropriate input terminal. PLCs do not provide power to the input or output terminals. The inputs and outputs must have power provided to them. The common (C) input terminal is connected to the neutral or grounded power conductor.

Motor starter coil 1M is connected to output terminal 17 and motor starter coil 2M is connected to output terminal 18. Also note that the normally closed overload contacts for starters 1M and 2M are connected in series with the appropriate starter. It is common practice by many industries to leave the overload contacts hard wired to the starter coil to ensure that the starter will de-energize in the event of an overload.

Some manufacturers of motor control equipment provide a second overload contact that is normally open. This contact can be used as an input to the PLC and placed in the logic of the circuit. In this example, however, it is assumed that the overload relay contains a single normally closed contact that will remain hard wired to the coil. The common output terminal is connected to the hot or ungrounded power conductor, and the other side of each coil is connected to the neutral or grounded power conductor.

CONVERTING THE SCHEMATIC

The next step is to change the control schematic or ladder diagram into a logic diagram that can be loaded into the PLC. There are some basic rules that should be followed when making this conversion:

- Each line of logic must end with a coil.
- Any contact labeled the same number as a coil is controlled by that coil.
- Each relay can have an infinite number of contacts and they can be assigned as normally open or normally closed.
- Any coil assigned the same number as an output will control that output.
- Any contact assigned the same number as an input is controlled by that input.
- The PLC assumes inputs to be low (no power applied) when the program is loaded into memory. When power is applied to an input, it will cause the contact assigned to that input to change state. A normally open contact will close and a normally closed contact will open.
- Any number of contacts can be assigned to the same input.

In the schematic shown in Figure 17-1, contacts controlled by relay CR are used throughout the circuit. It will be assumed that this relay will be assigned coil number 100. It will also be assumed that timer coil TR will be assigned coil number 200. The simplest way to convert a ladder diagram into a logic circuit is to make the changes in stages. The first step will be to draw a logic diagram that will control the operation of motor starter coil 1M. Because coil

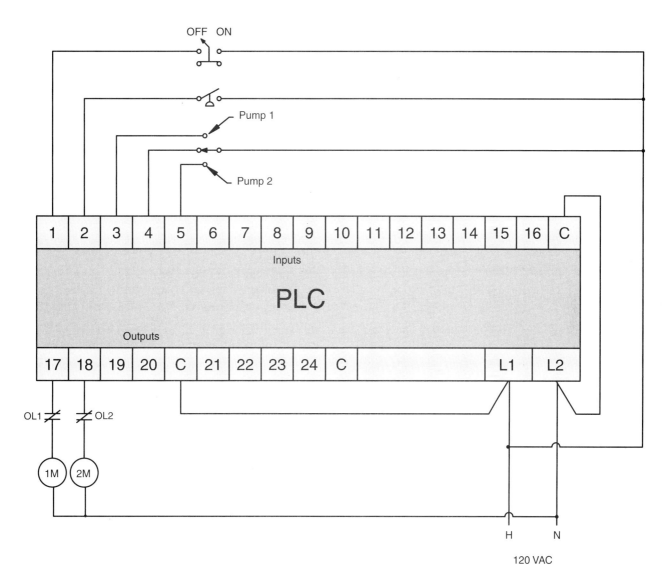

FIGURE 17-2 Input and output devices are connected to the proper terminals. (*Delmar/Cengage Learning*)

1M is connected to output 17, coil number 17 will be used for coil 1M. The circuit shown in Figure 17-3 will fulfill the first basic step of the logic. Note that the on–off switch is connected to input 1. Therefore, contact 1 will be controlled by the on–off switch. When the switch is turned on, power will be provided to input terminal 1 and the normally open contact labeled 1 will close. The pressure switch is connected to input 2 and the auto terminal of the selector switch is connected to input 4. Because the control relay is to be assigned coil number 100, the CR_1 contact is labeled 100. The normally

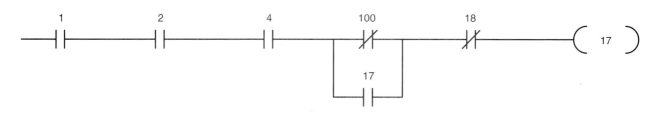

FIGURE 17-3 The first line of logic controls the coil of 1M starter. (*Delmar/Cengage Learning*)

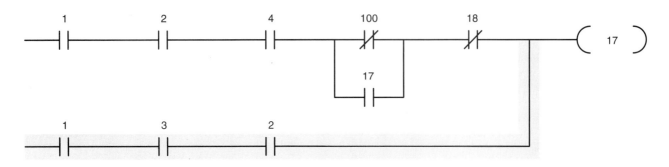

FIGURE 17-4 The pump 1 bypass circuit is added. (*Delmar/Cengage Learning*)

closed $2M_3$ contact is labeled 18 because 2M motor starter coil is connected to output terminal 18. Note that the normally closed overload contact (OL1) is not shown in the logic diagram because it is hard wired to the starter coil. The $1M_2$ contact connected in parallel with the normally closed CR_1 contact is labeled 17 because output 17 controls the operation of 1M starter coil.

The next step will to be to add the logic that permits the pump 1 terminal of the selector switch to bypass the automatic control circuit. The pump 1 terminal of the selector switch is connected to input terminal 3. The pressure switch must still control the operation of the pump if the collector switch is set in the pump 1 position. Therefore, another contact labeled 2 will be connected in series with contact 3. Another consideration is that the on–off switch controls power to the rest of the

circuit. There are several ways of accomplishing this logic, depending on the type of PLC used, but in this example it will be accomplished by inserting a normally open contact controlled by input 1 in series with each line of logic. One advantage of PLCs is that any input can be assigned any number of contacts. This amendment to the circuit is shown in Figure 17-4.

The logic to control motor starter coil 2M is developed in the same way as the logic for controlling starter coil 1M. The new logic is shown in Figure 17-5. The pump 2 terminal of the selector switch is connected to input 5 of the PLC. The bypass control for pump 2 is added to the circuit in Figure 17-6.

In the circuit shown in Figure 17-1, the coil of CR relay is controlled by an off-delay timer with a delay of 1 second. This timer is used to ensure that

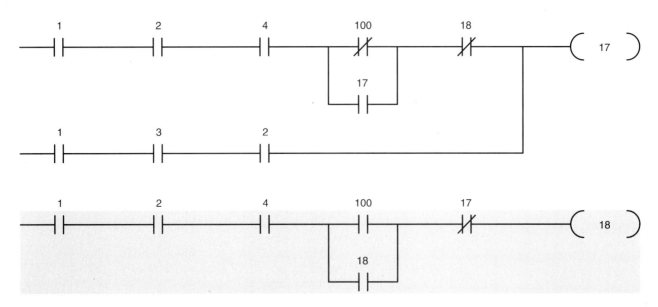

FIGURE 17-5 The logic for control of starter coil 2M is added to the circuit. (*Delmar/Cengage Learning*)

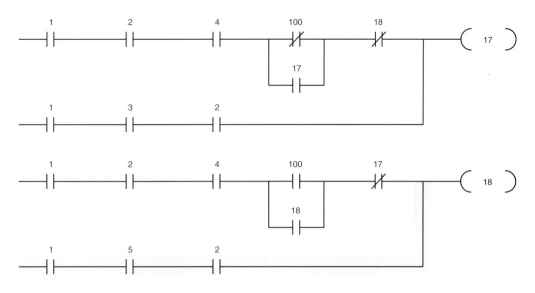

FIGURE 17-6   The pump 2 bypass circuit is added. (*Delmar/Cengage Learning*)

there is no problem with a contact race. A contact race occurs when it is possible for one contact to open before another closes or for one contact to close before another opens. It is assumed that the PLC in this example contains on-delay timers only.

It is therefore necessary to change the logic to make an off-delay timer. This was discussed in Chapter 16. To make this change, a second control relay labeled 101 will be required. The complete circuit is shown in Figure 17-7.

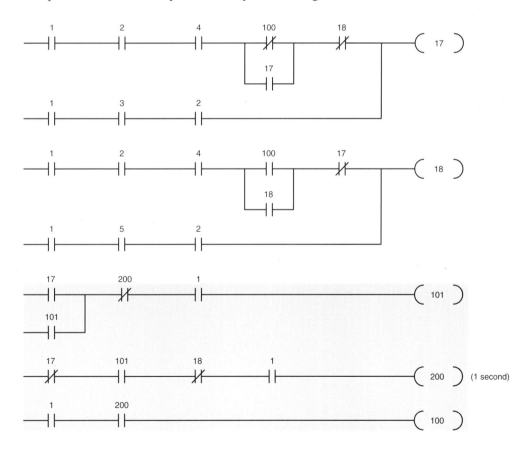

FIGURE 17-7   The complete logic circuit. (*Delmar/Cengage Learning*)

All answers should be written in complete sentences, calculations should be shown in detail, and *Code* references should be cited when appropriate.

1. In the circuit discussed in this chapter, to which input terminal is the pressure switch connected? _____

   _____

2. What coil numbers can be used as internal relays in the programmable logic controller discussed in this chapter? _____

   _____

3. Which output of the PLC discussed in this chapter controls the operation of motor starter coil 2M? _____

   _____

4. Refer to the circuit shown in Figure 17-1. Is the pressure switch normally open, normally closed, normally open held closed, or normally closed held open? _____

   _____

5. Does the PLC supply power to operate the devices connected to the output terminals?

   _____

   _____

6. The circuit shown in Figure 17-1 contains four coils: 1M, 2M, CR, and TR. Why are there only two coils connected to the output terminals of the PLC? _____

   _____

   _____

7. In this example, the normally closed overload contacts were not included in the logic diagram because they were hard wired to the motor starter coils. If these normally closed overload contacts had been connected to inputs 6 and 7 of the PLC, would they have been programmed as normally open or normally closed? Explain your answer. _____

   _____

   _____

   _____

   _____

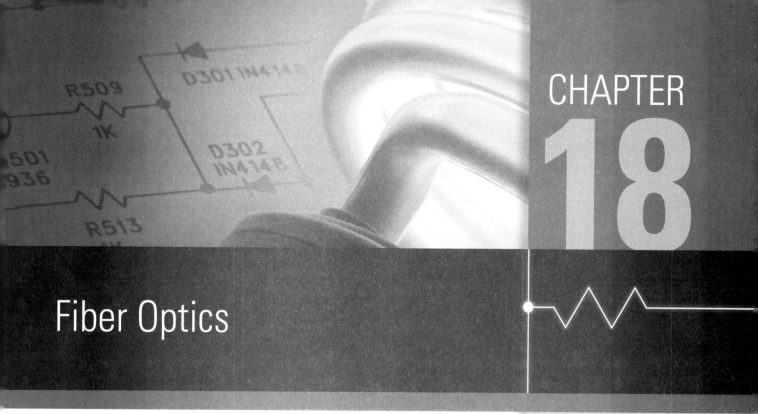

Fiber Optics

OBJECTIVES

After studying this chapter, the student should be able to

- discuss advantages of fiber optic cable over copper conductors.
- discuss the construction of fiber optic cable.
- discuss refraction.
- explain how light is transmitted through fiber optic cable.
- discuss fiber optic transmitters.
- discuss fiber optic receivers.
- list different types of fiber optical cable connectors.
- discuss concerns in making fiber optic connections.

⎍⎍⎍ FIBER OPTICS

Fiber optic cable is becoming increasingly popular for data transmission in industrial environments. Fiber optic cable has several advantages over copper wire for transmission of data. Copper wire is very susceptible to electromagnetic interference caused by electrical devices that draw large amounts of current, such as motors, transformers, and variable frequency drives. Fiber optic cable is totally immune to electromagnetic interference.

Also, the data transmission rate for fiber optic cable is much higher than for copper. Twisted-pair copper cable is generally limited to a data transmission rate of about 1 mbps (million bits per second). Coaxial cable can carry about 10 mbps. Some special coaxial cable can handle 400 mbps. Fiber optic cable can typically handle 8000 mbps, and laboratory tests have shown that rates as high as 200,000 mbps are possible.

Due to the high frequency of light, fiber optic cable has a very wide bandwidth as compared with copper wire. The bandwidth of fiber optic cable is about a million times that of copper wire. Fiber optic cables are much smaller and lighter in weight than copper cables. A single fiber is approximately 0.001 in. (1 micrometer [formerly "micron"]) in diameter and can carry five times more information than a telephone cable containing 900 pairs of 22 AWG twisted conductors. Single and duplex fiber optic cables are shown in Figure 18-1.

Cable Construction

Fiber optic cables are composed of three sections: the core, the cladding, and the sheath, Figure 18-2. The core is composed of either glass or plastic. Glass has a higher bit rate of transmission or bandwidth than plastic, and it has less line loss than plastic. Glass fibers are also able to withstand higher temperatures and are less affected by corrosive atmospheres and environments. Plastic core fibers are more flexible and can be bent to a tighter radius than glass. Plastic fibers are stronger and can be cut, spliced, and terminated with less difficulty than glass.

The cladding or clad surrounds the core and is made of glass or plastic also. The clad serves two basic functions. It protects the core from the surrounding environment and it increases the size and strength of the cable itself. Increasing the size of the cable makes it easier to handle. The core and cladding are considered the fiber optic.

The sheath is a polyurethane jacket that surrounds the cable. The sheath protects the fiber optics from the environment. Fiber optic cables may be packaged as a single fiber, fiber pairs, or several thousand fibers.

How Fiber Optic Cable Works

Light travels in a straight line. Fiber optic cable, however, makes it possible to bend light around corners and conduct it to any desired location,

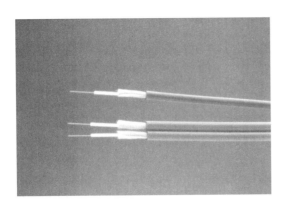

FIGURE 18-1 Single and duplex fiber optic cables. (*Delmar/Cengage Learning*)

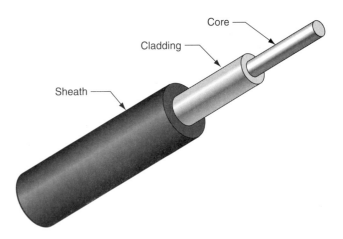

FIGURE 18-2 Fiber optic cable. (*Delmar/Cengage Learning*)

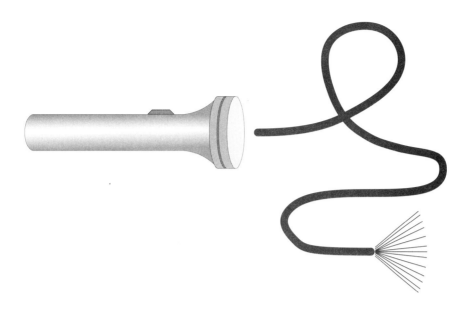

FIGURE 18-3   Fiber optic cable permits light to be bent. (*Delmar/Cengage Learning*)

Figure 18-3. The reason that light can travel through an optical fiber is because of refraction. Imagine that you are standing on the shore of a clear mountain lake on a calm, windless day. If you looked out over the surface of the lake, you would probably see the sun, clouds, and trees reflected on the surface of the water. If you looked directly at the water at your feet, you would no longer see reflections of the clouds or trees, but you would see down into the water. This is an example of refraction instead of reflection. The angle at which you stopped seeing the reflection of clouds and trees and started seeing down into the water is called the critical angle or acceptance angle. The critical angle occurs because air and water have a different optical property. The optical property is the speed at which light can travel through a material. Optical property is generally expressed as a term called index of refraction (IR), or η. The index of refraction is the speed at which light travels through a material. It is determined by comparing the speed of light traveling through vacuum to the speed of light traveling through a particular material.

$$\eta = \frac{\text{Speed of light in vacuum}}{\text{Speed of light in material}}$$

In glass optical fibers, the index of refraction is approximately 1.46 to 1.51.

When an optical fiber cable is connected to a light source, Figure 18-4, light strikes the cable at many

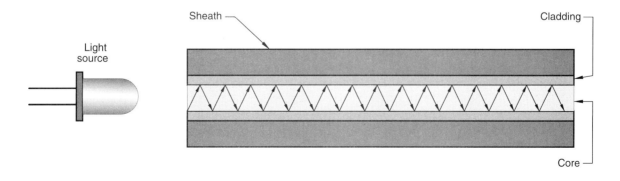

FIGURE 18-4   Light propagates through the core as a result of refraction. (*Delmar/Cengage Learning*)

different angles. Some photons strike at an angle that cannot be refracted and are lost through the cladding and absorbed by the sheath. Photons that can be refracted bounce down the core to the receiving device. This bouncing action of the photons causes a condition known as modal dispersion. Because photons enter the cable at different angles, some bounce more times than others before they reach the end of the cable, causing them to arrive later than photons that bounce fewer times. This causes a variance in the phase of the light reaching the source. Modal dispersion can be greatly reduced by using fiber cables called single-mode cables. Single-mode cables have a diameter of 1 to 2 micrometers. The cladding also affects modal dispersion. If the cladding thickness is kept to within three times the wavelength of the light, modal dispersion is eliminated.

Another type of cable that is much larger than single-mode cable is multimode cable. Multimode cable ranges in thickness from about 5 to 1000 micrometers. Multimode cable can cause severe modal dispersion in long lengths of several thousand feet. For short runs, however, it is generally preferred because it is larger in size and easier to work with than single-mode cable. Multimode cable is also less expensive than single-mode cable, and for short runs the modal dispersion generally is negligible.

Another type of multimode cable called graded cable has a core made of concentric rings. The rings on the outside have a lower density than that of the rings beneath. This produces a sharper angle of refraction for the outer rings. This arrangement helps to eliminate modal dispersion.

Cable Losses

Fiber optic cables do suffer some losses or attenuation. No fiber optic cable is perfect, and some amount of light does escape through the cladding and is absorbed by the sheath. The greatest losses generally occur when cable is terminated or spliced. The ends of fiber optic cable must be clean and free of scratches, nicks, or uneven strands. It is generally recommended that the ends of fiber optic cable be polished when they are terminated. A special hot knife cutting tool is available for cutting fiber optic cable. A cable with multiple fiber optic cables is shown in Figure 18-5.

Transmitters

Several factors should be considered in selecting a transmitter or light source for a fiber optic system. One is the *wavelength* of the light source. Many fiber optic cables specify a range of wavelengths for best performance. The wavelength can be measured by the color of the emitted light.

Another consideration is the *spectral width*. Spectral width is a measure of the range of colors that are emitted by the light source. The spectral width affects the color distortion that occurs in the optic fiber.

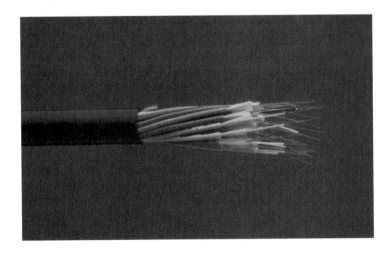

FIGURE 18-5 Multiple fiber optic cables. (*Courtesy of Optical Cable Corporation*)

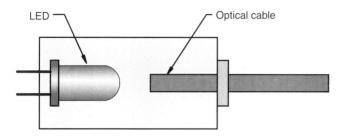

FIGURE 18-6 A transmitter is generally an LED or a laser diode. (*Delmar/Cengage Learning*)

The numerical aperture (NA) is a measure of the angle at which light is emitted from the source. If the NA of the source is too wide, it can overfill the NA of the optical fiber. If the NA of the source is too small, it will underfill the fiber. A low-NA light source helps reduce losses in both the optical fiber and at connection points.

Transmitter light sources are generally light-emitting diode (LED) or laser, Figure 18-6. LEDs are relatively inexpensive, operate with low power, and have a wide spectral width. They are generally used for short distances of about 7 kilometers, or 4.3 miles. LEDs have relatively low bandwidths of about 200 MHz or less. They can be used for data bit transmission rates of about 200 mbps or less. LEDs have wavelengths that range from about 850 to 1300 nanometers.

Laser diodes are expensive, require a large amount of operating power, and have a narrow spectral width. They can be used for extremely long distance transmission and can handle very high rates of data transmission. Laser diodes are generally used for telephone and cable television applications. Laser diodes operate at a wavelength of about 1300 nanometers.

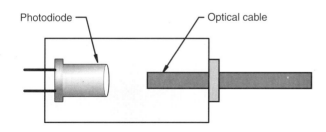

FIGURE 18-7 The receiver receives light from the optical cable. (*Delmar/Cengage Learning*)

Receivers

Receivers convert the light input signal into an electrical signal that can be used by the programmable controller or other devices. Receiver units generally consist of a photodiode, Figure 18-7. Photodiodes are preferred over other types of photo detection devices because of their speed of operation.

Transceivers

Transceivers house both a transmitter and receiver in the same package. Transceivers are often used as photodetection devices. Assume that half of the fiber optic fibers in a cable are connected to the transmitter and the other half are connected to the receiver. If a shiny object, such as a can on an assembly line, should pass in front of the cable, the light supplied by the transmitter would be reflected off the can back to the receiver, Figure 18-8. The output of the receiver could be connected to the input of a programmable controller that causes a counter to step each time a can is detected.

Another device that contains both a transmitter and receiver is call a *repeater*. A repeater is used to boost the signal when fiber optic cable is run long distances. Repeaters not only amplify the signal but also can reshape digital signals back to their original form. This ability of the repeater to reshape a digital signal back to its original form is one of the great advantages of digital-type signals over analog. The repeater "knows" what the original digital signal looked like, but it does not "know" what an original analog signal looked like. A great disadvantage of analog-type signals is that any distortion of the original signal or noise is amplified.

FIBER OPTIC CONNECTORS

One of the greatest problems with fiber optic systems is poor connections. There are two conditions that generally account for poor connections: bad alignment between cables, or devices and air gaps between cables or devices. An air gap changes the index of refraction, causing Fresnel reflection at the point of connection. An air gap can produce an

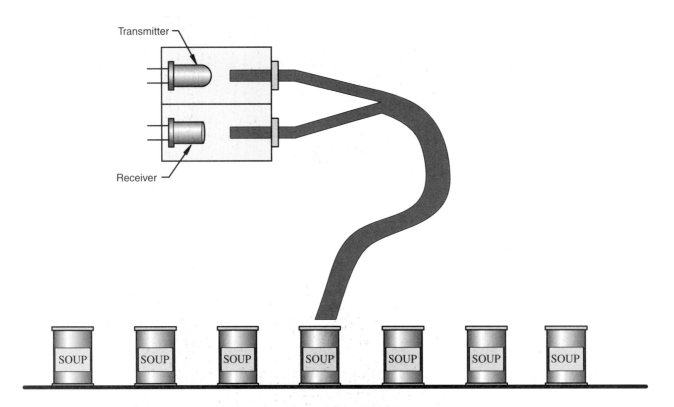

FIGURE 18-8 A transceiver contains both a transmitter and a receiver. (*Delmar/Cengage Learning*)

optical resonant cavity at the point of connection. This resonant cavity causes light to be reflected back to the transmitter, where it is bounced back again to be reflected back to the receiver.

If the fiber optic cable is not aligned correctly, part of the light signal will not be transmitted between the two cables and the device to which it is connected. Losses due to poor connection can be substantial. Losses are measured in decibels (dB).

$$dB = 10 \log_{10} \frac{\text{power out}}{\text{power in}}$$

Coupling Devices

There are three basic types of fiber optic connectors: the threaded, the bayonet, and the push-pull, as illustrated in Figure 18-9. The threaded type, shown in Figure 18-10, was one of the earliest to be introduced. The use of threaded connectors has decreased because of poor performance. One problem with this type of connector is how tight to make the connection. Also, there is nothing to control rotational alignment. If a threaded connector is disconnected and then reconnected, the two cables will probably not be in the same alignment. Threaded connectors typically have losses of 0.6 to 0.8 dB.

In the mid-1980s, the bayonet connector, shown in Figure 18-11, was introduced. This connector solved some of the basic problems of the threaded connector. The twist-lock action provides uniform tightness each time the connector is used, and rotational alignment is more constant. This connector typically has a loss of about 0.5 dB. Bayonet connectors are still widely used throughout industry. The push pull–type connector, Figure 18-12, has become very popular for connecting fiber optic cables because it offers excellent alignment and has less back reflection than other types of connectors. Push-pull connectors typically have a loss of about 0.2 dB.

Electrical connection devices, such as extension cords, generally have a male connector on one end and a female on the other. Fiber optic cables generally employ male connectors on both ends. Devices that they connect to have a female connection. When two fiber optic cables are to be joined together, a coupler with two female ends is employed, as illustrated in Figure 18-13.

FIGURE 18-9   Different types of optic connectors. (*Courtesy Tyco Electronics*)

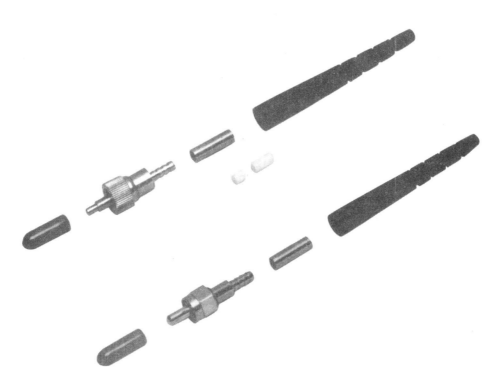

FIGURE 18-10   Threaded optic connectors. (*Courtesy Tyco Electronics*)

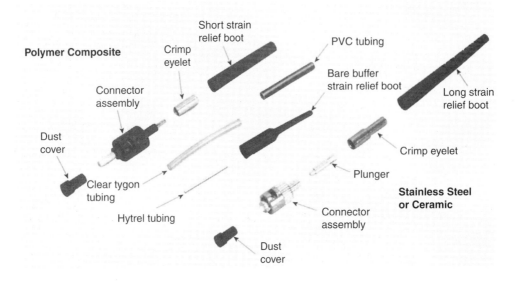

FIGURE 18-11 Bayonet-type optic connectors. (*Courtesy Tyco Electronics*)

Making Fiber Optic Connections

Attaching a connector to a fiber optic cable is different from connecting a male plug to an electrical extension cord. Fiber optic connections must be precise. The connections are epoxied and polished. Special crimp tools and dies are employed depending on the size of the cable. A microscope is generally used to examine the connection for possible problems. A kit containing the tools necessary for making fiber optic connections is shown in

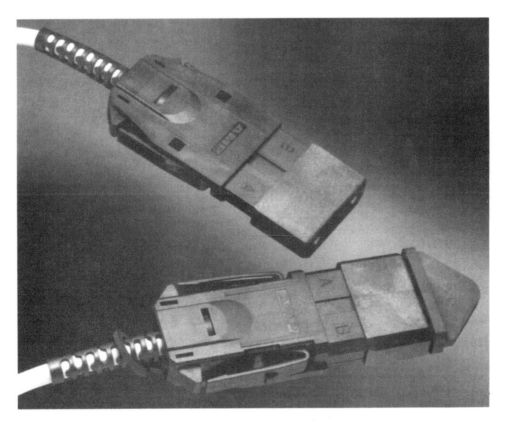

FIGURE 18-12 Push pull–type fiber optic connectors. (*Courtesy Tyco Electronics*)

FIGURE 18-13 Adapter for connecting two push-pull optic connectors. (*Courtesy Tyco Electronics*)

Figure 18-14. This kit contains a crimp tool and crimping dies, polishing bushings, polishing film, cable and fiber strippers, microscope with tripod, and an epoxy curing oven.

FIBER OPTIC LIGHTING

Fiber optic lighting has several advantages over conventional lighting.

- It can be employed in swimming pools and fountains without the shock hazard and electrical requirements of waterproof luminaires.

- It can be located in outdoor areas for accent lighting of buildings and walkways.

- It is supplied by a single light source; thus, the luminaires produce no heat and have no electrical connections, and the lamps never need replacement.

- It is ideal for applications in museums to accent valuable works of art because it does not produce heat or ultraviolet radiation, which is a major cause of color fading.

- The individual light sources do not produce an electromagnetic field that could interfere with sensitive electrical equipment. The only electrical connection is at the illuminator; thus, individual light sources do not produce an electromagnetic field.

Fiber optic lighting has a significant limitation: It cannot provide the amount of illumination that can be obtained with conventional light sources. Advances are being made in this technology, and in the near future it is likely that fiber optic cables will deliver the quantity of light necessary to compete with the common luminaire. Fiber optic cables could then be employed to provide illumination in hazardous locations without expensive explosion-proof luminaires.

Fiber optic lighting systems are composed of three primary sections: the illuminator, the cable, and the end fixtures, Figure 18-15. The illuminator is the only part of the system requiring electrical connections, permitting it to be located where it is convenient for maintenance and lamp (bulb) replacement.

FIGURE 18-14 Tool kit for making fiber optic connections. (*Courtesy Tyco Electronics*)

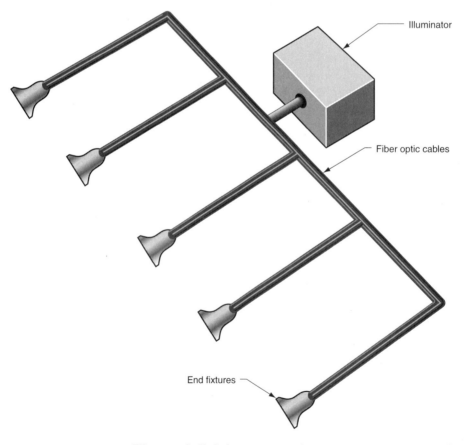

Illuminator

Fiber optic cables

End fixtures

FIGURE 18-15 Fiber optic lighting system. (*Delmar/Cengage Learning*)

All answers should be written in complete sentences, calculations should be shown in detail, and *Code* references should be cited when appropriate.

1. Name at least two advantages of fiber optic cable over copper wires. _____

   _____

   _____

2. Which type of fiber optic cable has a higher bit rate of transmission, glass or plastic?

   _____

3. Which type of fiber optic cable is more flexible and can be bent to a tighter radius, glass or plastic? _____

   _____

4. What is the index of refraction? _____

   _____

   _____

5. Name two devices that are generally used to transmit light in a fiber. _____

   _____

   _____

6. Name three types of coupling devices used to connect fiber optic cables. _____

   _____

   _____

7. Which fiber optic connector exhibits the least amount of loss? _____

   _____

   _____

8. Does fiber optic lighting transmit ultraviolet (UV) radiation? _____

   _____

   _____

9. What is the main disadvantage of fiber optic lighting systems? _____

   _____

   _____

10. What are the three sections that make up a fiber optic lighting system? _____

   _____

   _____

CHAPTER 19

Hazardous Locations

OBJECTIVES

After studying this chapter, the student should be able to

- discuss the different classes, divisions, and groups of hazardous locations.
- describe intrinsically safe circuits.
- discuss vertical and horizontal seals.
- describe the difference between explosionproof and enclosed and gasketed luminaires.
- discuss the installation requirements of pendant luminaires.
- list the conditions for the use of flexible cord in a hazardous location.

Hazardous locations are areas that exhibit a high risk of fire or explosion due to elements in the surrounding atmosphere or vicinity. The *NEC* divides hazardous locations into three classes. Class I locations are areas in which there are or may be high concentrations of flammable or explosive gases or vapors. Class II locations contain flammable or explosive dusts, and Class III locations are areas that contain combustible fibers.

Classes I and II are subdivided into several groups, each of which contain hazardous materials having similar properties and characteristics. The chart in Table 19-1 lists groups and typical atmospheres or hazards in each. The chart also indicates the typical

TABLE 19-1

Classification of hazardous atmospheres.

Group A	Group D
Acetylene (581°F, 420°C)	Acetone (869°F, 465°C)
	Acrylonitrile (898°F, 481°C)
Group B	Ammonia (928°F, 498°C)
Butadiene (788°F, 420°C)	Benzene (928°F, 498°C)
Ethylene oxide (804°F, 429°C)	Butane (550°F, 288°C)
Hydrogen (968°F, 520°C)	1-Butanol (650°F, 343°C)
Manufactured gases containing more than 30% hydrogen by volume	2-Butanol (781°F, 405°C)
Propylene oxide (840°F, 449°C)	N-butyl acetate (790°F, 421°C)
	Ethane (882°F, 472°C)
Group C	Ethanol (685°F, 363°C)
Acetadehyde (347°F, 175°C)	Ethyl dichloride (800°F, 427°C)
Cyclopropane (938°F, 503°C)	Gasoline (536 to 880°F, 280 to 471°C)
Diethyl ether (320°F, 160°C)	Heptane (399°F, 204°C)
Ethylene (842°F, 450°C)	Hexane (437°F, 225°C)
Unsymmetrical dimethyl hydrazine (480°F, 249°C)	Isoamyl alcohol (662°F, 350°C)
	Iosprene (428°F, 220°C)
	Methane (999°F, 630°C)
	Methanol (725°F, 385°C)
Group E	Methyl ethyl ketone (759°F, 404°C)
Atmospheres containing combustible metallic dust regardless of resistivity such as aluminum or magnesium, and other combustible dust having a resistivity less than 100 ohm-centimeters	Methyl isobutyl ketone (840°F, 440°C)
	2-Methyl-1-propanol (780°F, 416°C)
	2-Methyl-2-propanol (892°F, 478°C)
	Naptha (550°F, 288°C)
	Octane (403°F, 206°C)
	Pentane (470°F, 243°C)
Group F	1-Pentanol (572°F, 300°C)
Atmospheres containing carbonaceous dusts such as carbon black, coal, charcoal, or coke having a resistivity between 100 and 10^8 ohm-centimeters	Propane (842°F, 450°C)
	1-Propanol (775°F, 413°C)
	2-Propanol (750°F, 399°C)
	Propylene (851°F, 455°C)
	Styrene (914°F, 490°C)
Group G	Toluene (896°F, 480°C)
Atmospheres containing combustible dusts having a resistivity of 10^8 ohm-centimeters or greater	Vinyl acetate (756°F, 402°C)
	Vinyl chloride (882°F, 472°C)
	Xyenes (867 to 984°F, 464 to 529°C)

ignition temperature of these hazardous materials. Groups A, B, C, and D are found in Class I locations. Groups E, F, and G are found in Class II locations. There are no group listings for Class III locations.

In addition, hazardous locations are divided into two divisions that depend on the likelihood of the hazard being present. Division 1 locations are areas considered to be hazardous at any or all times during normal operations. Division 2 locations are areas that could become hazardous through a foreseeable accident. For example, an area in which gasoline is manufactured would be considered Class I, Division 1. An area in which maintenance is done on trucks that transport the gasoline would be Class I, Division 2.

EQUIPMENT APPROVAL

The *NEC* states in *500.8(B)* that equipment shall be approved not only for the class location but also for the specific type of atmosphere that will be present. In addition, equipment located in a Class I location cannot have an exposed surface that operates at a temperature greater than the ignition temperature of the surrounding gas or vapor. Equipment located in a Class II location cannot have a surface temperature greater than that specified in *500.8(D)(2)*, and equipment located in a Class III location cannot have a surface temperature greater than that specified in *503.5*.

Group A

Group A is an atmosphere that contains acetylene.

Group B

Group B contains flammable gas or flammable liquid-produced vapor having either a maximum experimental safe gap (MESG) value less than or equal to 0.45 mm or a minimum ignition current ratio (MIC) less than or equal to 0.40 mm.

Group C

Group C contains flammable gas or flammable liquid-produced vapor having either a maximum experimental safe gap value greater than 0.45 mm and less than or equal to 0.75 mm, or a minimum ignition current ratio greater than 0.40 mm and less than or equal to 0.80 mm.

Group D

Group D contains flammable gas or flammable-produced vapor having either a MESG value greater than 0.75 mm or a minimum current ratio greater than 0.80 mm. Typical materials and atmosphere ignition temperatures are shown in Table 19-1 for different groups.

Group E

Group E contains combustible metal dusts, including aluminum, magnesium, and their commercial alloys, or other combustible dusts whose particle size, abrasives, and conductivity present similar hazards in the use of electrical equipment.

Group F

Group F contains combustible carbonaceous dusts that have more than 8 percent total entrapped volatiles or that have been sensitized by other materials so that they present an explosion hazard.

Group G

Group G contains combustible dusts not included in Groups E or F, including flour, grain, wood, plastic, and chemicals.

INTRINSICALLY SAFE CIRCUITS AND EQUIPMENT

As stated in *504.10(B)*, equipment and associated apparatus that have been identified as intrinsically safe are permitted in any hazardous location. The equipment must be approved for the type of atmosphere in which it is to be used. Intrinsically safe equipment and circuits operate at low power levels. These circuits and equipment operate at a low enough power that, even under overload or fault conditions, they do not contain enough electrical or thermal energy to cause ignition of the surrounding atmosphere. Abnormal conditions are considered to be accidental damage to field wiring, failure of equipment, accidental application of overvoltage, and misadjustment of equipment. Intrinsically safe circuits must be physically separated from all other circuits that are not so considered. Seals must be used to prevent the passage of gas or vapor as they are in higher-voltage systems. The installation of intrinsically safe systems is covered in *504*.

EQUIPMENT

The *NEC* states in *500.8(B)* that the equipment used in a hazardous location must be approved not only for that location but also for the particular type of atmosphere in which the equipment is used. As stated in *501.10(A)(1)*, rigid metal conduit, threaded steel intermediate metal conduit, or type MI cable with approved termination fittings is required in a Class I, Division 1 location. In general, the use of rigid metal conduit or threaded steel intermediate metal conduit is also required in Class II locations

(*502.10*) and in Class III locations (*503.10*). In a Class I, Division 2 location, type MI, MC, MV, or TC cable may be used with approved fittings. A type SNM cable with an approved termination fitting is shown in Figure 19-1.

Class I, Division 1 locations are considered to be the most hazardous. Equipment used in a Class I, Division 1 location must be explosionproof. Explosionproof equipment is designed to withstand an internal explosion without permitting hot gases or vapors to escape to the outside atmosphere. This is accomplished by forcing the escaping gas to travel across large, flat surfaces or through screw threads before it exits to the outside atmosphere, Figure 19-2. This cools the hot gas or vapor below the ignition point of the surrounding atmosphere.

As a general rule, equipment installed in a Class II location must be dust-ignitionproof. The definition of a dust-ignitionproof enclosure is given in *500.2*. A dusttight enclosure is constructed so that dust cannot enter the enclosure.

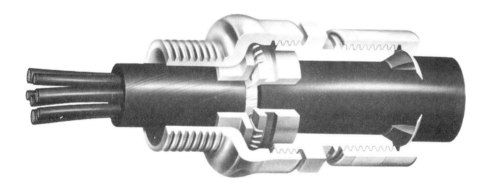

FIGURE 19-1 Type SNM cable with approved termination fittings: used in Class I, Groups A, B, C, D, Division 2; Class II, Group G, Division 2; and Class III. (*Courtesy Cooper Crouse-Hinds*)

FIGURE 19-2 Hot gas is cooled before leaving the enclosure. (*Delmar/Cengage Learning*)

Manufacturers of equipment intended for use in hazardous locations often design the equipment so that it can be used in more than one area. In Figure 19-3, several outlet boxes are shown. These outlet boxes can be used in Class I, Groups C and D; Class II, Groups E, F, and G; and Class III locations.

 **SEALS**

NEC 501.15 requires the use of conduit seals. Seals are used to minimize the passage of gas and to prevent the passage of flame through the conduit. In general, they are required within 18 in. (450 mm) of an explosionproof enclosure and in any conduit that exits from a more hazardous location to a less hazardous location, Figure 19-4. Seals are available in standard conduit sizes and can be installed vertically or horizontally. Seals designed to be installed in a vertical position are shown in Figure 19-5. Those designed to be installed in a horizontal position are shown in Figure 19-6.

When seals are installed, the conductors must be pulled through the seal and the system tested for shorts or grounds before the sealing compound is added. A cutaway view of a vertical seal is shown in

FIGURE 19-3 Explosionproof outlet boxes: used in Class I, Groups C, D; Class II, Groups E, F, G; and Class III. (*Courtesy Cooper Crouse-Hinds*)

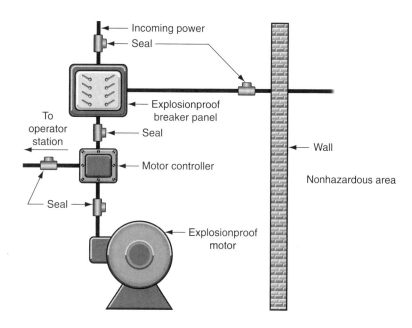

FIGURE 19-4   Seals are required in explosionproof installations. (*Delmar/Cengage Learning*)

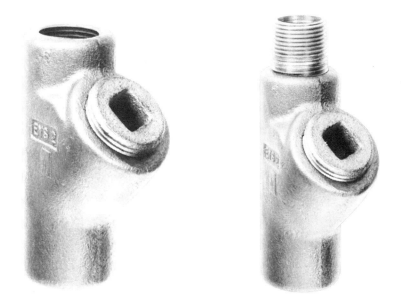

FIGURE 19-5   Seals designed to be installed in the vertical position: used in Class I, Groups A, B, C, D; and Class II, Groups E, F, G. (*Courtesy Cooper Crouse-Hinds*)

FIGURE 19-6   Seals designed to be installed in the horizontal position: Class I, Groups A, B, C, D; and Class II, Groups E, F, G. (*Courtesy Cooper Crouse-Hinds*)

Figure 19-7. To add the sealing compound, the large plug is removed. A fiber material is then packed around the inside bottom of the seal to form a dam. The dam prevents the sealing compound, before it becomes hard, from flowing down the conduit.

When horizontal seals are installed, two separate fiber dams must be used, as shown in Figure 19-8. Horizontal seals contain two separate plugs, a large one and a small one. When the seal is installed, the large plug is removed first to permit the installation

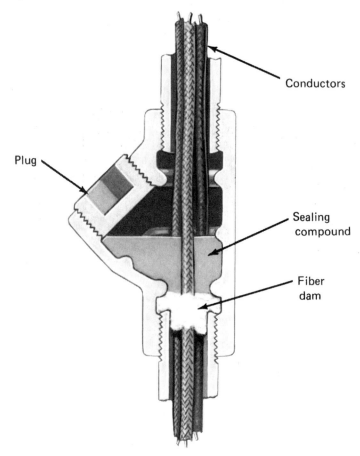

FIGURE 19-7   Cutaway view of a vertical seal. (*Courtesy Cooper Crouse-Hinds*)

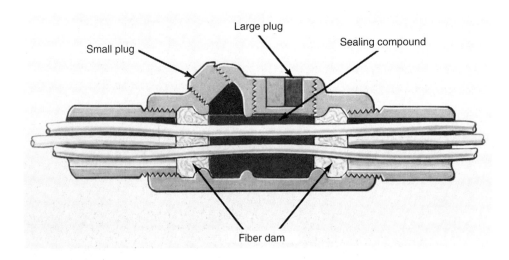

FIGURE 19-8   Two fiber dams must be used when installing horizontal seals.
(*Courtesy Cooper Crouse-Hinds*)

FIGURE 19-9 Vertical seals with drain plugs: used in Class I, Groups B, C, D; and Class II, Groups F, G. (*Courtesy Cooper Crouse-Hinds*)

of the fiber material. Once this has been accomplished, the plug is replaced and the liquid sealing compound is poured in through the opening provided by the smaller plug.

Condensation is sometimes a problem because it causes moisture to collect inside the conduit. To help prevent this problem, some vertical seals are designed with drain plugs, Figure 19-9. These seals are installed in low areas where long horizontal runs of conduit turn down. The seals are designed so that a hollow shaft extends through the sealing compound. The plugs can be removed periodically to drain moisture from the system. Some drain seal fittings contain *weep* holes to permit continuous draining.

CIRCUIT-BREAKER PANELBOARDS

When circuit breakers are installed in Class I locations, they must be explosionproof. The type of enclosure used is determined by the atmosphere in the area where the device is to be installed, the size of the breaker needed, and the number of breakers required. A single circuit breaker and enclosure is shown in Figure 19-10. This type of breaker can be obtained in 50-, 100-, and 225-ampere frame sizes.

FIGURE 19-10 Single circuit breaker in an explosionproof enclosure: used in Class I, Groups C, D; Class II, Groups E, F, G; and Class III. (*Courtesy Cooper Crouse-Hinds*)

A multiple circuit-breaker panelboard, suitable for Class II areas, is shown in Figure 19-11. This panelboard is considered to be dust-ignitionproof and can contain up to 24 single-pole breakers, 12 double-pole breakers, or eight 3-pole breakers. This panelboard, however, is not permitted in areas containing hazardous gas or vapors. A multiple

FIGURE 19-11 Dust-ignitionproof circuit-breaker panelboard: used in Class I, Groups C, D; Class II, Groups E, F, G; and Class III. (*Courtesy Cooper Crouse-Hinds*)

FIGURE 19-12 Explosionproof circuit-breaker panelboard: used in Class I, Groups C, D; Class II, Groups E, F, G; and Class III. (*Courtesy Cooper Crouse-Hinds*)

circuit-breaker panelboard that is permitted in Class I areas is shown in Figure 19-12.

LUMINAIRES

Luminaires used in hazardous locations can be obtained in many different types, styles, and sizes. The luminaires shown in Figure 19-13 are known as *enclosed* and *gasketed*. They are considered to be vaportight, but they are not explosionproof. The glass globe that covers the lamp is not designed to contain an internal explosion. For this reason, these luminaires are not permitted in a Class I, Division 1 location. They can be used in a Class I, Division 2 location, however.

An explosionproof luminaire is shown in Figure 19-14. This luminaire can be used in a Class I, Division 1 location. The glass globe is made of tempered glass, which can withstand an internal explosion. This type of luminaire can also be equipped with an inner globe of colored glass if desired. A

FIGURE 19-13 Enclosed and gasketed luminaires: used in Class I, Division 2. (*Courtesy Cooper Crouse-Hinds*)

FIGURE 19-14 Explosionproof incandescent luminaire. (*Courtesy Cooper Crouse-Hinds*)

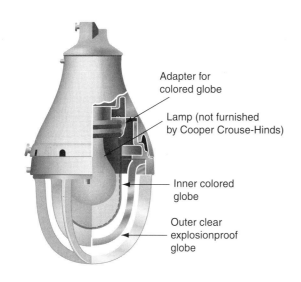

Adapter for colored globe

Lamp (not furnished by Cooper Crouse-Hinds)

Inner colored globe

Outer clear explosionproof globe

FIGURE 19-15 Cutaway view of an explosionproof incandescent luminaire. (*Courtesy Cooper Crouse-Hinds*)

cutaway view of this type of luminaire is shown in Figure 19-15.

Fluorescent-type luminaires can also be obtained for use in hazardous locations. The luminaire shown in Figure 19-16 can be used in Class I and Class II locations. These luminaires are also permitted in paint-spray areas and wet locations. They use heavy-duty glass tubes to cover the fluorescent lamps. The ballast is contained in an explosionproof enclosure.

T Ratings of Luminaires

Luminaires intended for use in hazardous locations have a temperature or "T" rating. The T rating indicates the maximum operating temperature of the luminaire. The operating temperature must be kept below the ignition temperature of the surrounding atmosphere. The chart in Table 19-2 lists the T number and the maximum operating temperature of a luminaire with that number. Luminaires listed for use in Division 1 locations are expected to be installed in areas where hazardous material is present at all times. For this reason, the T rating is established by measuring the temperature on the outer surface of the luminaire. Luminaires listed for use in Division 2 locations are not expected to be in the presence of hazardous material under normal conditions. The T rating of these luminaires is established by measuring the temperature at the hottest spot on the luminaire, which is the lamp itself.

Installation of Luminaires

NEC 501.130(A)(3) describes the method for installing pendant (hanging) luminaires in a hazardous location. In general, power is supplied through threaded rigid metal conduit or threaded steel intermediate conduit. If the stem is 12 in. (300 mm) long or less, no extra bracing is required. If the stem is longer than 12 in. (300 mm), however, lateral braces must be placed within 12 in. (300 mm) of the luminaire, or an explosionproof flexible coupling must be used within 12 in. (300 mm) of the junction box, Figure 19-17. An explosionproof flexible coupling used for this purpose is shown in Figure 19-18.

MOTOR CONTROLS

NEC 501.115 states that switches, circuit breakers, and make-and-break contacts of push buttons, relays, alarms, and so on, must be enclosed in an explosionproof enclosure when used in a Class I location. There is an exception, however, if the contacts are immersed in oil or hermetically sealed, or if the circuit does not contain sufficient energy to ignite the surrounding atmosphere. Explosionproof

FIGURE 19-16   Explosionproof fluorescent luminaire: used in Class I, Group C, D; Class II Groups E, F, G; paint-spray areas; and wet locations. (*Courtesy Cooper Crouse-Hinds*)

TABLE 19-2

T numbers and corresponding temperatures.

T Number	Temperature °F	Temperature °C	T Number	Temperature °F	Temperature °C
T1	842	450	T3A	356	180
T2	572	300	T3B	329	165
T2A	536	280	T3C	320	160
T2B	500	260	T4	275	135
T2C	446	230	T4A	248	120
T2D	419	215	T5	212	100
T3	392	200	T6	185	85

manual motor starters are shown in Figure 19-19. These starters can be used to control ac or dc motors. They contain heaters to provide running overcurrent protection for the motor. A single manual motor starter is shown in Figure 19-20. This particular starter can be used in atmospheres that contain hydrogen.

With use of explosionproof starters, it is often necessary to adjust the rating of the overload heater size. This is because the heater must be contained inside the explosionproof enclosure, which makes heat dissipation difficult. The chart in Figure 19-21 is used to select the proper overload heater size for the manual motor starter shown in Figure 19-20.

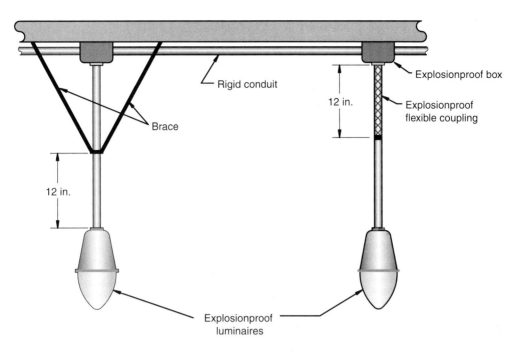

FIGURE 19-17 Installing pendant luminaires in a Class I, Division 1 location. (*Delmar/Cengage Learning*)

FIGURE 19-18 Explosionproof flexible coupling: used in Class I, Groups A, B, C, D; Class II, Groups E, F, G; and Class III. (*Courtesy Cooper Crouse-Hinds*)

FIGURE 19-19 Dual manual motor starter used in hazardous locations: used in Class I, Groups C, D; Class II, Groups E, F, G; and Class III. (*Courtesy Cooper Crouse-Hinds*)

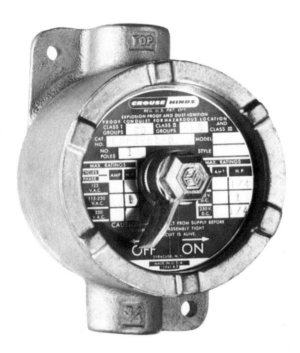

FIGURE 19-20 Single manual motor starter used in hazardous location: used in Class I, Groups B, C, D; Class II, Groups E, F, H; and Class III.
(*Courtesy Cooper Crouse-Hinds*)

FIGURE 19-22 Push buttons used in hazardous location: used in Class I, Groups C, D; Class II, Groups E, F, G; and Class III.
(*Courtesy Cooper Crouse-Hinds*)

HEATER TABLE			
FULL LOAD MOTOR CURRENT	**HEATER RATING**	**FULL LOAD MOTOR CURRENT**	**HEATER RATING**
.40– .43	.50	2.72– 2.95	3.40
.44– .48	.55	2.96– 3.27	3.70
.49– .53	.61	3.28– 3.59	4.10
.54– .58	.67	3.60– 3.99	4.50
.59– .64	.74	4.00– 4.39	5.00
.65– .71	.81	4.40– 4.79	5.50
.72– .78	.89	4.80– 5.26	6.00
.79– .87	.98	5.27– 5.83	6.60
.88– .95	1.10	5.84– 6.39	7.30
.96– 1.03	1.20	6.40– 7.03	8.00
1.04– 1.15	1.30	7.04– 7.74	8.80
1.16– 1.27	1.45	7.75– 8.46	9.70
1.28– 1.35	1.60	8.47– 9.35	10.60
1.36– 1.51	1.70	9.36– 10.30	11.70
1.52– 1.67	1.90	10.31– 11.35	12.90
1.68– 1.83	2.10	11.36– 12.47	14.20
1.84– 1.99	2.30	12.48– 13.67	15.60
2.00– 2.23	2.50	13.68– 15.12	17.10
2.24– 2.47	2.80	15.13– 16.00	18.60
2.48– 2.71	3.10		

Note: These heaters are for motors rated 40°C continuously. For motors rated 50° or 55°C, multiply full load motor current by 0.9 and use this value to select heaters. Symbol 0 (zero) may be used to indicate heater omitted.

FIGURE 19-21 Overload heater table for explosionproof manual starters.
(*Courtesy Cooper Crouse-Hinds*)

Most motor control circuits are either semiautomatic or automatic, and require the use of pilot devices such as the push buttons shown in Figure 19-22. All pilot devices, such as limit switches, flow switches, float switches, and so on, must be contained inside explosionproof enclosures when used in a Class I location. Semiautomatic and automatic controls require the use of magnetic starters, contactors, and relays. These contactors and/or starters also must be contained inside explosionproof enclosures. The motor starter shown in Figure 19-23 contains a circuit breaker and motor starter with overload relay.

FLEXIBLE CORDS AND RECEPTACLES

Flexible cords and attachment plugs are permitted in hazardous locations for the operation of portable lights or equipment, Figure 19-24 (*501.140, 502.140*, and *503.140*). When they are used, they must comply with the following conditions:

1. Be approved for extra-hard usage.

2. Contain a separate grounding conductor.

3. Be properly connected to terminals or supply conductors.

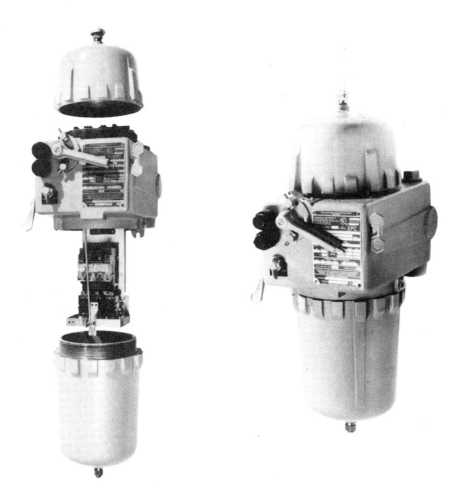

FIGURE 19-23   Explosionproof motor starter with circuit breaker: used in Class I, Groups C, D; Class II, Groups E, F, G; and Class III. (*Courtesy Cooper Crouse-Hinds*)

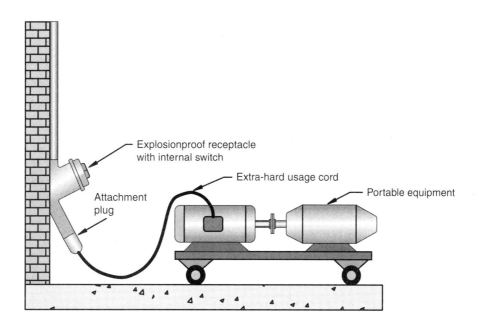

FIGURE 19-24   Portable equipment connected by a cord. (*Delmar/Cengage Learning*)

4. Be supported by clamps in such a manner that no tension is transmitted to the terminal connection.

5. Be supplied with seals to prevent the entrance of flammable vapors in a Class I location, flammable dust in a Class II location, or the entrance of fibers in a Class III location.

Plugs and receptacles must be approved for the location in which they are used (*501.145, 502.145, and 503.145*). Explosionproof receptacles are constructed with an internal switch that disconnects the power from the circuit before the attachment plug is removed. A cutaway view of this type of receptacle is shown in Figure 19-25. This plug and receptacle are constructed in such a manner that the attachment plug can be inserted into or removed from the receptacle only when the disconnect switch is in the off position. This is done to prevent the possibility of an arc being produced outside of the explosionproof enclosure when the plug is connected or disconnected. Once the plug has been inserted into the receptacle, the switch is turned on by twisting the plug in a clockwise direction. An explosionproof receptacle and attachment plug are shown in Figure 19-26.

FIGURE 19-25 Cutaway view of explosionproof receptacle. (*Courtesy Cooper Crouse-Hinds*)

FIGURE 19-26 Explosionproof attachment plug and receptacle: used in Class I, Groups B, C, D; Class II, Groups F, G; and Class III. (*Courtesy Cooper Crouse-Hinds*)

HAZARDOUS AREAS

Commercial Garages

NEC Article 511 deals with locations where motor vehicles are serviced or repaired. These vehicles include automobiles, trucks, buses, and tractors. In general, the floor area of a garage, up to a level of 18 in. (450 mm) above the floor, is considered to be a Class I, Division 2 location, Figure 19-27. The only exception to this is if a mechanical ventilation system provides sufficient airflow to produce at least four changes of air per hour. A pit area below the floor is considered to be a Class I, Division 1 location unless a mechanical ventilation system provides at least six changes of air per hour. In this instance, the pit area is considered to be Class I, Division 2.

Aircraft Hangars

NEC 513.1 defines an aircraft hangar as a location used to store or service aircraft that contains gasoline, jet fuel, or other flammable vapors. Areas used to store aircraft that have never contained flammable fuel or that have been drained and purged are not included in this definition.

In general, any pit located below floor level is considered to be a Class I, Division 1 location, Figure 19-28. Any area up to a level of 18 in. (450 mm) above the floor, any area within 5 ft (1.5 m) of the engine or engines, any area within 5 ft (1.5 m) of fuel tanks, and any area extending 5 ft (1.5 m) above the surface of the wings or engines is considered to be a Class I, Division 2 location.

Gasoline-Dispensing and Service Stations

NEC 514.2 describes the class and division for different areas where gasoline is stored or dispensed. In general, any space below grade level, within 10 ft (3 m) of an underground tank fill pipe, is considered to be Class I, Division 1, Figure 19-29. Any area within a radius of 10 ft (3 m) of an underground tank fill line and up to 18 in. (450 mm) above grade level is considered to be Class I, Division 2. Any area within 3 ft (900 mm), extending in all directions, of a gasoline tank vent is considered to be Class I, Division 1. Any space between 3 ft and 5 ft (900 mm and 1.5 m), extending in all directions, of a

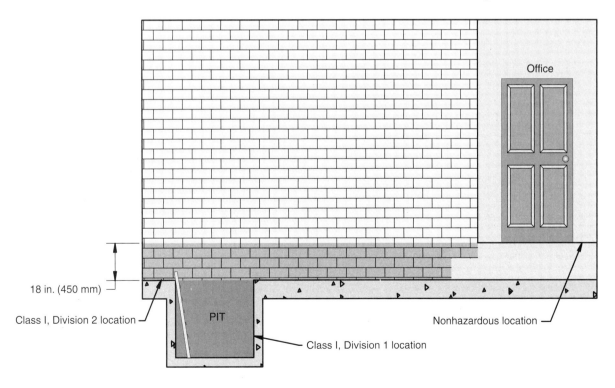

FIGURE 19-27 Commercial garage. (*Delmar/Cengage Learning*)

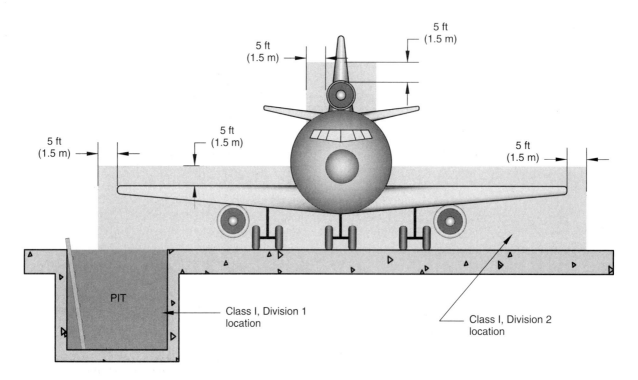

5 ft
(1.5 m)

5 ft
(1.5 m)

5 ft
(1.5 m)

5 ft
(1.5 m)

5 ft
(1.5 m)

PIT

Class I, Division 1
location

Class I, Division 2
location

FIGURE 19-28   Aircraft hangar. (*Delmar/Cengage Learning*)

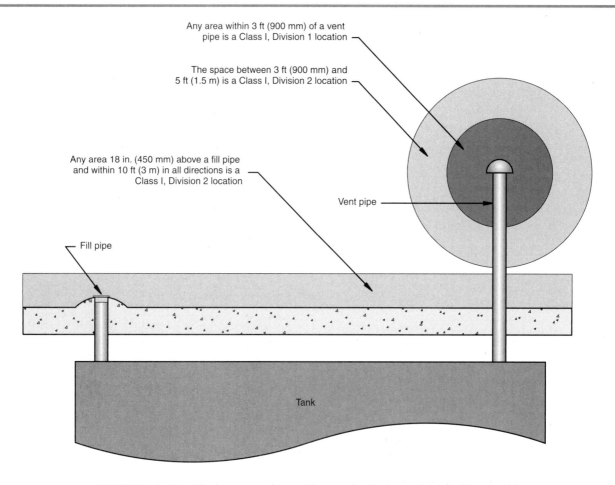

Any area within 3 ft (900 mm) of a vent
pipe is a Class I, Division 1 location

The space between 3 ft (900 mm) and
5 ft (1.5 m) is a Class I, Division 2 location

Any area 18 in. (450 mm) above a fill pipe
and within 10 ft (3 m) in all directions is a
Class I, Division 2 location

Vent pipe

Fill pipe

Tank

FIGURE 19-29   Underground gasoline tank. (*Delmar/Cengage Learning*)

gasoline tank vent is considered to be a Class I, Division 2 location.

Any space below grade level, within 20 ft of an outside gasoline-dispensing pump, is considered to be Class I, Division 1. Any space inside the pump enclosure, up to a height of 4 ft above the base, and any space within a nozzle boot are considered to be Class I, Division 1, Figure 19-30. Class I, Division 2 locations are as follows:

1. Any area 18 in. (450 mm) above grade level, within 20 ft (6 m) of an outside dispensing pump.

2. Any space, within the interior of the pump enclosure, isolated by a solid partition.

3. Any space within the pump enclosure higher than 4 ft (1.2 m).

4. Any space within 18 in. (450 mm), in all directions, of the pump enclosure.

Spray and Dipping Processes

NEC 516.3 gives the location classification for areas that contain hazardous concentrations of flammable vapor or dusts produced by spraying or dipping. Class I or Class II, Division 1 locations are as follows:

1. The interior of spray booths or rooms, Figure 19-31.

2. The interior of exhaust ducts.

3. Any area within the direct path of spray operation.

4. Any area within 5 ft (1.5 m) in any direction from the vapor source, extending from the source to the floor. The source is considered to be the surface of the dip tank, the wet surface of the drain board, or the surface of the dipped object, Figure 19-32.

5. Any pit within 25 ft (7.5 m) of the vapor source.

Class I or Class II, Division 2 locations are as follows:

1. For open spraying, the space extending 20 ft (6 m) horizontally and 10 ft (3 m) vertically from the Division 1 location, Figure 19-33.

2. For open-front spray booths with a closed top and closed sides, any space within 3 ft (900 mm)

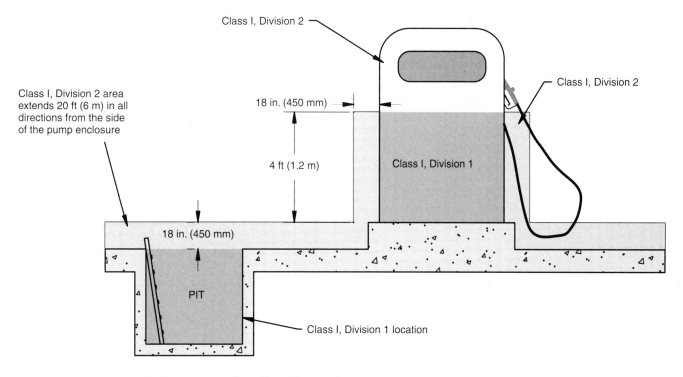

FIGURE 19-30 Gasoline-dispensing pump. (*Delmar/Cengage Learning*)

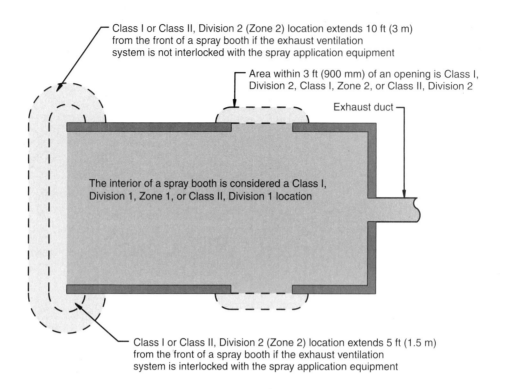

Class I or Class II, Division 2 (Zone 2) location extends 10 ft (3 m) from the front of a spray booth if the exhaust ventilation system is not interlocked with the spray application equipment

Area within 3 ft (900 mm) of an opening is Class I, Division 2, Class I, Zone 2, or Class II, Division 2

Exhaust duct

The interior of a spray booth is considered a Class I, Division 1, Zone 1, or Class II, Division 1 location

Class I or Class II, Division 2 (Zone 2) location extends 5 ft (1.5 m) from the front of a spray booth if the exhaust ventilation system is interlocked with the spray application equipment

FIGURE 19-31 Spray booth. (*Delmar/Cengage Learning*)

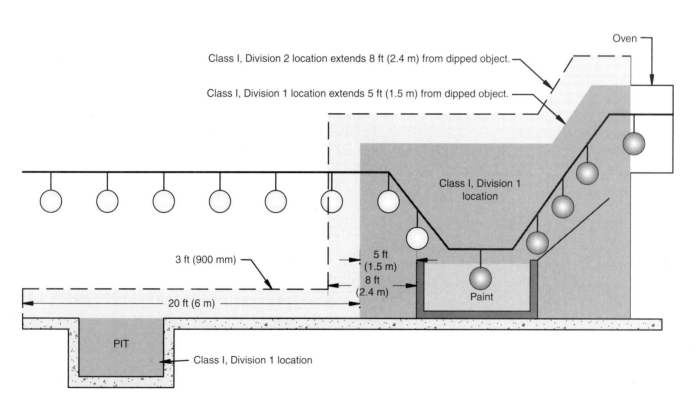

Class I, Division 2 location extends 8 ft (2.4 m) from dipped object.

Class I, Division 1 location extends 5 ft (1.5 m) from dipped object.

Oven

Class I, Division 1 location

3 ft (900 mm)

5 ft (1.5 m)

8 ft (2.4 m)

Paint

20 ft (6 m)

PIT

Class I, Division 1 location

FIGURE 19-32 Dip vat. (*Delmar/Cengage Learning*)

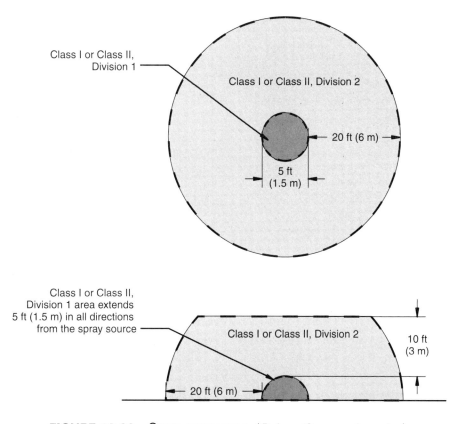

FIGURE 19-33   Open spray area. (*Delmar/Cengage Learning*)

in any direction of an opening other than the front. The amount of hazardous area extending around the front opening is determined by the type of ventilation system. If the ventilation system is interlocked with the spray equipment in such a manner that the spray equipment cannot be used when the ventilation system is not in operation, the Class I or Class II, Division 2 area is considered to be 5 ft in all directions from the front opening of the booth. If the ventilation system is not interlocked with the spray equipment, the hazardous area is considered to be 10 ft (3 m) in all directions from the front opening.

3. For spray booths with an open top, the hazardous area is considered to be 3 ft (900 mm) above the booth and 3 ft (900 mm) in any direction from an opening.

4. For completely enclosed spray booths or rooms, an area within 3 ft (900 mm) in any direction of any opening is considered to be Class I or Class II, Division 2.

5. For dip tanks and drain boards, any space between 5 ft and 8 ft (1.5 m and 2.5 m) of the vapor source, and any space between 5 ft and 25 ft (1.5 m and 7.5 m) horizontally of the vapor source, extending to a height of 3 ft (900 mm) above the floor, is considered to be a Class I or Class II, Division 2 location.

EXPLOSIONPROOF EQUIPMENT

Great care must be exercised when installing or maintaining explosionproof equipment. The improper installation or maintenance of this equipment can completely negate the system's integrity. The mating surfaces of explosionproof boxes are ground flat to provide a very close fit. A screwdriver gouge or deep scratch can provide the exit point for hot gases to escape and ignite the surrounding atmosphere, Figure 19-34.

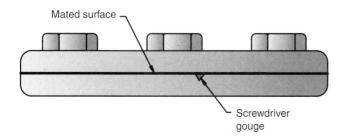

FIGURE 19-34 A screwdriver gouge or scratch can destroy the integrity of the enclosure by providing an exit point for hot gas. (*Delmar/Cengage Learning*)

The ground surfaces of explosionproof enclosures should be cleaned with solvent to remove dirt particles before the parts are bolted together. If it is necessary to remove a substance that cannot be cleaned with solvent, use a fine steel wool; *never* use sandpaper or a sharp scraping instrument such as a putty knife. All bolts must be in place and tight. If one bolt is missing, the enclosure may not be able to prevent ignition of the surrounding atmosphere in the event of an internal explosion.

Seals must be installed between explosion-proof enclosures to prevent a condition known as pressure piling. Pressure piling occurs when the hot gases produced by an internal explosion travel through a conduit to an adjacent enclosure, Figure 19-35.

These hot gases add to the volume of gas already present in the second enclosure. If the gases in the second enclosure are ignited by the hot gas produced by the explosion in the first enclosure, the pressure produced in the second enclosure can be as great as three times what would be normally expected. A properly installed seal, however, prevents the passage of hot gas from one enclosure to the other, Figure 19-36.

Seals are especially important when the conduit leaves a hazardous location and enters a nonhazardous location, Figure 19-4. The equipment and enclosures in a nonhazardous location are not designed to withstand any type of internal explosion and would be completely destroyed.

Some types of explosionproof enclosures are designed to be drilled and tapped in the field. When this is done, at least five full threads should be engaged between the conduit and the enclosure, Figure 19-37.

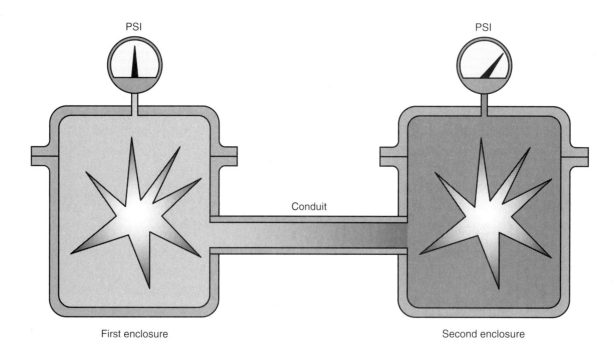

FIGURE 19-35 Pressure piling occurs when the hot gases of one enclosure are forced into the second. This produces an increase of pressure in the second. (*Delmar/Cengage Learning*)

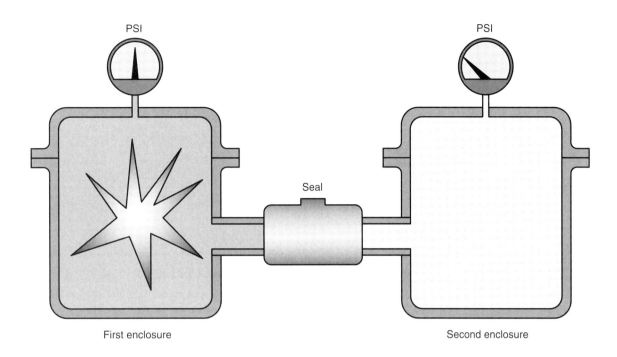

FIGURE 19-36 A seal prevents the passage of hot gas from one enclosure to the other.
(*Delmar/Cengage Learning*)

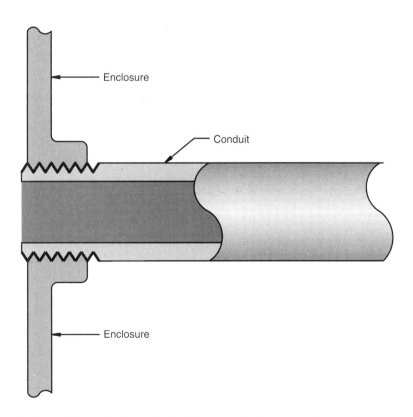

FIGURE 19-37 At least five full threads shall be made between the enclosure and the conduit.
(*Delmar/Cengage Learning*)

REVIEW QUESTIONS

All answers should be written in complete sentences, calculations should be shown in detail, and *Code* references should be cited when appropriate.

1. What would be the class, division, and group of an area in which acetylene gas was manufactured? _____

_____

2. What would be the class, division, and group of an area in which gasoline was manufactured? _____

_____

3. What would be the class, division, and group of an area in which flour was manufactured? _____

_____

4. What would be the class, division, and group of an area in which coal was stored? ___

_____

_____

5. What class is used for areas in which combustible fibers are woven into cloth? _____

_____

6. What is the maximum operating temperature of equipment to be used in a Group B location? _____

_____

7. What are intrinsically safe circuits? _____

_____

8. What is the maximum length of a stem used for pendant lighting before bracing is required? _____

_____

9. Why are seals used in explosionproof wiring systems? _____

_____

_____

10. Are "enclosed and gasketed" luminaires permitted in a Class II, Division 1 location?

_____

_____

_____

11. Name five conditions that must be met before flexible cords can be used in a hazardous location. _____

_____

12. In a commercial garage, what is the classification of any area less than 18 in. (450 mm) above the floor (other than a pit)? _____

_____

_____

13. To what height above an aircraft wing does a Class I, Division 2 location extend? ___

_____

_____

14. The interior of a gasoline-dispensing pump enclosure is considered to be Class I, Division 1 up to what height? _____

_____

_____

15. An open-front spray booth has a closed top and sides. The booth is equipped with a ventilating system that is interlocked with the spray equipment in such a manner that the spray equipment will not operate when the ventilation system is not in operation. What is the classification of the area 8 ft (2.5 m) in front of the opening of the spray booth? _____

_____

_____

CHAPTER 20

Harmonics

OBJECTIVES

After studying this chapter, the student should be able to

- describe a harmonic.
- discuss the problems concerning harmonics.
- identify the characteristics of different harmonics.
- perform a test to determine whether harmonic problems exist.
- discuss methods of dealing with harmonic problems.

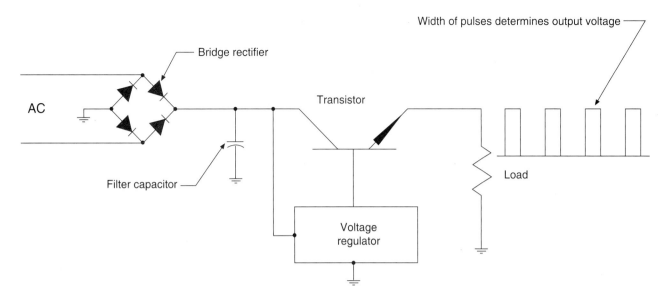

FIGURE 20-1   Pulse-width modulation regulates the output voltage by varying the time the transistor conducts as compared to the time it is off. (*Delmar/Cengage Learning*)

Harmonics are voltages or currents that operate at a frequency that is a multiple of the fundamental power frequency. If the fundamental power frequency is 60 hertz, for example, the second harmonic would be 120 hertz, the third harmonic would be 180 hertz, and so on. Harmonics are produced by nonlinear loads that draw current in pulses rather than in a continuous manner. Harmonics on single-phase power lines are generally caused by devices such as computer power supplies, electronic ballasts in fluorescent lights, triac light dimmers, and so on. Three-phase harmonics are generally produced by variable-frequency drives for ac motors and electronic drives for dc motors. A good example of a pulsating load is one that converts accurrent into dc and then regulates the dc voltage by pulse-width modulation, Figure 20-1. Many regulated power supplies operate in this manner. The bridge rectifier in Figure 20-1 changes the ac into pulsating dc. A filter capacitor is used to smooth the pulsations. The transistor turns on and off to supply power to the load. The amount of time the transistor is turned on as compared with the time it is turned off determines the output dc voltage. Each time the transistor turns on, it causes the capacitor to begin discharging. When the transistor turns off, the capacitor will begin to charge again. Current is drawn from the ac line each time the capacitor charges. These pulsations of current

produced by the charging capacitor can cause the ac sine wave to become distorted. These distorted current and voltage waveforms flow back into the other parts of the power system, Figure 20-2.

HARMONIC EFFECTS

Harmonics can have very detrimental effects on electrical equipment. Some common symptoms of harmonics are overheated conductors and transformers and circuit breakers that seem to trip when they should not. Harmonics are classified by name, frequency, and sequence. The name refers to whether the harmonic is the second, third, fourth, and so on, of the fundamental frequency. The frequency refers to the operating frequency of the harmonic. The second harmonic operates at 120 hertz, the third at 180 hertz, the fourth at 240 hertz, and so on. The sequence refers to the phasor rotation with respect to the fundamental waveform. In an induction motor, a positive sequence harmonic would rotate in the same direction as the fundamental frequency. A negative sequence harmonic would rotate in the opposite direction of the fundamental frequency. Harmonics called *triplens* have a zero sequence. Triplens are the odd multiples of the third harmonic (3rd, 9th, 15th, 21st, and so on). The sequence of the first nine harmonics is shown in Table 20-1.

TABLE 20-1									
Name, frequency, and sequence of the first nine harmonics.									
Name	**Fund.**	**2nd**	**3rd**	**4th**	**5th**	**6th**	**7th**	**8th**	**9th**
Frequency	60	120	180	240	300	360	420	480	540
Sequence	+	−	0	+	−	0	+	−	0

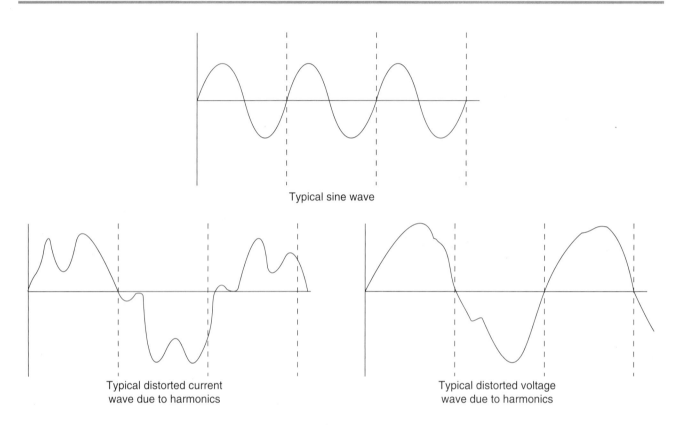

Typical sine wave

Typical distorted current wave due to harmonics

Typical distorted voltage wave due to harmonics

FIGURE 20-2 Harmonics cause an ac sine wave to become distorted. (*Delmar/Cengage Learning*)

Harmonics with a positive sequence generally cause overheating of conductors and transformers and circuit breakers. Negative sequence harmonics can cause the same heating problems as positive harmonics plus additional problems with motors. Because the phasor rotation of a negative harmonic is opposite that of the fundamental frequency, it will tend to weaken the rotating magnetic field of an induction motor, causing it to produce less torque. The reduction of torque causes the motor to operate below normal speed. The reduction in speed results in excessive motor current and overheating.

Although triplens do not have a phasor rotation, they can cause a great deal of trouble in a 3-phase, 4-wire system, such as a 208/120-volt or 480/277-volt system. In a common 208/120-volt

wye-connected system, the primary is generally connected in delta and the secondary is connected in wye, Figure 20-3.

Single-phase loads that operate on 120 volts are connected between any phase conductor and the neutral conductor. The neutral current will be the vector sum of the phase currents. In a balanced 3-phase circuit (all phases having equal current), the neutral current will be zero. Although single-phase loads tend to cause an unbalanced condition, the vector sum of the currents will generally cause the neutral conductor to carry less current than any of the phase conductors. This is true for loads that are linear and draw a continuous sine wave current. When pulsating (nonlinear) currents are connected to a 3-phase, 4-wire system, triplen harmonic frequencies disrupt the normal phasor relationship of

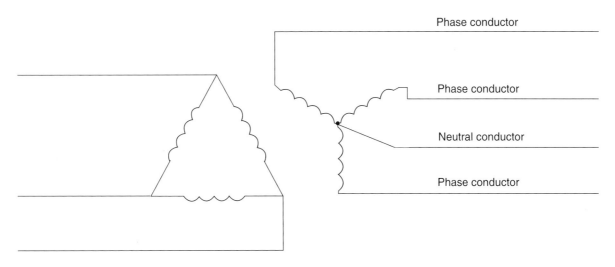

FIGURE 20-3 In a 3-phase, 4-wire connected system, the center of the wye-connection secondary is tapped to form a neutral conductor. (*Delmar/Cengage Learning*)

the phase currents and can cause the phase currents to add in the neutral conductor instead of cancel. Because the neutral conductor is not protected by a fuse or circuit breaker, there is real danger of excessive heating in the neutral conductor.

Harmonic currents are also reflected in the delta primary winding where they circulate and cause overheating. Other heating problems are caused by eddy current and hysteresis losses. Transformers are typically designed for 60-hertz operation. Higher harmonic frequencies produce greater core losses than the transformer is designed to handle. Transformers that are connected to circuits that produce harmonics must sometimes be derated or replaced with transformers that are specially designed to operate with harmonic frequencies.

Transformers are not the only electrical component to be affected by harmonic currents. Emergency and standby generators can be affected in the same way as transformers. This is especially true for standby generators used to power data processing equipment in the event of a power failure. Some harmonic frequencies can even distort the zero crossing of the waveform produced by the generator.

CIRCUIT-BREAKER PROBLEMS

Thermomagnetic circuit breakers use a bimetallic trip mechanism that is sensitive to the heat produced by the circuit current. These circuit breakers are designed to respond to the heating effect of the true RMS current value. If the current becomes too great, the bimetallic mechanism trips the breaker open. Harmonic currents cause a distortion of the RMS value, which can cause the breaker to trip when it should not, or not to trip when it should. Thermomagnetic circuit breakers, however, are generally better protection against harmonic currents than electronic circuit breakers. Electronic breakers sense the peak value of current. The peaks of harmonic currents are generally higher than the fundamental sine wave, Figure 20-4. Although the peaks of harmonic currents are generally higher than the fundamental frequency, they can be lower. In some cases, electronic breakers may trip at low currents and in other cases they may not trip at all.

BUS DUCTS AND PANELBOARD PROBLEMS

Triplen harmonic currents can also cause problems with neutral bus ducts and connecting lugs. A neutral bus is sized to carry the rated phase current. Because triplen harmonics can cause the neutral current to be higher than the phase current, it is possible for the neutral bus to become overloaded.

Electrical panelboards and bus ducts are designed to carry currents that operate at 60 hertz. Harmonic

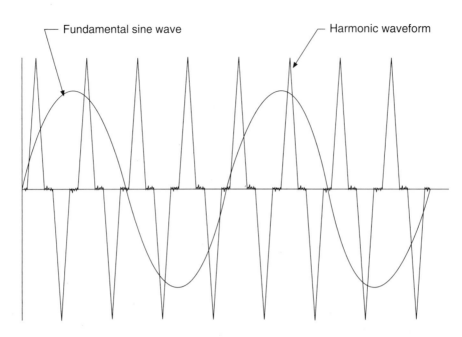

Fundamental sine wave

Harmonic waveform

FIGURE 20-4 Harmonic waveforms generally have higher peak values than those of the fundamental waveforms. (*Delmar/Cengage Learning*)

currents produce magnetic fields that operate at higher frequencies. If these fields should become mechanically resonant with the panelboard or bus duct enclosures, the panelboards and bus ducts can vibrate and produce buzzing sounds at the harmonic frequency.

Telecommunications equipment is often affected by harmonic currents. Telecommunication cable is often run close to power lines. To minimize interference, communication cables are run as far from phase conductors as possible and as close to the neutral conductor as possible. Harmonic currents in the neutral conductor induce high-frequency currents into the communication cable. These high-frequency currents can be heard as a high-pitched buzzing sound on telephone lines.

DETERMINING HARMONIC PROBLEMS ON SINGLE-PHASE SYSTEMS

There are several steps that can be followed in determining whether there is a problem with harmonics. One step is to do a survey of the equipment. This is especially important in determining whether there is a problem with harmonics in a single-phase system.

1. Make an equipment check. Personal computers, printers, and fluorescent lights with electronic ballast are known to produce harmonics. Any piece of equipment that draws current in pulses can produce harmonics.

2. Review maintenance records to see whether there have been problems with circuit breakers tripping for no apparent reason.

3. Check transformers for overheating. If the cooling vents are unobstructed and the transformer is operating excessively hot, harmonics could be the problem. Check transformer currents with an ammeter capable of indicating a true RMS current value. Make sure that the voltage and current ratings of the transformer have not been exceeded.

It is necessary to use an ammeter that responds to true RMS current when making this check. Some ammeters respond to the average value, not the RMS value. Meters that respond to the true RMS value generally state this on the meter. Meters that respond to the average value are generally less expensive and do not state that they are RMS meters. A clamp-type ammeter that responds to a true RMS current is shown in Figure 20-5.

FIGURE 20-5   True RMS ammeter.
(Courtesy Fluke Corporation)

Meters that respond to the average value use a rectifier to convert the alternating current into direct current. This value must be increased by a factor of 1.111 to change the average reading into the RMS value for a sine wave current. True RMS-responding meters calculate the heating effect of the current. The chart in Figure 20-6 shows some of the differences between average indicating meters and true RMS meters. In a distorted waveform, the true RMS value of current will no longer be the average value multiplied by 1.111, Figure 20-7. The distorted waveform generally causes the average value to be as much as 50 percent less than the RMS value.

Another method of determining whether a harmonic problem exists in a single-phase system is to make two separate current checks. One check is made using an ammeter that indicates the true RMS value and the other is made using a meter that indicates the average value, Figure 20-8. In this example, it is assumed that the true RMS ammeter indicates a value of 36.8 amperes and the average ammeter indicates a value of 24.8 amperes. Determine the ratio of the two measurements by dividing the average value by the true RMS value.

$$\text{Ratio} = \frac{\text{Average}}{\text{RMS}}$$

$$= \frac{24.8}{36.8}$$

$$= 0.674$$

Ammeter type	Sine wave response	Square wave response	Distorted wave response
Average responding	Correct	Approx. 10% high	As much as 50% low
True RMS responding	Correct	Correct	Correct

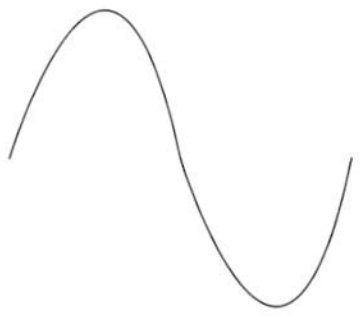

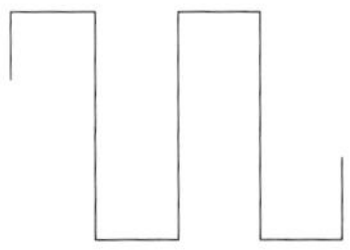

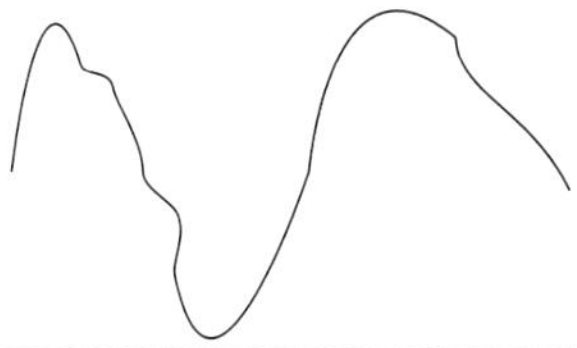

FIGURE 20-6   Comparison of average responding and true RMS responding ammeters.
(Delmar/Cengage Learning)

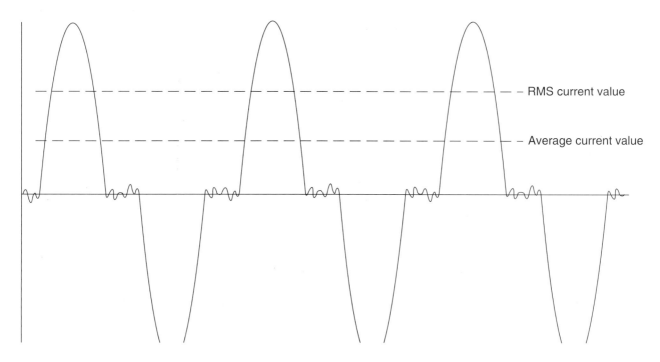

FIGURE 20-7 Average current values are generally greater than the true RMS value in a distorted waveform. (*Delmar/Cengage Learning*)

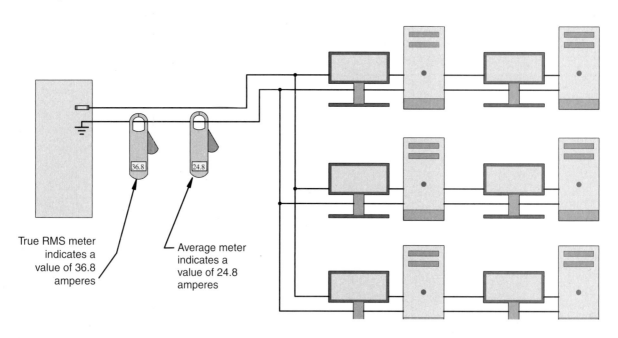

True RMS meter indicates a value of 36.8 amperes

Average meter indicates a value of 24.8 amperes

FIGURE 20-8 Determining the harmonic problems using two ammeters. (*Delmar/Cengage Learning*)

A ratio of 1 would indicate no harmonic distortion. A ratio of 0.5 would indicate extreme harmonic distortion. This method does not reveal the name or sequence of the harmonic distortion, but it does give an indication that there is a problem with harmonics.

The most accurate method for determining whether there is a harmonics problem is to use a harmonic analyzer. The harmonic analyzer will determine the name, sequence, and amount of harmonic distortion present in the system. A harmonic analyzer is shown in Figure 20-9.

FIGURE 20-9   Harmonic analyzer.
(*Courtesy Fluke Corporation*)

	TABLE 20-2	
Measuring phase and neutral currents in a 3-phase, 4-wire wye-connected system.		
Conductor	True RMS Responding Ammeter	Average Responding Ammeter
Phase 1	365	292
Phase 2	396	308
Phase 3	387	316
Neutral	488	478

DETERMINING HARMONIC PROBLEMS ON 3-PHASE SYSTEMS

Determining whether a problem with harmonics exists in a 3-phase system is similar to determining the problem in a single-phase system. Because harmonic problems in a 3-phase system generally occur in a wye-connected 4-wire system, this example will assume a delta-connected primary and wye-connected secondary with a center-tapped neutral as shown in Figure 20-3. To test for harmonic distortion in a 3-phase, 4-wire system, measure all phase currents and the neutral current with both a true RMS-indicating ammeter and an average indicating ammeter. It will be assumed that the 3-phase system being tested is supplied by a 200-kVA transformer, and that the current values shown in Table 20-2 are recorded. The current values indicate that a problem with harmonics does exist in the system. Note the higher current

measurements made with the true RMS-indicating ammeter, and also the fact that the neutral current is higher than any phase current.

DEALING WITH HARMONIC PROBLEMS

After it has been determined that harmonic problems exist, something must be done to deal with the problems. It is generally not practical to remove the equipment causing the harmonic distortion, so other methods must be employed. It is a good idea to consult a power quality expert to determine the exact nature and amount of harmonic distortion present. Some general procedures for dealing with harmonics are as follows:

1. In a 3-phase, 4-wire system, reduce the 60-hertz part of the neutral current by balancing the current on the phase conductors. If all phases have equal current, the neutral current would be zero.

2. If triplen harmonics are present on the neutral conductor, add harmonic filters at the load. These filters can help reduce the amount of harmonics on the line.

3. Pull extra neutral conductors. The ideal situation would be to use a separate neutral for each phase instead of using a shared neutral.

4. Install a larger neutral conductor. If it is impractical to supply a separate neutral conductor for each phase, increase the size of the common neutral.

5. Derate or reduce the amount of load on the transformer. Harmonic problems generally involve overheating of the transformer. In many instances, it is necessary to derate the transformer to a point that it can handle the extra current caused by the harmonic distortion. When this is done, it is generally necessary to add a second transformer and divide the load between the two.

DETERMINING TRANSFORMER HARMONIC DERATING FACTOR

Probably the most practical and straightforward method for determining the derating factor for a transformer is recommended by the Computer & Business Equipment Manufacturers Association. To use this method, two ampere measurements must be made. One is the true RMS current of the phases and the second is the instantaneous peak phase current. The instantaneous peak current can be determined with an oscilloscope connected to a current probe or with an ammeter capable of measuring the peak value. Many of the digital clamp-on ammeters are capable of measuring the average, true RMS and peak values of current. For this example, it will be assumed that peak current values are measured for the 200-kVA transformer discussed previously. These values are added to the previous data obtained with the true RMS- and average indicating ammeters, Table 20-3. The formula for determining the transformer harmonic derating factor (THDF) is

$$THDF = \frac{(1.414)\ (RMS\ phase\ current)}{Instantaneous\ peak\ phase\ current}$$

This formula will produce a derating factor somewhere between 0 and 1.0. Because the

TABLE 20-3

Peak currents are added to chart.

Conductor	True RMS Responding Ammeter	Average Responding Ammeter	Instantaneous Peak Current
Phase 1	365	292	716
Phase 2	396	308	794
Phase 3	387	316	737
Neutral	488	478	957

instantaneous peak value of current is equal to the RMS value multiplied by 1.414, if the current waveforms are sinusoidal (no harmonic distortion), the formula will produce a derating factor of 1.0. Once the derating factor is determined, multiply the derating factor by the kVA capacity of the transformer. The product will be the maximum load that should be placed on the transformer.

If the phase currents are unequal, find an average value by adding the currents together and dividing by 3.

$$Phase\ (RMS) = \frac{365 + 396 + 387}{3}$$
$$= 382.7$$

$$Phase\ (peak) = \frac{716 + 794 + 737}{3}$$
$$= 749$$

$$THDF = \frac{(1.414)\ (382.7)}{749}$$
$$= 0.722$$

The 200-kVA transformer in this example should be derated to 144.4 kVA (200 kVA multiplied by 0.722).

All answers should be written in complete sentences, calculations should be shown in detail, and *Code* references should be cited when appropriate.

1. What is the frequency of the second harmonic? _____

   _____

   _____

   _____

   _____

2. Which of the following are considered triplen harmonics: 3rd, 6th, 9th, 12th, 15th, and 18th? _____

   _____

   _____

   _____

   _____

3. Would a positive rotating harmonic or a negative rotating harmonic be more harmful to an induction motor? Explain your answer. _____

   _____

   _____

   _____

   _____

4. What instrument should be used to determine what harmonics are present in a power system?

   _____

   _____

   _____

5. A 22.5-kVA single-phase transformer is tested with a true RMS ammeter and an ammeter that indicates the peak value. The true RMS reading is 94 amperes. The peak reading is 204 amperes. Should this transformer be derated and if so by how much? _____

   _____

   _____

   _____

Electrical Specifications

The electrical contractor shall furnish and install all electrical materials, equipment, and electrical wiring in accordance with the plans and specifications. All work shall be done in a neat and workmanlike manner, and shall conform in a satisfactory way to the architectural features of the structure. It is further required that all electrical work, including the installation of equipment, shall be done in accordance with the standards and provisions of the *NEC* and modern trade practice.

MATERIALS

All electrical materials shall be new, and of the make, type, and description specified. All electrical installations, circuits, controls, and equipment shall be in proper operating condition before the contract is deemed to be completed.

LIGHTING

All lighting circuits located within the structure are to be rated at 120 volts. These circuits are to be run from power panelboards as shown on the plans and further described in the specifications. All lighting circuits are to be rated at 20 amperes, except for the lighting in the manufacturing area. This lighting will have a rating of 50 amperes per circuit and will be suspended from 50-ampere trolley busway as specified elsewhere.

RECEPTACLES AND SWITCHES

All receptacles located throughout the structure are to be rated at 125 volts and supplied from the lighting panelboards as indicated. Receptacle circuits are to be rated at 20 amperes, and the receptacles are to be of the grounding type. Special-purpose receptacles are to be installed as indicated and as further described in the specifications. All wall switches shall be ac general-use snap switches.

CONDUITS

Conduits smaller than a size ¾ (21) are *not* to be installed in the building.

CONDUCTORS

All conductors used within the building that are larger than 12 AWG shall be stranded. No conductor smaller than 12 AWG shall be used except for the luminaire wiring. This wiring may consist of 14 AWG or 16 AWG conductors. Lighting and receptacle circuit conductors shall have type TW insulation or higher. Conductors used for feeders and other purposes may be types TW, THW, or THWN, unless specified otherwise.

TROLLEY BUSWAY RUNS FOR LIGHTING

All lighting within the manufacturing area shall be suspended from 50-ampere, 2-wire, 300-volt trolley busway. The busway shall be suspended from messenger wires run along the lower part of the roof truss work. The busway shall be supplied from feed-in boxes as indicated on the plans. One terminal-type trolley, with an outlet box and ground strap and equipped with metal wheels, shall be furnished to suspend each luminaire. Two heavy-duty weight supports shall be provided for each lighting unit. Runs shall be supported by messenger cables, rod-hanger mounted, and hung by steel clip-over hangers for each 10-ft (3-m) section of busway. The installed runs shall be rigidly constructed of standard busway sections joined electrically and mechanically by self-locking, telescope-type couplings. Coupling junctures shall permit unimpeded passage of current collecting trolleys. Current tapoff devices shall contain means for feeding electrical loads and also means for connecting equipment grounding conductors.

TROLLEY BUSWAY RUNS FOR ELECTRIC TOOLS

Assembly lines as indicated on the plans shall be provided with 100-ampere, 208-volt, 3-phase, 3-wire, industrial-type trolley busway suspended by rod or strap-type hangers from the overhead structure at a height of 8 ft (2.5 m) from the floor. The final assembled runs shall be rigid in construction and neat and symmetrical in appearance. Hangers shall be spaced at not more than 5-ft (1.5-m) intervals along the runs. A feed-in adapter is to be furnished for each individual run, and the assembly is to be complete with bus connectors for electrically and mechanically joining the bus bars. End plates with bumper attachments shall be used to terminate duct runs and to prevent damage to trolleys. Trolley drop-out devices are to be provided in sections of the standard runs as indicated on the plans. The drop-out device shall have provisions for locking in a closed position and be so constructed that trolleys are easily accessible for insertion or removal. Trolleys are to be provided with metal wheels and with a provision for grounding the equipment. One trolley shall be furnished for each 15 ft (4.5 m), or fraction thereof, of trolley busway. The trolleys shall be box-type tool hangers with 30-ampere fusible pullouts, grounding receptacle, and cord clamp.

LIGHTING AND POWER PANELBOARDS

Lighting and power panelboards are to be located as shown on the plans. These panelboards are to be fed from dry-type transformers located as indicated and having the voltage and kilovolt-ampere ratings as shown in a later section of these specifications.

OUTLET BOXES AND FITTINGS

All outlet boxes and conduit fittings shall be of the type and size required by the *NEC*.

DRY-TYPE TRANSFORMERS

All panelboards shall be fed from dry-type transformers. The transformers shall be 3-phase with 480-volt primaries and 3-phase, 4-wire secondaries rated at 208Y/120 volts. The neutral point of all distribution transformer secondaries shall be effectively grounded at the transformer. The transformers shall be mounted along the walls of the manufacturing area of the plant as indicated on the plans and shall supply power to the panelboards through conduit runs containing conductors of the size and type specified. However, the 75- and 175-kVA transformers listed shall be floor-mounted. The transformers shall be supplied with power from the ventilated feeder busway designated as Feeder No. 1, as indicated on the plans and further described in these specifications. The dry-type transformers shall be provided with Class B or Class H insulation and shall be installed in accordance with *Article 450* of the *NEC*. The transformers shall have ratings as listed.

Cable tap boxes and accessories shall be used to provide the necessary tapoff from the feeder busway No. 1 to supply the transformer primaries. A suitable enclosed fusible disconnect switch shall be installed between the cable tap box and the transformers.

VENTILATED FEEDER BUSWAY NO. 1

A system of ventilated feeder busways shall be installed as indicated on the plans. The duct shall be rated at 1000 amperes and shall be installed in an edgewise manner. The duct shall be provided with end closers, edgewise elbows, and tees where necessary to turn corners and to otherwise complete the system as shown on the plans. This feeder busway shall be installed in standard 10-ft (3-m) lengths where possible. The busway shall be listed by the UL, and all busway sections shall be so labeled. A cable tap box and fusible disconnect switch of appropriate size and rating shall be installed at each transformer location. The necessary conduit and fittings shall be supplied to connect the transformer through the cable tap box to the safety switch.

FEEDER BUSWAY NO. 2

A second system of ventilated feeder busway shall be installed as indicated on the plans and will be known as Feeder No. 2. This system shall be installed in a flatwise manner and shall have a current rating of 1600 amperes. This system shall start at the low-voltage section of the unit substation

(to be described later in these specifications) and will extend to the approximate center of the manufacturing area as indicated. At this point, a tee section shall be installed and the feeder shall then extend in northerly and southerly directions to points where the extreme northerly and southerly plug-in busway runs are located as indicated on the plans. Circuit-breaker cubicles, containing two 225-ampere circuit breakers, shall be provided to connect and protect the plug-in busway indicated on the plans. Five of the double circuit-breaker cubicles shall be provided and installed on the other side of the feeder busway No. 2. The plug-in busway shall extend in both directions, east and west, from these cubicles.

PLUG-IN BUSWAY

Plug-in busway shall be installed as shown on the plans. The busway shall have a current-carrying capacity of 225 amperes and shall contain three busbars. The entire system shall be supported at a height of 12 ft (3.7 m) above the floor by standard clamp hangers and shall have maximum strength and rigidity.

Busbars shall be fabricated from pure copper with a conductivity of 98 percent and shall be silver-plated along their entire length to ensure a good electrical contact at every connection point and at every plug-in point. Takeoff plug-in openings shall be spaced at convenient intervals along both sides of the busway. These plug-in openings shall be divided equally and alternately on both sides of the busway. The plug-in busway runs shall be provided with end closers at the ends of the runs. One plug-in unit of the proper size and rating shall be furnished for each takeoff point to operate individual machines as further described in these specifications.

MOTOR BRANCH CIRCUITS AND FEEDERS

Motor branch circuits and feeders shall be installed as shown on the plans. These circuits and feeders shall consist of a fusible plug inserted in the busway system and heavy-duty type SJ flexible rubber, 4-wire cord running to each individual machine.

SCHEDULE OF ELECTRIC PANELBOARDS FOR THE INDUSTRIAL BUILDING

Panelboard No.	Location	Mains	Voltage Rating	No. of Circuits	Breaker Ratings	Poles	Purpose
P-1	Basement N. Corridor	Breaker 100 A	208/120 V 3 φ, 4 W	19 2 5	20 A 20 A 20 A	1 2 1	Lighting and Receptacles Spares
P-2	1st Floor N. Corridor	Breaker 100 A	208/120 V 3 φ, 4 W	24 2 0	20 A 20 A	1 2	Lighting and Receptacles Spares
P-3	2nd Floor N. Corridor	Breaker 100 A	208/120 V 3 φ, 4 W	24 2 0	20 A 20 A	1 2	Lighting and Receptacles Spares
P-4	Basement S. Corridor	Breaker 100 A	208/120 V 3 φ, 4 W	24 2 0	20 A 20 A	1 2	Lighting and Receptacles Spares
P-5	1st Floor S. Corridor	Breaker 100 A	208/120 V 3 φ, 4 W	23 2 1	20 A 20 A 20 A	1 2 1	Lighting and Receptacles Spares
P-6	2nd Floor S. Corridor	Breaker 100 A	208/120 V 3 φ, 4 W	22 2 2	20 A 20 A 20 A	1 2 1	Lighting and Receptacles Spares
P-7	Mfg. Area N. Wall E.	Breaker 100 A	208/120 V 3 φ, 4 W	5 7 2	50 A 20 A 20 A	1 1 1	Lighting and Receptacles Spares
P-8	Mfg. Area N. Wall W.	Breaker 100 A	208/120 V 3 φ, 4 W	5 7 2	50 A 20 A 20 A	1 1 1	Lighting and Receptacles Spares
P-9	Mfg. Area S. Wall E.	Breaker 100 A	208/120 V 3 φ, 4 W	5 7 2	50 A 20 A 20 A	1 1 1	Lighting and Receptacles Spares
P-10	Mfg. Area S. Wall W.	Breaker 100 A	208/120 V 3 φ, 4 W	5 7 2	50 A 20 A 20 A	1 1 1	Lighting and Receptacles Spares
P-11	Mfg. Area East Wall	Lugs only 225 A	208 V 3 φ, 3 W	6	20 A	3	Blowers and Ventilators
P-12	Boiler Room	Breaker 100 A	208/120 V 3 φ, 4 W	10 4	20 A 20 A	1 1	Lighting and Receptacles Spares
P-13	Boiler Room	Lugs only 225 A	208 V 3 φ, 3 W	6	20 A	3	Oil Burners and Pumps
P-14	Mfg. Area East Wall	Lugs only 400 A	208 V 3 φ, 3 W	3 2 1	175 A 70 A 40 A	3 3 3	Chillers Fan Coil Units Fan Coil Units
P-15	Mfg. Area West Wall	Lugs only 600 A	208 V 3 φ, 3 W	5	100 A	3	Trolley Busway and Elevator

Note: Where a 2-pole circuit breaker is used, the space required is the same as for two single-pole breakers.

The rubber cord shall have 3-phase conductors and one green equipment grounding conductor. The cord shall be supported under tension both horizontally and vertically by strain relief and cable grips and accessories. One 80-pound safety spring shall be used at each bus plug to start the horizontal cable run. Two cable grips are to be used at the point where the cable changes from the horizontal to the vertical drop to the machine. One 80-pound safety spring shall be used in connection with these two grips. One cable grip shall be provided at the lower end of the vertical cable run as well as one grip. Each rubber cord run shall be level and plump and shall present a neat and symmetrical appearance.

SCHEDULE OF 3-PHASE DRY-TYPE TRANSFORMERS

Transformer No.	kVA Rating	Primary Voltage	Secondary Voltage	Supplies Panelboard No.
T-A	50	480	208/120Y	P-1 P-2 P-3
T-B	50	480	208/120Y	P-4 P-5 P-6
T-C	50	480	208/120Y	P-7 P-8
T-D	50	480	208/120Y	P-9 P-10
T-E	100	480	208/120Y	P-14
T-F	75	480	208/120Y	P-11 P-12 P-13
T-G	175	480	208/120Y	P-15

MOTORS AND CONTROLLERS

The machines and machine tools indicated for the machine layout of the manufacturing area of the plant are to be installed by another contractor. The motors and controllers used to drive and control these machines, however, shall be connected to the motor branch circuits and feeders with rubber cord drops, and shall be tested for satisfactory electrical operation by this contractor. All machines listed for the plant are to have built-in drives and controls and are laid out as shown on Sheet E-2 of the plans and as described in these specifications.

PRECIPITATION UNITS

Precipitation units shall be furnished and installed at the rear of each vertical boring mill, turret lathe, and cylindrical grinder. All wiring and connections necessary to complete the installation shall be furnished and the work involved shall be done by the electrical contractor. This work shall include furnishing and installing the transformers necessary to provide 230 and/or 115 volts needed for the precipitation units.

THE SYNCHRONOUS CONDENSERS

Two synchronous condensers rated at 350 kVAR, 480 volts, 60 hertz, and 600 rpm, with direct-connected exciters and an automatic control panelboard, shall be furnished and installed as indicated on the plans. The synchronous condensers and auxiliary equipment shall be installed and connected to feeder busway No. 2 by means of cable tap boxes and conduit runs to each control panelboard. Two cables in parallel will constitute each of the three phases required. All electrical wiring and installation work in connection with the control panelboards, the exciters, and the synchronous condensers shall be done by the electrical contractor. The machines shall be tested and placed in satisfactory operating condition.

THE ROOF BLOWERS

Four ventilating blowers shall be installed on the roof of the manufacturing structure. These blowers shall be connected to panelboard No. P-11. Two additional ventilating blowers shall be installed on the roof of the office structure as indicated on the plans. Conduits and conductors for these blowers shall also be installed and connected to panelboard No. P-11. All blowers are to be controlled by line starters with push-button control. The electrical contractor shall install, on either side of panelboard No. P-11, a section of 6-in. (150-mm) square busway or wireway, from which line starters will be connected by the necessary conduit nipples and fittings needed to feed the six blower circuits.

ELEVATOR POWER SUPPLY

The elevator installation is not to be done by the electrical contractor. However, a supply circuit for the elevator operation shall be installed. This circuit is to be a 3-phase, 208-volt circuit and shall be installed from panelboard No. P-15. The circuit shall be run to the elevator penthouse.

SCHEDULE OF MACHINES TO BE CONNECTED FROM BUSWAY SYSTEM IN THE MANUFACTURING AREA

Code No.	Type of Machine	No. Used	Type and Number of Motors	Type of Control	Hp	Amperes, Full Load	Plug-in Fusible Switch Rating	Time-Delay Fuse Size
MA	Engine Lathes	20	1-Squirrel Cage	Line Starter For. and Rev.	5.0	7.6	30	15
MB	Turret Lathes	10	1-Squirrel Cage	Line Starter	7.5	11.0	30	20
MC	Vertical Drills	12	1 -Squirrel Cage	Line Starter	1.0	1.8	30	15
MD	Multispindle Drills	8	1-Four Speed, Two Winding	Reduced Voltage and Cam. Sw.	10.0	14.0	30	30
ME	Milling Machines	6	3-Squirrel Cage	Primary Resistor	10.0 1.0 1.0	14.0 1.8 1.8	30	30
MF	Shapers	6	1-Squirrel Cage	Line Starter	7.5	11.0	30	20
MG	Vertical Boring Mills	5	1-DC Shunt 3-Squirrel Cage	Electronic Control Line Starters	25.0 3.0 3.0 3.0	34.0 4.8 4.8 4.8	100	90
MH	Planers	3	1-DC Shunt	Electronic Control	25.0	34.0	60	60
MI	Power Hacksaws	6	1-Squirrel Cage	Line Starter	3.0	4.8	30	15
MJ	Band Saws	4	1-Squirrel Cage	Line Starter	5.0	7.6	30	15
MK	Surface Grinders	6	1-Squirrel Cage	Line Starter	10.0	14.0	30	30
ML	Cylindrical Grinders	10	1-Squirrel Cage	Line Starter	7.5	11.0	30	20
MN	Punch Presses	10	1-Wound Rotor	Secondary Resistor	10.0	14.0	30	30
MO	Special Machines	5	1-Squirrel Cage	Line Starter	5.0	7.6	30	15

SCHEDULE OF RECEPTACLE OUTLETS

Panelboard No.	No. Duplex Grounding Receptacles	No. 208 V Receptacles 2-Pole and Gr.	Total No. Circuits in Panelboard	No. Circuits Reserved for Light	No. Circuits Reserved for Receptacles	Spare Circuits Left in Panelboard
P-1	15	2	28	16	3	5
P-2	24	2	28	16	8	0
P-3	24	2	28	16	8	0
P-4	15	2	28	19	5	0
P-5	24	2	28	17	6	1
P-6	24	2	28	14	8	2
P-7	15		14	5	7	2
P-8	15		14	5	7	2
P-9	15		14	5	7	2
P-10	15		14	5	7	2
P-12	12		14	6	4	4

AIR-CONDITIONING EQUIPMENT

All air-conditioning equipment shall be supplied with power from panelboard No. P-14 as described elsewhere in the specifications. The electrical contractor shall furnish and install three liquid chillers as indicated on the plans. These chillers are to be connected with conduits to panelboard P-14. Ten fan coil units shall be installed by the electrical contractor and shall be connected by conduits to panelboard No. P-14. The installation shall be arranged to consist of three circuits, two of which shall have four fan coil units connected to each circuit. The two fan coil units installed for cooling the office structure shall be connected together on a single circuit. All piping between the liquid chillers and the fan coil units shall be done by another contractor.

PAGING SYSTEM

A paging system shall be installed as indicated on the plans. It shall be provided with four audible industrial chime signals in the office structure. Two of the paging signals located in the plant shall be single-stroke bells. The twenty-call unit shall be installed at the telephone switchboard near the reception desk.

CLOCK AND PROGRAM SYSTEM

A clock system shall be installed in the office section of the building. The clocks shall be controlled by a flush-mounted, automatic-reset control panelboard located behind the reception desk.

A two-circuit program instrument shall be installed in the same location and shall work in conjunction with the clock system. All installation details shall be in accordance with the installation instructions furnished by the manufacturer. Two 8-in. (200-mm) pilot clocks shall be furnished and installed.

FIRE ALARM SYSTEM

A system of fire alarm boxes and sirens (horns) shall be installed in the building. The system shall be a supervised, closed-circuit, city-connected system designed for coded operation. The system shall be installed in accordance with the installation instructions furnished by the manufacturer. Eight station boxes and seven fire sirens (horns) shall be furnished and installed in the plant at the points indicated on the plans. A surface-mounted control panelboard shall be installed in the janitor's room. The stations

in the office structure shall be flush-mounted. The stations shall be surface-mounted in the plant. Fire sirens shall be installed in the office portion, with megaphone horns installed elsewhere.

THE UNIT SUBSTATION

A unit substation shall be installed as indicated on the plans. The substation shall consist of a high-voltage section, an air-cooled transformer, and a low-voltage section. A pothead on the high-voltage section shall be furnished and connected to incoming power. The high-voltage section shall be provided with a load-break, air-interrupter switch, and also shall be able to accommodate high-voltage current and potential transformers for metering purposes. Conduits must be provided to carry the low-voltage secondaries of the instrument transformers to the location indicated. Suitable test blocks and meter trims are to be furnished to accommodate the watt-hour meter.

The transformer section shall be provided with a dry-type, 3-phase transformer rated at 1500 kVA with a 4160-volt primary and a 480-volt secondary. The transformer shall be air-cooled and have class B insulation. The low-voltage section of the unit substation shall be provided with two draw-out type circuit breakers rated at 1000 and 1600 amperes at 480 volts. Space shall be reserved for one future breaker. The 1000-ampere breaker shall connect to feeder busway No. 1, and the 1600-ampere breaker shall connect to feeder busway No. 2.

HIGH-VOLTAGE METERING FACILITIES

The electrical contractor shall install facilities for the high-voltage metering of the power used in the industrial building. Two size $\frac{3}{4}$ (21) conduits shall be run from the high-voltage section of the unit substation to the point indicated on the side wall of the loading platform alcove at the east end or rear of the building. A steel cabinet with a lock and keys shall be installed at the point at which the two conduits terminate. The cabinet shall measure 30 in. by 30 in. (750 mm by 750 mm) and shall have a depth of 20 in. (525 mm). A double meter socket trough shall be installed directly above the cabinet and shall be connected to the cabinet by means of size 1 (27) conduit nipples. The left-hand meter socket shall be wired to receive a standard two-element watt-hour meter. This meter will be set in place by employees of the power company. The right-hand meter socket shall be wired to receive a standard two-element reactive kilovolt-ampere-hour meter. This meter will be installed by the electrical contractor. All of the wiring in the conduit runs from the current and potential transformers located in the high-voltage section of the unit substation to the meter cabinet, and meter sockets shall be wired with 12 AWG conductors. An autotransformer designed for use with a reactive kilovolt-ampere-hour meter shall be installed in the cabinet and all necessary connections made. The current and potential transformers located in the high-voltage compartment are to be furnished as an integral part of the unit substation.

TELEPHONE RACEWAYS

A system of telephone raceways shall be installed according to the following schedule. Floor boxes or wall boxes shall be provided as listed and as indicated on the plans. When completed, tagged fish wires shall be left in all of the conduit runs to facilitate the installation of cables by the telephone company. A junction or pull box shall be provided in the janitor's room at the point of entrance of the conduit run from the telephone pole. A weep hole, $\frac{3}{8}$ in. in diameter, shall be drilled at the lowest point in the elbow at the pole and a dry well constructed to receive any drainage. A second pull box shall be located in the conduit run from the boiler room to the cafeteria telephone outlets. No individual phone run shall be greater than 150 ft (45 m) long, nor shall any run contain more than the equivalent of two quarter bends.

SCHEDULE OF CONDUIT RUNS FOR TELEPHONE EQUIPMENT

From Terminal Cabinet To	Phone Outlets on Run	Floor Box	Wall Box	Conduit Sizes	
Outside Pole	—	—	—	3	(78)
PBX Board	—	—	—	1½	(41)
Dial Switch Unit	—	—	—	1½	(41)
Rectifier	—	—	—	1½	(41)
Drafting and Engineering	3	—	3	1	(27)
General Offices and Engineering	3	2	1	1	(27)
Exec. Suite No. 1	3	3	—	1	(27)
Exec. Suite No. 2 and No. 6	3	3	—	1	(27)
Exec. Suite No. 3	3	2	1	1	(27)
Exec. Suite No. 4 and Receptionist	3	2	1	1	(27)
Exec. Suite No. 5	3	3	—	1	(27)
Public Phones	2	—	1	1	(27)
Cafeteria, Custodian, Boiler Room	3	—	3	1	(27)
Shipping Dept. and Manufacturing Area	3	—	3	1	(27)
Receiving Dept. and Manufacturing Area	3	—	3	1	(27)

Code Index

Note: Page numbers in **bold** reference non-text material.

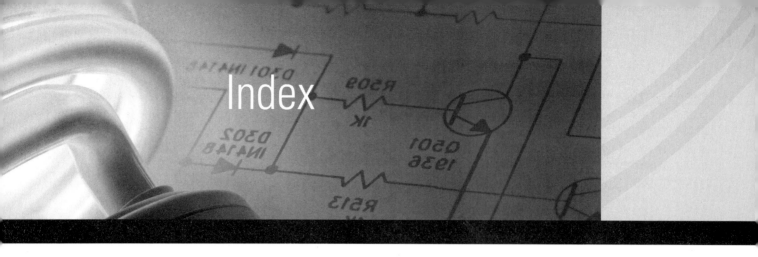

Index

Note: Page numbers in **bold** reference non-text material.